# VISUAL INFORMATION PROCESSING

# VISUAL INFORMATION PROCESSING

Edited by Sal Soraci Jr. and Kimiyo Murata-Soraci

PRAEGER

Westport, Connecticut
London

**Library of Congress Cataloging-in-Publication Data**

Visual information processing / edited by Sal Soraci Jr. and Kimiyo Murata-Soraci.
p. cm.
Includes bibliographical references and index.
ISBN 1-56750-685-2 (alk. paper)
1. Visual perception. 2. Cognition. 3. Psychology, Comparative. 4. Cognition in animals. I. Soraci, Sal. II. Murata-Soraci, Kimiyo.

BF241.V567 2003
152.14—dc21 2002029331

British Library Cataloguing in Publication Data is available.



Library of Congress Catalog Card Number: 2002029331
ISBN: 1-56750-685-2

First published in 2003

Praeger Publishers, 88 Post Road West, Westport, CT 06881
An imprint of Greenwood Publishing Group, Inc.
www.praeger.com

Printed in the United States of America

The paper used in this book complies with the Permanent Paper Standard issued by the National Information Standards Organization (Z39.48-1984).

10 9 8 7 6 5 4 3 2 1

# Contents

# Preface

In the present book, we have chosen to focus on visual information processing and intellectual functioning for conceptual and methodological reasons. Much of the evolutionary path of higher primate species has involved the development of sophisticated visual systems that interact with complex, higher-order cognitive processes. Key questions in cognitive science address the manner in which the environment is represented by the organism, and thus relate to how knowledge about the world is gleaned, with implications for theories of action and decision making. Finally, it has become apparent that the distinction between perceptual and cognitive processes is not always a clear one, and that these processes interact in critical ways in underlying complex behavioral repertoires.

The contributors' chapters in this volume cover three areas. Human visual processing, animal visual processing, and philosophical approaches to visual processing are all represented. Consistent with the emphasis on individual differences, both typical and atypical development are explored. In Part I, on human visual processing, Cronin-Golomb and Gilmore examine deficits in visual cognition in Alzheimer's disease, and potential intervention strategies. Oross and Woods investigate perceptual abilities in individuals with intellectual disabilities, and suggest that certain types of visual cueing have implications for the development of training methods. Carlin, Soraci, and Strawbridge employ structural manipulations of the visual array and demonstrate the facilitation of visual search and detection process in individuals with and without mental retardation. Dube, Lombard, Farren, Flusser, Balsamo, Fowler, and Tomanari apply stimulus control methods to modify observing behaviors and stimulus overselectivity in individuals with mental retardation. Gardner,

Karmel, and Flory explore visual attention in neonates, emphasizing arousal modulation and influences on information-processing skills and development. Hodapp and Ly focus on visual processing strengths in individuals with Down syndrome, especially as they inform etiology-related interventions involving reading skills. Young and Wasserman examine the effects of visual display variability on discrimination performances and the implications for enhancing our understanding of conceptual skills in human and animal populations.

In Part II, on animal visual processing, Katz and Cook delineate visual search processes in pigeons and discuss how assessing problem solving in avian species can contribute in important ways to our understanding of strategic behavior in humans. In the following chapter, da Silva Barros, de Faria Galvão, and McIlvane take a behavioral approach to assessing the perceptual and cognitive processes of the capuchin monkey, and discuss the relevance of their research to facilitating the effectiveness of programmed instructional methods in determining the cognitive capabilities of children with developmental disabilities.

The final part of the book includes three chapters on philosophical approaches to visualism. We believe these chapters complement and supplement the empirically based contributions in the first two parts. Ihde posits that the evidentiary basis for the sciences can be characterized as one that is essentially in visual form and he discusses the development of techniques such as imaging technologies to support his premise. Walters forwards the need for investigators of visual information processing to interact in an interdisciplinary manner with philosophers in addressing the relationships between visual, auditory, and what he refers to as "listening" domains. Murata-Soraci discusses the importance of what she refers to as the "alterity in memory," emphasizing the importance of a reconceptualization of the conventional measurement of time and space for memory studies, and a consideration of the ethical dimension of such studies.

We hope that this volume as a whole contributes to enhancing an understanding of intellectual functioning in the many domains in which it is examined. We believe that these chapters have import both for basic science and for the development of applications, facilitating theory and practice.

**Part I**

# Human Visual Processing

Chapter 1

# Visual Factors in Cognitive Dysfunction and Enhancement in Alzheimer's Disease

*Alice Cronin-Golomb and Grover C. Gilmore*

Most aspects of visual cognition are impaired in Alzheimer's disease (AD), including the abilities to recognize and discriminate objects, faces, and patterns. Deficits in visual cognition arise not only from pathological changes in high-order association areas of the brain in this disorder, but also from defective input from lower-level visual processing areas. Impairments in basic vision are prevalent in AD, and such impairments can strongly predict deficits in visual cognition in this disorder.

Links between visual deficits and cognitive performance have mostly been correlative. More recently, investigators have attempted to establish the causal relation between visual and cognitive deficits. Although it is agreed that top-down or feedback processing is important to understanding visual dysfunction in AD, the impact of the feed-forward process—that is, of basic visual capacity upon higher-order visual cognition—has received little study. Further, most studies to date have focused on how poor vision predicts poor cognition but have not at all explored how enhanced vision might lead to enhanced cognition. If basic visual capacities can be enhanced, it is reasonable to predict that multiple aspects of visual cognition also may be enhanced, including reading, face discrimination, face and object recognition, and pattern completion.

The goal of this chapter is to describe deficits in basic vision and visual cognition in AD and to discuss the relation between these domains. We will emphasize contrast sensitivity and backward masking as basic visual capacities that have been investigated in some depth in our laboratories. These visual capacities will be related to a variety of aspects of visual cognition that encompass the broad categories of object recognition and spatial localization. Finally, we will mention new work applying visual interventions to everyday function to improve the quality of life in demented individuals.

## DEFICITS IN VISUAL COGNITION IN AD

Clinical observations and research reports indicate that AD leads to severe impairments in visual cognition (Appel, Kertesz, & Fishman, 1982; Bäckman & Herlitz, 1990; Becker, Lopez, & Boller, 1995; Cronin-Golomb, 2001; Cronin-Golomb & Amick, 2001; Cronin-Golomb, Corkin, & Rosen, 1993a; Cronin-Golomb, Corkin, & Growdon, 1995; Cummings, Houlihan, & Hill, 1986; Della Sala, Muggia, Spinnler, & Zuffi, 1995; Henderson, Mack, & Williams, 1989; Hodges, Salmon, & Butters, 1993; Kurylo, Corkin, Dolan, Rizzo, Parker, & Growdon, 1994a; Kurylo, Corkin, & Growdon, 1994b; Mendez, Mendez, Martin, Smyth, & Whitehouse, 1990a; Mendez, Turner, Gilmore, Remler, & Tomsak, 1990b; Mendola, Cronin-Golomb, Corkin, & Growdon, 1995; Nebes, Martin, & Horn, 1984; Ogden 1990; Ricker, Keenan, & Jacobson, 1994). Visual cognition comprises the broadly defined capacities of *object recognition,* which includes the ability to recognize and discriminate objects, faces, letters and words, and other patterns; and *spatial localization,* which includes the ability to orient oneself to aspects of the environment (egocentric localization) and to relate spatially those aspects of the environment external to the self (allocentric localization). Converging evidence points to dysfunction of two main cortical pathways for visual processing, one of which is specialized for object recognition (the ventral, or occipito-temporal pathway) and the other of which is specialized for spatial localization (the dorsal, or occipito-parietal pathway) (Ungerleider & Mishkin, 1982). Behavioral and physiological evidence for dysfunction of the temporal and parietal lobes in AD is provided by poor performance on tests of both object recognition and spatial localization (Butter, Trobe, Foster, and Berent, 1996; Haxby et al., 1991; Haxby, Horwitz, Ungerleider, Maisog, Pietrini, & Grady, 1994; Kurylo, Corkin, Rizzo, Growdon, 1996; Mendez et al., 1990a, 1990b), and is supported by studies of the distribution of AD neuropathological changes in those cortical areas (Arnold, Hyman, Flory, Damasio, & Van Hoesen, 1991; Braak, Braak, & Kalus, 1989; Lewis, Campbell, Terry, & Morrison, 1987; Pearson, Esiri, Hiorns, Wilcock, & Powell, 1985). The visual-spatial capabilities of AD patients with Balint's syndrome especially highlight dysfunction of the dorsal pathway (Mendez et al., 1990b).

## DEFICITS IN VISION IN AD

Whereas impairments in visual cognition are well established in AD, a consensus that this disorder is also associated with significant impairments in basic visual processes has only recently begun to emerge. Although AD patients are less likely than healthy elderly individuals to report vision problems to their physicians (McCormick, Kukull, Van Belle, Bowen, Teri, & Larson, 1994), deficits in basic vision may be quite prevalent, with dysfunction in a single capacity (backward pattern masking) occurring in 58 percent of a large sample of patients (Mendola et al., 1995). The pathological changes characteristic of AD, including amyloid plaques and neurofibrillary tangles, affect the visual processing streams not only at the level of the temporal and parietal cortices, but also at earlier levels. The ventral and dorsal visual processing streams are dominated, respectively, by the P- (parvocellular, or color-opponent) and M- (magnocellular, or broadband) pathways that extend from the retina to the visual association cortex. Although the pathways are not strictly isolated from each other, each is relatively specialized for certain visual functions. Those functions dependent on the P-pathway, such as color discrimination and contrast sensitivity for a range of spatial frequencies, are impaired in many individuals with AD (Cronin-Golomb, 2001; Cronin-Golomb, Corkin, Rizzo, Cohen, Growdon, & Banks, 1991; Cronin-Golomb, Sugiura, Corkin, & Growdon, 1993; Gilmore, Koss, Wenk & Whitehouse, 1993; Kurylo, Corkin, Dolan, Rizzo, Parker, & Growdon, 1994a; Mendola et al., 1995; Neargarde, Stone, Cronon-Golomb & Oross, in press). P-pathway dysfunction is supported neuropathologically by findings of the presence of senile plaques in parvocellular regions of the lateral geniculate nucleus (Leuba & Saini, 1995). Further, AD pathological changes occur in the specific striate area (IVcB) that receives color information from the lateral geniculate nucleus (Beach & McGeer, 1988).

Visual capacities dependent on the M-pathway may be performed normally by some AD patients. For example, flicker perception (Cronin-Golomb et al., 1991; Mendola et al., 1995) and motion perception at some speeds (Kurylo, Corkin, Dolan, et al., 1994; Mendola et al., 1995) may be normal in AD. Impaired performance on some tasks of motion perception (Gilmore, Wenk, Naylor, & Koss , 1994; Trick & Silverman, 1991) may reflect relatively high signal strength or cognitive demands of those tasks (Cronin-Golomb, 1995). Some investigators, however, report M-pathway dysfunction in AD using passive visual stimulation, thereby minimizing cognitive demands. Mentis and colleagues (1996) described abnormal AD response to high-frequency pattern flash and to apparent motion, suggestive of dysfunction of striate cortex and area MT (a motion-sensitive area), respectively. The dissociation between normal motion detection (unconscious) and abnormal motion perception (conscious) found in one study (Silverman, Tran, Zimmerman, & Feldon, 1994) underscores the idea that the source of most visual dysfunction in AD occurs beyond striate cortex. Taken together with recent neuropathological findings of impaired capacity for oxidative metabolism in the dorsal lateral geniculate nucleus as indicated by reduction of cytochrome oxidase levels (Wong-Riley et al., 1997), and a high density of AD senile plaques in primary visual cortex relative to other brain areas (Kuljis & Tikoo, 1997; Wong-Riley et al., 1997) as well as disruption of feedback projections from extrastriate (area V2) to striate cortex (V1) in AD patients with prominent visual symptoms (Hof, Vogt, Bouras, &

Morrison, 1997), these behavioral studies show that numerous visual capacities dependent on either the P- or M-pathways are dysfunctional in AD. The AD visual profile is not seen in normal elderly adults (Cronin-Golomb et al., 1991) or individuals with Parkinson's disease, another age-related neurological disorder (Kurylo, Corkin, & Growdon, 1992), but does appear in adults with Down syndrome, a disorder that shares with AD a number of neuropathological and behavioral characteristics (Rocco, Cronin-Golomb, & Lai, 1997).

The visual deficits that are observed in many individuals with AD are likely to be of cortical origin. Neuro-ophthalmological examination and electrophysiological testing often yield normal results in samples of AD patients who showed significant impairments on tests of basic vision, indicating that defects in the anterior visual structures up to and including the optic radiations did not account for the observed visual impairments (Aguglia, Gambarelli, Farnarier, & Quattrone, 1991; Celesia, Villa, Brigell, Rubboli, Bolcioni, & Fiori, 1993; Martinelli et al., 1996; Rizzo et al. 1992; Wright, Drasdo, & Harding, 1987; Wright, Harding, & Orwin, 1984). Although some investigators find evidence of degeneration of retinal ganglion cells in AD autopsy tissue (Blanks, Torigoe, Hinton, & Blanks, 1991; Blanks, Schmidt, Torigoe, Porrello, Hinton, & Blanks, 1996; Blanks, Torigoe, Hinton, & Blanks, 1996; Hinton, Sadun, Blanks, & Miller, 1986), others do not (Curcio & Drucker, 1993; Davies, McCoubrie, McDonald, & Jobst, 1995). In living patients, retinal nerve-fiber layer-changes with AD are not found reliably (Hedges, Galves, Spiegelman, Barbas, Peli, & Yardley, 1996; Tsai et al., 1991). The distribution of neuropathological changes in the AD brain implies that the disease may cause retrograde degeneration within the visual system, from association to primary visual cortex (Arnold, Hyman, Flory, Damasio, & Van Hoesen, 1991; Braak, Braak, & Kalus, 1989; Lewis, Campbell, Terry, & Morrison, 1987; Pearson, Esiri, Hiorns, Wilcock, & Powell, 1985; Rebeck & Hyman, 1993) and also along cortico-geniculate and cortico-tectal routes (Leuba & Saini, 1995), rather than originating in lower-level regions of the visual system.

## RELATION BETWEEN BASIC VISION AND VISUAL COGNITION IN AD

The significance of the appearance of deficits in basic vision lies in the fact that they are likely to impact negatively upon visual cognition. High-order visual processes build on, and so depend on, the integrity of visual processes at earlier stages. A small number of behavioral studies have shown correlations between visual and cognitive performance in AD. One reported that central visual acuity correlated inversely with dementia severity as assessed with a brief mental status examination (Uhlmann, Larson, Koepsell, Rees, & Duckert, 1991). Most studies, however, report normal central acuity in AD (Cronin-Golomb et al., 1991; Kiyosawa et al., 1989; Levine, Lee, & Fisher, 1993; Mendez, Mendez, Martin, Smyth, & Whitehouse, 1990; Rizzo et al., 1992, Sadun &

Bassi, 1990). A second study (Mendez, Mendez, 1990) found that "visual symptoms" were more strongly correlated with performance on tests of object recognition than with performance on tests of spatial localization. Although the results of this study are difficult to interpret because the basic visual deficits were not quantified, they suggest that disorders of basic visual function may have a greater impact on object recognition than spatial localization, a finding that we have reported as well (Cronin-Golomb, Corkin, & Growdon, 1995).

In our study, we found two visual capacities, backward pattern masking and contrast sensitivity function, that were strong predictors of deficits in visual cognition in a large sample of AD patients ($N = 72$). The percentage of variance on certain cognitive tests predicted by backward masking ranged from 27 to 51, and contrast sensitivity at low spatial frequencies accounted for up to 33 percent of variance in performance. Pattern masking and low spatial frequency contrast sensitivity were also the tests on which deficits in AD were the most prevalent (Mendola et al., 1995). Because of the prevalence of deficit on these vision tests and their predictive value for cognitive function, we have targeted these two visual capacities as potential points of intervention for cognitive studies.

Comparison studies suggest that object recognition is more impaired than spatial localization in AD (Butter, Trobe, Foster, & Berent, 1996; Kurylo, Corkin, Rizzo, & Growdon, 1996), reflecting the greater density of AD neuropathology in inferotemporal—relative to posterior—parietal cortex (Arnold et al., 1991; Bouros, Hof, Giannakopoulas, Michel, & Morrison, 1994) and the greater density of senile plaques in parvocellular than in magnocellular regions of the lateral geniculate nucleus (Leuba & Saini, 1995). Our correlative study of vision and cognition in AD indicated that pattern masking and contrast sensitivity deficits accounted for substantially more performance variance on tests of object recognition than on tests of spatial localization (Cronin-Golomb et al., 1995), a result also reported by Mendez, Mendez et al. (1990) in their study relating visual symptoms to cognitive performance. Nevertheless, we acknowledge the disruption of the dorsal as well as the ventral visual processing stream across the AD population as a whole (Grady et al., 1993), and note that spatial localization may be especially impaired in a subgroup of patients, including those with symptoms of Balint's syndrome or other such prominent visual symptoms (Hof & Bouras, 1991; Hof, Bouras, Constantinidís, & Morrison, 1989, 1990; Kiyosawa et al., 1989; Pietrini et al., 1996; Trick, Barris, & Bickler-Bluth, 1989).

## DIRECT COGNITIVE CONSEQUENCES OF VISUAL DYSFUNCTION IN AD

Our previous studies of the relation between deficits in basic vision and cognition in AD were correlative, as dictated by the methods of data collection. More recently, both the Boston and Cleveland laboratories have begun to address this issue experimentally. We chose as our first visual-system intervention the experimental manipulation of contrast sensitivity. We found that enhancement of contrast eliminated a deficit in picture recognition by AD patients (Gilmore, 1995; Gilmore, Turner, et al., 1995; Lustig et al., 1995). Likewise, under normal contrast conditions, the letter-reading speed of AD patients was slower than that of healthy age-matched control participants, but under very high contrast conditions the groups' performance was equivalent (Gilmore, Thomas, Klitz, Persanyi, & Tomsak, 1996). Conversely, decreasing contrast in order to simulate vision in normal old age or AD resulted in impairment of performance of healthy young individuals on a variety of cognitive tasks, such as object naming and pattern completion. Not only did the young adults' performance worsen, but it also achieved the characteristics of AD performance, including an increase in the number of perceptual errors on object naming (Spinks, Gilmore & Thomas, 1996). Finally, we found that enhancing contrast sensitivity at low spatial frequencies normalizes performance on a task of face discrimination (Brown et al., 1996; Cronin-Golomb et al., 2000). The results of these studies are detailed below, following a brief description of the study participants and general procedures.

## Participants

The AD patients for our studies were outpatients attending the Alzheimer clinics of the Alzheimer Disease Centers (ADCs) of Boston and Cleveland, including the Boston Medical Center, the Bedford (MA) Veterans Affairs Medical Center, Case Western Reserve University, and the University Hospitals of Cleveland, as well as area day centers. Elderly control participants were spouses or siblings of the patients or other volunteers participating in research at Boston University and Case Western Reserve University. Young adults were participants from the university participant pools.

Young and elderly control participants received detailed health history screening. Additionally, elderly control participants were screened for dementia with the Mini-Mental State Exam (MMSE) (Folstein, Folstein, & McHugh, 1975). Persons with scores of 25 and above with normal health history who met all inclusion and exclusion criteria were included in the normal elderly adult group. The healthy elderly adults were matched to the AD patients for age, and all three groups were matched for education level and other demographic variables.

Participants with AD were over the age of fifty and met the clinical criteria for mild to moderate dementia of the Alzheimer type as specified by the NINCDS-ADRDA work group and the Health and Human Services Task Force on Alzheimer's Disease (McKhann, Drachman, Folstein, Katzman, Price, & Stadlan, 1984). Specifically, patients carried the diagnosis of probable Alzheimer's disease as determined by clinical examination, documented by the MMSE and confirmed by neuropsychological tests. The evaluation demonstrated deficits in two or more areas of cognition, progressive worsening of memory and other cognitive functions, no disturbance of consciousness, onset between ages forty and ninety, and absence of evidence for other causes of dementia such as may be provided by imaging scan and laboratory tests. Accuracy of diagnosis using these criteria may be 80 percent or better relative to autopsy findings (Lanska & Schoenberg, 1993). Mild dementia was defined by MMSE scores of 21 and above, and moderate dementia by MMSE scores of 10–20 (stratified by scores of 10–15 and 16–20). The Alzheimer Disease Centers use neuropsychological testing to support the diagnosis of probable AD, including numerous individual measures of memory (immediate and delayed), language, visuospatial and executive function as well as standard batteries such as that recommended by the Consortium to Establish a Registry for Alzheimer's Disease (CERAD) (Mirra, Hart & Terry, 1993; Morris, Mohs, Rogers, Fillenbaum, & Heyman, 1988).

Inclusion and exclusion criteria for all groups applied to all of our studies. Participants were required to be native speakers of English with eight or more years of formal education, in good health, and living at home rather than in an institution. Exclusion criteria included coexisting cancer, serious cardiac disease, other serious chronic medical illness, prior intracranial surgery, history of traumatic brain injury, mental retardation, use of large doses of psychoactive medications, psychiatric or neurological diagnoses other than AD, history of alcoholism or other drug abuse, history of treatment with electroshock therapy, or history of eye disease or other abnormalities. We also excluded individuals whose corrected binocular acuity was poorer than 20/40. This criterion did not result in the exclusion of more AD than control participants because central acuity appears to be normal in AD (Cronin-Golomb et al., 1991; Kiyosawa et al., 1989; Levine et al., 1993; Mendez et al., 1990; Rizzo et al., 1992, Sadun & Bassi, 1990). It should be noted, however, that variability in acuity levels, even within the restricted range of 20/20 to 20/40, can affect performance on tests of contrast sensitivity and hence further visual processing (Neargarder et al., in press).

Our inclusion and exclusion criteria are quite strict and include vision considerations not incumbent upon most AD studies. Because disorders of the eye and optic nerve are common in normal aging, we typically enroll fewer AD and elderly control participants for vision studies than for studies of cognition. Typical vision impairments with normal age and AD occur at the ocular level, including cataract, glaucoma, and age-related macular degeneration. Any of these conditions would make a potential participant ineligible for any study of vision in AD, resulting in a reduced pool of potential participants relative to the size of the pool available in studies of cognition. We estimate that our participant pools of AD patients and healthy elderly adults is reduced by at least 10 percent for our vision studies relative to studies of cognition, depending on the sample and the tests involved. However, because the presence of visual behavioral symptomatology (as opposed to ocular abnormalities) was not a factor in selection for our studies, which was based solely on consecutive appearance of patients in the clinic, we believe that our study participants were representative of most AD patients and healthy elderly adults.

## Heterogeneity of AD patients' presentation

AD is a heterogeneous disease in terms of factors such as age, family history, rate of progression, and presenting or predominant behavioral symptomatology. We did not restrict the age range of our participants in order to capture possible subgroups with clusters of visual symptoms that may shed light on underlying pathology. For example, in the Mendola et al. study (1995), we identified a subgroup of fourteen AD patients who performed exceptionally poorly on the backward masking task; these patients were significantly younger and had a shorter duration of AD than other patients who took the test. We hypothesized that the impairment in this subgroup may have reflected pathological change in the left occipito-temporal cortex (Petersen, Fox, Snyder, & Raichle, 1990) and was consistent with the suggestion of increased severity, possibly biased toward left hemisphere, language-related functions, for early-onset AD (Brandt, Mellits, Rovner, Gordon, Selnes, & Folstein, 1989; Chui, Teng, Henderson, & Moy, 1985; Faber-Langendoen, Morris, Knesevich, LaBarge, Miller, & Berg, 1988; Filley, Kelly, & Heaton, 1986; Koss et al., 1996; Selzer & Sherwin, 1983). Other early-onset subgroups may show prominent visuoperceptual disorders, such as visual agnosia and impaired visuospatial abilities, early in the disease course (e.g., Reid et al., 1996). Such subgroups may respond differently from other AD patients to experimental manipulations for effecting cognitive change.

Some individuals with probable AD present with visual complaints such as hallucinations, difficulty reading, identifying familiar objects or persons, driving, manipulating objects, dressing, or judging distances. At the Cleveland Center twenty out of one hundred and sixty AD patients were experiencing visual hallucinations and had more behavioral symptoms (Lerner et al., 1994). There is evidence that AD patients with prominent visual symptoms show more deficits in the visual–spatial domain (implicating the dorsal, occipito-parietal pathway devoted to spatial localization) (Butter et al., 1996; Hof & Bouras, 1991; Hof et al., 1989, 1990; Kiyosawa et al., 1989; Pietrini et al., 1996; Trick et al., 1989) but do not differ from patients without visual symptoms in form processing (dependent on the ventral, occipito-temporal object-recognition pathway) (Butter et al., 1996). These observations underscore the importance of separating our participants according to their visual complaints to give us sufficient sensitivity in identifying the visual processing deficits associated with the disease.

Apolipoprotein epsilon genotype may be related to visual and cognitive performance. Genotype is defined by the load of the epsilon 4 (E4) allele, the presence of which describes the risk of developing AD (Roses, 1997). We expect that genotype may predict the success of the visual manipulations on cognitive enhancement, based on studies linking E4 load to cognition and functional abilities in AD (e.g., Henderson et al., 1995) and nondemented elderly adults (Albert, Gurland, Maestre, Jacobs, Stern, & Mayeux, 1995; Blesa, Adroer, Santacruz, Ascaso, Tolosa, & Oliva, 1996; Bondi et al., 1995). There are no data from which to infer that E4 load will likewise predict visual performance. However, because visual deficits in AD result from neuropathological changes that are similar in kind, if not in site, to the changes driving cognitive dysfunction, we tentatively predict a correlation of epsilon 4 load with visual as well as cognitive deficits. In our current work, we are investigating the role of E4 genotype in visual and cognitive performance in AD.

### A note on procedures

The cognitive dysfunction of AD is reflected in impairments in memory, attention, spatial function, language, and other specific domains (Cronin-Golomb, 2001; Cronin-Golomb & Amick, 2001; Cronin-Golomb, Corkin, et al., 1993). In order for us to understand visual changes in this disorder, participants in our studies must be able to meet the cognitive demands of the vision tasks. Although many tests provide sensitive measures of thresholds, they may coincidentally require sustained attention, good working memory, and other feats of cognition that may be beyond what can be asked of a demented individual. Our experience and that of other investigators dictates that we seek out or design vision tests that most individuals with AD understand how to perform and are able to perform at an optimal level. Often, forced-choice designs are preferable to designs such as method-of-adjustment, which are subject to variability in even a normal observer's judgment of threshold (e.g., Higgins, Jaffe, Coletta, Caruso, & de Monasterio, 1984); such variability may be exacerbated in AD. Forced-choice procedures with few response alternatives, long stimulus-presentation times, and other aspects of design geared toward a reduction in task complexity help to maximize the ability to perform reliably and at an optimal level.

## CONTRAST SENSITIVITY

Deficits in spatial frequency contrast sensitivity in AD have been reported in a number of investigations (Lakshminarayanan, Lagrave, Kean, Dick, & Shankle, 1996; Nissen, Corkin, Buonanno, Growdon, Wray, & Bauer, 1985), including our studies (Cronin-Golomb et al., 1991; Cronin-Golomb et al., 1995; Gilmore & Levy, 1991; Mendola et al., 1995; Neargarder et al., in press). Decrements in contrast sensitivity in humans occur with lesions of the occipital, temporal, and parietal cortices (Bodis-Wollner, 1972, 1976; Bodis-Wollner & Diamond, 1976; Bulens, Meerwaldt, VanderWildt, & Keemink, 1989; Kobayashi, Mikuno, Ishikawa, & Tasaki, 1985; Nissen et al., 1985). The widespread pathology of the extrastriate visual, temporal, and parietal cortices in AD, as well as the noted reduction in oxidative capacity in both blobs (low spatial-frequency processing) and interblobs (high spatial-frequency processing) in the primary visual cortex (Wong-Riley et al., 1997) presumably result in the observed changes in contrast sensitivity throughout the frequency range.

Differences in the cognitive demands or other aspects of the measures used in the various studies are likely to underlie differences in individual participants' ability to perform reliably. We found that on a contrast sensitivity test in which participants were instructed to remember the relation between two consecutive screen displays and tones, mildly demented AD patients were impaired, and more demented patients were unable to perform the task (Cronin-Golomb et al., 1991). In comparing two tests of contrast sensitivity that appeared to make low cognitive demands, we found that fewer participants were able to perform the computerized task than the task in a chart format, presumably because of differences in task duration and motor requirements (Gilmore and Levy, 1991).

However, two key findings strongly suggest that impaired performance on contrast sensitivity is not primarily a result of general cognitive decline, but instead indicates genuine perceptual dysfunction. First, we have reported that contrast sensitivity performance across the frequency range does not necessarily correlate with dementia severity (Cronin-Golomb et al., 1991; Mendola et al., 1995; Neargarder et al., in press). Second, we have found an increased prevalence of deficits on measures of low—relative to high—spatial frequencies in AD patients (Mendola et al., 1995), pointing to an unequal vulnerability to the AD process across spatial frequencies. The correlation between contrast sensitivity performance and dementia severity as assessed with the Information, Memory and Concentration subtest of the Blessed Dementia Scale was significant at the two lowest frequencies ($r = 0.73$ and 0.81 at 0.5 and 1.0 cpd, respectively) but not at the three higher frequencies, in our study of nine patients with AD (Cronin-Golomb et al., 1991) and in our larger study of sixteen patients with AD ($r = 0.61$ and 0.69 at 0.5 and 1.0 cpd, respectively) (Mendola et al., 1995).

We interpret the correlation with dementia severity to indicate that the measures of low spatial frequency contrast sensitivity are sensitive to cognitive changes over time, whereas performance at higher frequencies is relatively stable over time regardless of changes in dementia severity. This interpretation is supported by the findings of the Cleveland laboratory. Although elderly adults have poorer contrast sensitivity at high spatial frequencies than do young adults (Cronin-Golomb et al., 1991; Owsley, Sekuler, & Siemsen, 1983), we reported that spatial contrast sensitivity was stable for healthy elderly adults over a one-year interval (Gilmore & Whitehouse, 1995). AD patients experienced a drop in sensitivity over as little as six months in their response to low spatial frequencies that were presented abruptly. This finding demonstrates that the visual response capacity of AD patients does change in a relatively short period of time in the range of spatial frequencies that have been implicated in accurate face and object perception. Further, the change in response to abruptly occurring visual events implicates the M-cell stream of processing as a locus of the processing deficit.

It may be argued that poor spatial-contrast sensitivity creates weak visual signals that are ineffectively processed by the higher order visual areas. Indeed, Regan, Raymond, Grinsburg and Murray (1981) argued that a contrast-sensing deficit could masquerade as a higher-order disability, such as object recognition. Work by Owsley and her colleagues has substantiated this point. They demonstrated that a portion of the problem that elderly adults have with face perception can be linked to their low contrast sensitivity (Owsley, Sekuler & Boldt, 1981). Furthermore, the detection and identification of signs and other objects is related directly to the contrast sensitivity of adults (Owsley & Sloane, 1987).

The relatively greater loss of spatial contrast sensitivity by AD patients (Cronin-Golomb et al., 1991, 1995; Gilmore & Levy, 1991; Gilmore & Whitehouse, 1995; Lakshminarayanan et al., 1996; Mendola et al., 1995; Neargarder et al., in press; Nissen et al., 1985; Sadun, Borchert, DeVita, Hinton, & Bassi, 1987; Wright et al., 1987) suggests that they may have problems with reading and recognition tasks independent of their memory-related deficits. Difficulty with reading and the recognition of objects is a common complaint of persons who have been diagnosed with AD. Whereas AD patients are able to read aloud accurately (Crawford, Hart, & Nelson, 1990; Cummings, Houlihan, & Hill, 1986), their reading speed (Nebes, Martin, & Horn, 1984) and comprehension (Cummings et al., 1986) are poor.

Likewise, AD patients perform poorly at confrontation naming tasks (Appel, Kertesz, & Fishman, 1982). Analysis of the linguistic content of the naming errors has led a number of investigators to conclude that semantic processing problems are the major source of the deficit (e.g. Bayles & Kaszniak, 1987; Bowles, Obler, & Albert, 1987). However, there is evidence that AD patients are quite sensitive to the perceptual quality of a stimulus. The closer the representation of a stimulus is to its real world attributes, the better the recognition performance of AD patients. For example, AD patients do well with real three-dimensional objects and colored pictures but have difficulty with line drawings (Bisiach, 1966; Kirshner, Webb, & Kelly, 1984). The literature on confrontation naming was conducted under the then-prevailing assumption that AD did not involve a primary visual deficit. Recent empirical evidence reviewed in this application has called that assumption into question. Since AD patients require a stronger signal to detect an object, it may be argued that the presentation of normal contrast stimuli results in a weak proximal representation in the visual system of the AD patients. That is, one source of the naming problems exhibited by AD patients is the weak signal carried by their visual system.

In sum, because of the AD patients' good acuity, it has been assumed that their poor performance on reading, recognition, and confrontation naming tasks is due to a linguistic and/or memory deficit rather than a visual perception problem. We have challenged this position, arguing that the very poor spatial-contrast sensitivity of the AD patients can be expected by itself to lead to poor performance on any task or endeavor involving the perception of visual stimuli.

Our position is supported by our recent studies. The data from the first set of studies are shown in Table 1.1. We found that both healthy young and elderly adults did poorly on an object-naming task (Gilmore, Turner, et al., 1995; Lustig et al., 1995) and a pattern recognition test (Spinks et al., 1996; Spinks, Gilmore, Thomas, & Hinman, 1998) when they were presented with pictures or patterns that had been degraded to simulate the visual experience of AD patients. The degree of degradation was dictated by the empirically observed contrast-sensitivity deficit of AD patients. That is, the simple contrast degradation of the stimuli led the healthy adults to exhibit Alzheimer-like test behavior. Furthermore, the presentation of contrast-enhanced line drawings boosted the accuracy of the AD patients to the same level as the elderly adults who viewed the normal contrast stimuli (Gilmore, Turner, et al., 1996). Thus, the appropriate adjustment of the stimulus contrast eliminated the AD participants' object-naming deficit. This is an important finding because it suggests that the universally poor performance of AD patients in confrontation tasks may have a strong basis in their visual disorder rather than in a cognitive dysfunction.

In the second study, AD patients read briefly presented letters and sentences as quickly as healthy elderly adults when the stimuli had a sufficiently high contrast (Gilmore, Thomas, et al., 1996; Gilmore, Townsend, Lustig, & Morrison, 1998). Compensating for the weak contrast sensitivity of the AD patients by presenting very strong stimuli eliminated their letter-reading speed deficit. Thus, the studies with AD patients clearly illustrate the powerful impact of a contrast sensitivity deficit on higher order tasks and demonstrate that a manipulation of contrast can be used to improve cognitive performance.

We have also conducted an examination of the effect of impaired contrast sensitivity on face discrimination in AD (Brown et al. 1996; Cronin-Golomb et al., 2000). Face discrimination is dependent upon normal contrast sensitivity at low spatial frequencies expressed in cycles per face (cpf) (Costen, Parker, & Craw, 1994, 1996; Owsley et al., 1981). The ability to discriminate faces often is affected in AD and may seriously affect interpersonal relations for patients and their caregivers. A possible source of this impairment is the decline in contrast sensitivity, expressed in cycles per degree (cpd), seen throughout the frequency range in AD but most prevalent at low frequencies. We and others have previously shown that contrast sensitivity is abnormal in AD, especially at the low frequencies. We tested the hypothesis that face discrimination in AD is compromised by the contrast-sensitivity deficit at low frequencies specifically. The participants included eighteen patients with AD and eighteen healthy elderly adults matched to the AD groups for age, education, and central visual acuity (median 20/25 binocular). Probable AD was diagnosed according to standard guidelines, and patients were mildly to moderately demented.

**Table 1.1**
Accuracy (%) in Picture Naming Task by Healthy Young and Elderly Adults and Persons Diagnosed with Probable Alzheimer's Disease (AD) under Three Levels of Stimulus Contrast

| Group | Enhanced | Normal | Degraded |
|---|---|---|---|
| Young | 97 | 93 | 84 |
| Elderly | 94 | 88 | 69 |
| AD | 87 | 80 | --* |

*Participants with AD were not tested in the Degraded condition.

The use of images enhanced by computer at particular cpf can result in nonlinear distortions in the interactions of the various frequencies in the display and in the visual system. In order to overcome this problem, we presented normal images from the Benton Facial Recognition Test (Benton & Van Allen, 1968) at three sizes to enhance 3.5, 11, and 27 cpf, using the natural cpd filter response of the visual system. The three sets of trials were matched for difficulty. All faces were presented at 61 centimeters. Participants were instructed to match a target face with the identical face or faces of a six-choice array.

Enlarging the stimulus brought the high cpf to peak cpd response, and reducing the size of the stimulus brought the low cpf to peak cpd response. Peak response for face discrimination is approximately between 3 and 6 cpd (Costen et al., 1994; Peli, Lee, Trempe, Buzney, 1994). The first set of trials was scaled to a width of 1.25 centimeter so that the visual system was most sensitive to low frequencies (3.5 cpf). The second set was scaled to 3.8 centimeters for greatest sensitivity to middle frequencies (11 cpf). The third set was scaled to 9.7 centimeters for greatest sensitivity to high frequencies (27 cpf). The visual filter was calibrated using Vistech contrast-sensitivity charts. As has been reported in previous studies, we found that contrast sensitivity of the AD patients was significantly impaired relative to the elderly control group at the two lowest spatial frequencies of 1.5 and 3.0 cpd.

When middle or high spatial frequencies (11 and 27 cpf) of the faces were enhanced, the patients performed the matching task significantly more poorly than did the control group. However, when low spatial frequencies were enhanced (3.5 cpf), the AD and control groups performed equivalently. The results indicate that AD patients benefit from low-frequency enhancement of contrast sensitivity, which serves to counteract the low-frequency loss that is often seen in this disease.

To explain why a reduction in face size led to better performance in the Boston study, the Cleveland group developed a demonstration using a band-pass filter on pictures of faces that were 4 centimeters wide and 1.25 centimeters wide. The filter passed the frequencies between 3 and 6 cpd. The outline of the face and its features could be seen clearly in the small image but only dimly in the larger image. Thus, an observer who was relying on the 3–6 cpd range would have very little data to use in the larger display. The "face information" evident in the small display is shifted into a lower frequency range in the large display (approximately .9–1.9 cpd). The control participants, who were reasonably sensitive in this range of spatial frequencies, could still use those data. However, because the patients with AD were lower in sensitivity in that range, they could not use that information and would be expected to perform more poorly.

To argue that the visual deficits of both healthy elderly adults and AD patients may influence their high-order information processing is consistent with current theories of information processing. Murdock (1989) and Theios and Amrhein (1989 a,b) have incorporated the quality of stimulus encoding in their recognition–memory and picture–word processing models, respectively. It is clear that manipulations of physical characteristics of the stimulus, such as its intensity (contrast, luminance) (Loftus & Ruthruff, 1994), size or spatial-frequency content (Theios & Amrhein, 1989a), or its duration (Busey & Loftus, 1994; Loftus, 1974; Loftus & Ruthruff, 1994; Yonelinas, Hockley, & Murdock, 1992), can have direct impacts on the processing speed and accuracy of young observers whose visual systems are intact. The degradation of a stimulus may have both a main effect on performance by slowing sensory acquisition and encoding processes and an interactive effect with higher-order decision and processing stages (den Heyer & Benson, 1988; Norris, 1984). Such effects may account for the strong interactions between age, sensory status and cognitive abilities that have been reported for several sensory modalities (Baltes & Lindenberger, 1997; Lindenberger & Baltes, 1994; Salthouse, Hancock, Meinz, & Hambrick, 1996; Stevens, Cruz, Marks, & Lakatos, 1998). It is suggested here that if manipulations of the distal stimuli can influence information processing, then alterations of the proximal stimulus by a deficient visual system can create similar processing burdens.

Finally, we recently completed our first intervention studies relating visual contrast to everyday functioning in demented individuals. The interventions consisted of increasing light intensity and enhancing visual stimulation (i.e., providing tableware with maximal visual contrast) during evening meals. In the first study (Koss & Gilmore, 1998), two dependent variables were studied: daily amount of food intake and frequency of agitated behaviors. Agitated behavior in patients with AD and related disorders has been reported to worsen at night or near sunset. This pattern of behavior, described as "sundowning," shows considerable interpatient variability, with a prevalence of 12 percent to 50 percent. Little is known about relevant patient-related factors. Though sundowning may be related to the circadian timing system, an alternate explanation may be that agitation is triggered by a decrease in ambient light and that the patients' maladaptive behavior may be understood as impaired visual apprehension of their surroundings. We observed thirteen residents in a high-functioning long-term care dementia unit for three consecutive periods of twenty-one days each, divided into baseline, intervention and post-intervention. Daily amount of food intake and frequency of agitated behaviors were extracted from the residents' charts, as recorded by nursing staff. The amount of food ingested at dinner was significantly different across baseline, intervention, and post-intervention ($p < .01$), with more food consumed during intervention. No significant differences were found for amounts of food ingested during lunch, where no intervention was implemented. Findings concerning agitation were even more striking. Most agitated behaviors occurred during the evening shift, after the last meal. The frequency of agitated behaviors decreased dramatically during the intervention phase ($p = 0.01$).

In the second study, it was found that enhanced color contrast of tableware (bright red plates and cups substituted for institutional white) resulted in significant increases in the volume of food and liquid ingested relative to baseline in nine severely demented inpatients. The volume ingested declined again in the post-intervention (white) phase. The pattern of performance held for each of the nine participants (Dunne & Neargarder, 1999; Dunne et al., submitted).

These preliminary findings are encouraging. They indicate that simple, easy to implement environmental interventions may have an impact on food intake and agitation in long-term care. The findings suggest that at least some of the behavioral disturbances in demented patients may be attributable to visual functions that have received relatively little attention in the research and clinical literature.

## BACKWARD MASKING

Besides the following studies, which showed a high prevalence of deficient masking (Cronin-Golomb et al., 1991, 1995; Mendola et al., 1995), there has been little work on backward masking in AD. All studies that have examined pattern masking in AD have reported a deficit (Coyne, Liss, & Geckler, 1984; Miller, 1977; Schlotterer, Moscovitch, & Crapper-McLachlan, 1984). The masking deficit for letter stimuli is supported by evidence that AD patients require longer exposure time than healthy elderly adults to identify briefly presented words (Keane, Gabrieli, Fennema, Growdon, & Corkin, 1991) and form degraded or incomplete percepts when comparing pairs of letters (Allen, Namazi, Patterson, Crozier, & Groth, 1992). Masking involves both retinal and cortical processes (Bowen & Wilson, 1994), with cortical processing demonstrated by orientation and spatial-frequency selectivity (Bennett & Cortese, 1996; Bowen & Wilson, 1994), modulation by voluntary visual attention (Green, Nuechterlein, & Mintz, 1994; Ramachandran & Cobb, 1995) and by degree of semantic relatedness of the target and mask (Curry, Hung, Wilder, & Julesz, 1995). The mechanism is as yet poorly understood, with some investigators arguing for impairments in the dorsal processing stream (Husain, Shapiro, Martin, & Kennard, 1997; Saccuzzo, Cadenhead, & Braff, 1996) and others showing, in macaques, firing of neurons in the inferior temporal cortex and in the superior temporal sulcus, implicating the ventral processing stream (Rolls and Tovee, 1994; Rolls, Tovee, Purcell, Stewart, & Azzopardi, 1994). When the target or mask consists of patterned elements, including letters, occipito-temporal cortex is activated (Petersen et al., 1990). In AD, masking performance worsens with increased dementia severity, presumably reflecting progressive dysfunction of the devoted visual-processing pathways (Cronin-Golomb et al., 1991, 1995).

Of all tests of basic visual capacities that we have conducted, backward pattern masking stands out as the capacity that is most subject to impairment in AD, with the prevalence of significant deficit found to be 58 percent in a sample of fifty-six patients with AD compared with twenty-seven healthy elderly adults (Mendola et al., 1995). Pattern masking is also noteworthy for being the best predictor of impairments in visual cognition in AD of eight tests of basic vision, explaining 25 to 50 percent of variance in performance on tests of word reading, object recognition, pattern completion, complex figure copying, and arrangements of pictures in a logical storyline (Cronin-Golomb et al., 1995).

The correlation between test performance and dementia severity was significant in our study of thirty-one patients with AD (Cronin-Golomb et al., 1991), but was not significant in our larger study of fifty-five patients with AD (Mendola et al., 1995). The discrepancy in results may be attributable to the presence in the latter study of the subgroup of fourteen patients with exceptionally impaired masking performance who were in the same range of dementia severity as the rest of the AD sample, again pointing to the value of identifying subgroups of AD patients in terms of visual performance.

In our study that included young adults as well as healthy elderly and AD groups (Cronin-Golomb et al., 1991), using the same tests as in the Mendola et al. study above, we found that the young adults outperformed the elderly adults ($p = .06$). It has been well established that elderly adults are more susceptible than are young adults to interference from visual stimuli that briefly follow a target visual event (e.g., Walsh, Till, & Williams, 1978). Older adults require a longer period of time between the target and mask stimuli in order to accurately identify the target. This finding has been interpreted commonly as strong evidence that older adults are slow in information processing. Recent research on masking with healthy adults by our laboratory in Cleveland has called the cognitive-slowing interpretation into doubt. We argue that the age-related deficit in backward masking tasks may be related to a change in the sen-

**Figure 1.1**
Target Identification Accuracy in Masking Task by Three Groups of Healthy Adults Stratified by Age

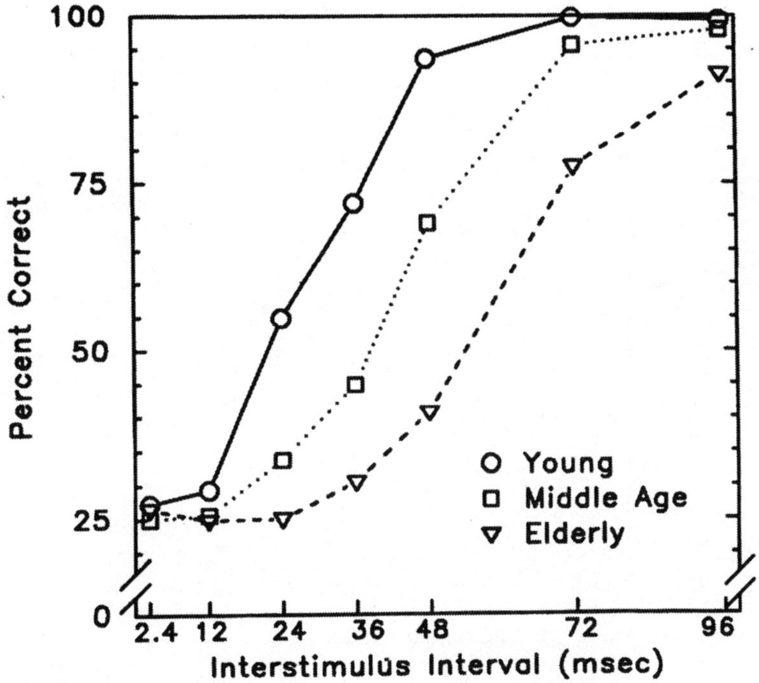


sory response capability of the visual system in older adults and not necessarily a direct change in cognitive processing speed.

In two experiments, we investigated the role of target luminance on the performance of healthy adults (Gilmore, Seoane, Thomas, & Xue, 1995). In the first experiment, fifty-seven persons from nineteen to seventy-five years of age participated in a target masking study that was modeled closely on the study by Walsh, Till, & Williams (1978). Figure 1.1 shows the target identification accuracies of young, middle-aged, and elderly adults at a range of target and mask interstimulus intervals (ISI). This finding closely replicates the results of Walsh et al. (1978) and demonstrates that middle-aged adults yield intermediate performance levels. The performance functions in Figure 1.1 show that the older participants required more time—a longer ISI—to achieve a high level of accuracy in the target identification task.

To examine the relation between age and critical processing time more directly, a critical ISI was calculated for each observer. The critical ISI was the amount of time needed between the target and mask for the participant to achieve 75 percent accuracy. Figure 1.2 presents the critical ISIs for each participant and illustrates that this dependent variable was highly correlated (*r* [55]

**Figure 1.2**
Age and the Critical Interstimulus Interval (ISI) for Achieving 75% Accuracy for Each Participant

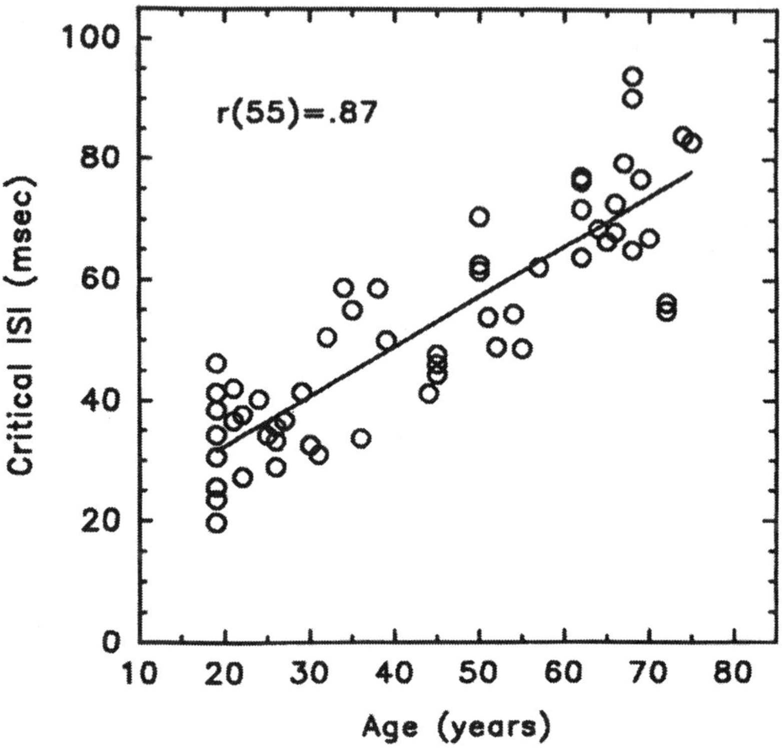


= .87) with the age of the participant. Thus, the factors that dictate the increase in critical ISI change throughout the adult life span.

It has been well established that light sensitivity declines starting in the third decade of life (McFarland, Domey, Warren, & Ward, 1960; McFarland & Fisher, 1955). Indeed, the relation between maximum light sensitivity and adult age is $r = .895$ (McFarland & Fisher, 1955). We hypothesized that the performance of adults in the masking task may be mediated by their sensitivity to light and not a higher-order slowing in cognitive operations per se. To examine this hypothesis, a second experiment was conducted in which a sample of twelve young and twelve elderly adults were asked to identify a target with a mask following at 48 milliseconds. The latter value was chosen on the basis of Experiment 1 to yield an appropriate range of performance from young and elderly participants. An adaptive staircase procedure was used to determine the target luminance required by each participant to achieve 75 percent target accuracy. Contrast of the target of course does change under this procedure. However, because the back-

**Figure 1.3**
Critical Luminance Level for Achieving 75% Accuracy for Each Participant

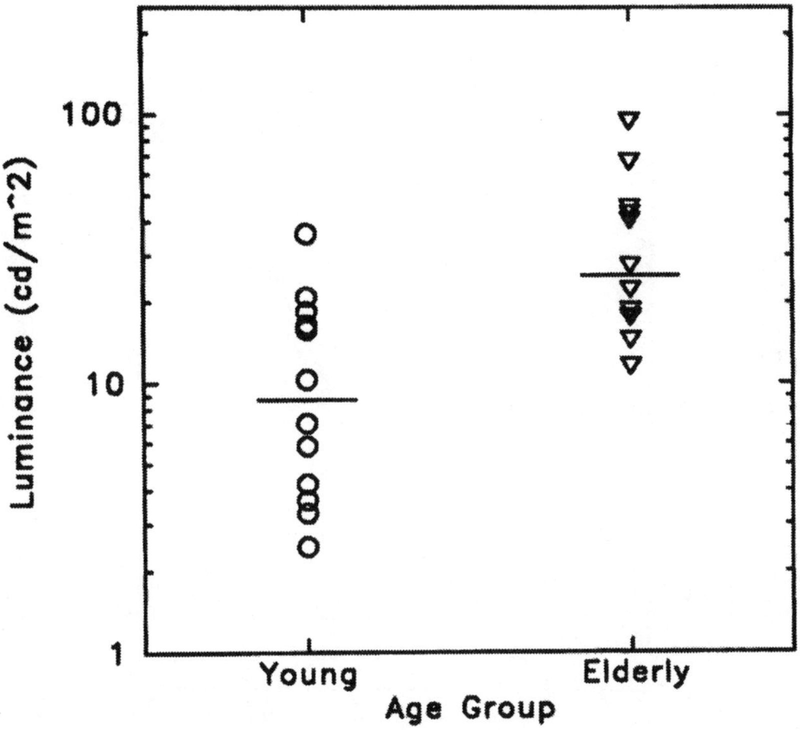


ground was kept at a constant very low level of luminance, the contrast variation was within 96 percent and 99 percent. Thus, the more relevant stimulus characteristic of the changing target was its luminance.

The critical target luminances of each participant are given in Figure 1.3. It is important to note that the criterion performance was achieved by each participant. Second, while there is noteworthy overlap in critical luminance for the two groups, the older adults did require a higher level of target luminance to achieve the criterion accuracy.

The second phase of Experiment 2 was to establish the masking function for each participant using their individually determined critical luminance level. In Figure 1.4 the performance functions of the two age groups are shown. It is clear that the adjustment of the target luminance eliminated the age effects that were so apparent in Figure 1.1. This study demonstrates that the age-related change in target processing in a backward masking task can be accounted for entirely by the participants' response to target luminance. When the latter variable is adjusted, no age differences in processing speed remain. Thus, rather

**Figure 1.4**
Target Identification Accuracy in Backward Masking Task by Young and Elderly Groups of Healthy Adults for Whom the Target Luminance Was Individually Determined

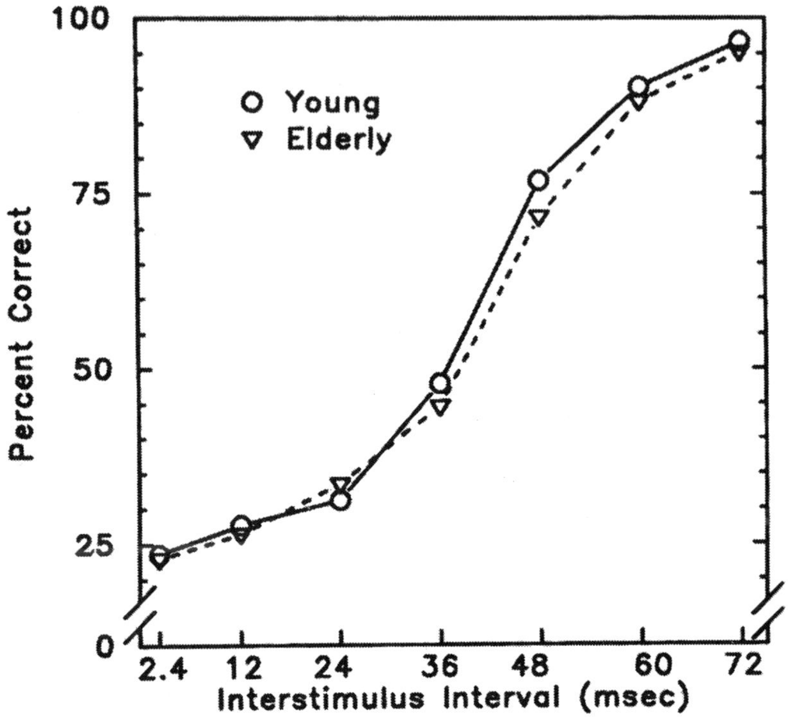


than demonstrating the existence of cognitive slowing in older adults, it may be argued that backward masking tasks reveal the impact of sensory deficits on higher-order tasks. This result is similar in interpretation to that of Raz, Millman, & Moberg (1990), who found in an auditory masking experiment that increased stimulus persistence rather than cognitive slowing accounted for age-related differences in masking performance.

A final relevant point for the present project is demonstrated in Figure 1.5. For the small samples of young and elderly participants, critical target luminance in Experiment 2 was correlated with contrast sensitivity through a range of spatial frequencies. The peak correlation for each group was between the critical target luminance and contrast sensitivity for 1-cpd displays. While the samples were small, a cautious observation may be made that the response of the low spatial-frequency channels is the best predictor of performance in a target masking study. This finding is consistent with the report of Mendola et al.

**Figure 1.5**
Log Contrast Sensitivity for 1 cpd and Log Critical Luminance for Target Identification in Experiment 2

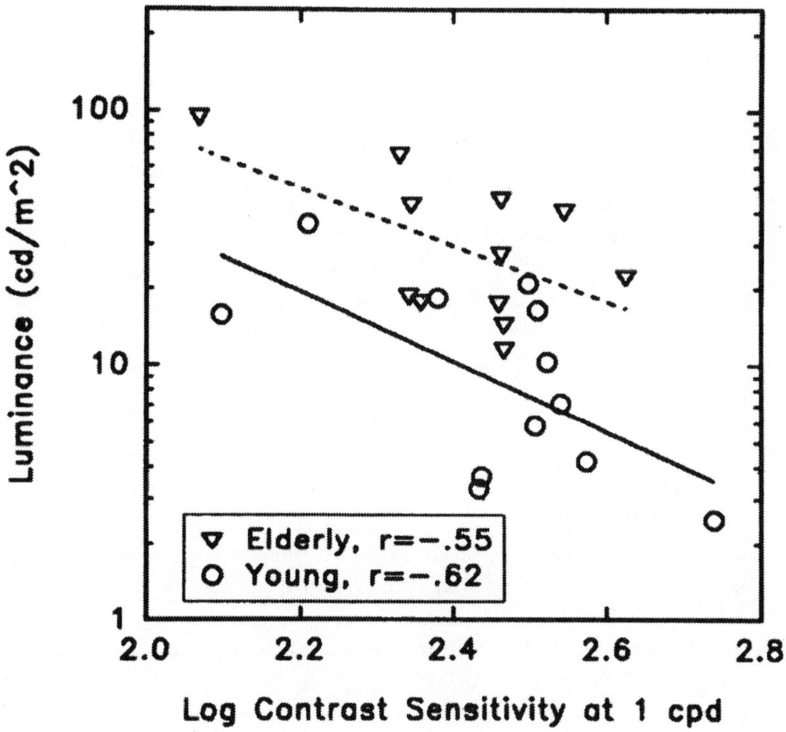


(1995) that AD patients who are quite impaired in low spatial-frequency contrast sensitivity also do poorly in a backward masking task.

To date we have tested three persons with probable AD with the adaptive luminance procedure of the second experiment. The key finding of this pilot test was that each of the participants was able to perform the task. That is, they were able to achieve the criterion accuracy level for a fixed ISI using the adaptive luminance procedure. Furthermore, when the adaptive target luminance was used in the second phase of the experiment to establish a masking function, each AD participant yielded performance functions that overlapped that of the healthy adults. The pilot data support the notion that the masking problem of AD patients can be linked directly to a sensory-processing deficit rather than a higher-order information-processing problem. This preliminary finding also suggests that the cognitive tasks that we have shown to be highly correlated with masking performance may themselves be profoundly affected by the sensory-processing deficit. Work in our laboratories is now addressing this hypothesis.

## SUMMARY

Taken together, the studies by our laboratories and others demonstrate that manipulations of basic visual variables may help to ameliorate some of the cognitive dysfunction characteristic of AD. Specifically, we have determined that contrast sensitivity enhancement may improve performance on tests of reading, face discrimination, object recognition, and object naming in AD, as well as improving everyday function in the areas of food intake and agitated behavior. We have found that backward masking performance is a strong predictor of reading ability as well as object naming and recognition in AD, and have preliminary evidence to suggest that a deficiency in luminance sensitivity may be one source of impaired masking performance.

New information about the sensory and perceptual deficits of healthy elderly adults and persons diagnosed with AD is the first step toward the development of nonpharmacological methods of visual intervention by which their cognitive performance can be improved. Clinicians and researchers may be able to enhance cognitive performance through interventions aimed at restoring deficient visual capacities. For example, enhancing contrast sensitivity at specific spatial frequencies as well as enhancing stimulus strength and persistence by identifying optimal stimulus size, distance, luminance, and duration, may normalize performance of AD patients on cognitive tasks such as reading, face discrimination, memory, and object naming, and may improve aspects of everyday behavior such as amount of food intake and frequency of agitation. By bridging the findings of clinical deficits to innovative and effective intervention strategies, we have the hope of obtaining new insights into improving cognition and hence the quality of life of normal elderly adults and especially individuals with AD.

## NOTE

We thank Nassim Behi, Elisabeth Koss, Sarah Morrison, and Nathan Weisz for their comments on an earlier draft of this chapter.

## REFERENCES

Aguglia, U., Gambarelli, D., Farnarier, G., & Quattrone, A. (1991). Different susceptibilities of the geniculate and extrageniculate visual pathways to human Creutzfeldt-Jakob disease (a combined neurophysiological and neuropathological study). *Electroencephalography and Clinical Neurophysiology, 78,* 413–423.

Albert, S.M., Gurland, B., Maestre, G., Jacobs, D.M., Stern, Y., & Mayeux, R. (1995). APOE genotype influences functional status among elderly without dementia. *American Journal of Medical Genetics, 60,* 583–587.

Allen, P.A., Namazi, K.H., Patterson, M.B., Crozier, L.C., & Groth, K.E. (1992). Impact of adult age and Alzheimer's disease on levels of neural noise for letter matching. *Journal of Gerontology, 47,*344–349.

Appel, J., Kertesz, A., & Fisman, M. (1982). A study of language functioning in Alzheimer patients. *Brain and Language, 17,* 73–91.

Arnold, S.E., Hyman, B.T., Flory, J., Damasio, A.R., & Van Hoesen, G.W. (1991). The topographical and neuroanatomical distribution of neurofibrillary tangles and neuritic plaques in the cerebral cortex of patients with Alzheimer's disease. *Cerebral Cortex, 1,* 103–116.

Bäckman, L., & Herlitz, A. (1990). The relationship between prior knowledge and face recognition memory in normal aging and Alzheimer's disease. *Journal of Gerontology, 45,* 94–100.

Baltes, P.B. & Lindenberger, U. (1997). Emergence of a powerful connection between sensory and cognitive function across the adult lifespan: A new window to the study of cognitive aging? *Psychology and Aging, 2,* 12–21.

Bayles, K.A., & Kaszniak, A.W. (1987). *Communication and cognition in normal aging and dementia.* New York: Little, Brown.

Beach, T.G., & McGeer, E.G. (1988). Lamina-specific arrangement of astrocytic gliosis and senile plaques in Alzheimer's disease. *Brain Research, 463,* 357–361.

Becker, J., Lopez, O., & Boller, F. (1995). Understanding impaired analysis of faces by patients with probable Alzheimer's disease. *Cortex, 31,* 129–137.

Bennett, P.J., & Cortese, F. (1996). Masking of spatial frequency in visual memory depends on distal, not retinal, frequency. *Vision Research, 36,* 233–238.

Benton, A.L., & Van Allen, M.W. (1968). Impairment in facial recognition in patients with cerebral disease. *Cortex, 4,* 344–358.

Bisiach, E. (1966). Perceptual factors in the pathogenesis of anomia. *Cortex, 2,* 90–95.

Blanks, J.C., Torigoe, Y., Hinton, D.R., & Blanks, R.H.I. (1991). Retinal degeneration in the macula of patients with Alzheimer's disease. *Annals of the New York Academy of Sciences, 640,* 44–46.

Blanks, J.C., Schmidt, S.Y., Torigoe, Y., Porrello, K.V., Hinton, D.R., & Blanks, R.H.I. (1996). Retinal pathology in Alzheimer's disease. II. Regional neuron loss and glial changes in GCL. *Neurobiology of Aging, 17,* 385–395.

Blanks, J.C., Torigoe, Y., Hinton, D.R., & Blanks, R.H.I. (1996). Retinal pathology in Alzheimer's disease. I. Ganglion cell loss in foveal/parafoveal retina. *Neurobiology of Aging, 17,* 377–384.

Blesa, R., Adroer, R., Santacruz, P., Ascaso, C., Tolosa, E., & Oliva, R. (1996). High apolipoprotein epsilon E4 allele frequency in age-related memory decline. *Annals of Neurology, 39,* 548–551.

Bodis-Wollner, I. (1972). Visual acuity in patients with cerebral lesions. *Science, 178,* 769–771.

Bodis-Wollner, I. (1976). Vulnerability of spatial frequency channels. *Nature, 261,* 309–311.

Bodis-Wollner, I., & Diamond, S.P. (1976). The measure of spatial contrast sensitivity in cases of blurred vision associated with cerebral lesions. *Brain, 99,* 695–710.

Bondi, M.W., Salmon, D.P., Monsch, A.U., Galasko, D., Butters, N., Klauber, M.R., Thal, L.J., & Saitoh, T. (1995). Episodic memory changes are associated with the APOE-epsilon 4 allele in nondemented older adults. *Neurology, 45,* 2203–2206.

Bouros, C., Hof, P.R., Giannakopoulas, P., Michel, J.P., & Morrison, J.H. (1994). Regional distribution of neurofibrillary tangles and senile plaques in the cerebral cortex of elderly patients: A quantitative evaluation of a one-year autopsy population from a geriatric hospital. *Cerebral Cortex, 4,* 138–150.

Bowen, R.W., & Wilson, H.R. (1994). A two-process analysis of pattern masking. *Vision Research, 34,* 645–657.

Bowles, N.L, Obler, L. K., & Albert, M. (1987). Naming errors in healthy aging and dementia of the Alzheimer type. *Cortex, 23,* 519–524.

Braak, H., Braak, E., & Kalus, P. (1989). Alzheimer's disease: Areal and laminar pathology in the occipital isocortex. *Acta Neuropathologica, 77,* 494–506.

Brandt, J., Mellits, E.D., Rovner, B., Gordon, B., Selnes, O.A., & Folstein, M. (1989). Relation of age of onset and duration of illness to cognitive functioning in Alzheimer's disease. *Neuropsychiatry, Neuropsychology, and Behavioural Neurology, 2,* 93–101.

Bulens, C., Meerwaldt, J.D., Van Der Wildt, G.J., & Keemink, C.J. (1989). Spatial contrast sensitivity in unilateral cerebral ischaemic lesions involving the posterior visual pathway. *Brain, 112,* 507–520.

Busey, T.A., & Loftus, G.R. (1994). Sensory and cognitive components of visual information acquisition. *Psychological Review, 101,* 446–469.

Butter, C.M., Trobe, J.D., Foster, N.L., & Berent, S. (1996). Visual–spatial deficits explain visual symptoms in Alzheimer's disease. *American Journal of Ophthalmology, 122,* 97–105.

Celesia, G.G., Villa, A.E.P., Brigell, M., Rubboli, G., Bolcioni, G., & Fiori, M.G. (1993). An electrophysiological study of visual processing in Alzheimer's disease. *Electroencephalography and Clinical Neurophysiology, 87,* 97–104.

Chui, H.C., Teng, E.L., Henderson, V.W., & Moy, A.C. (1985). Clinical subtypes of dementia of the Alzheimer type. *Neurology, 35,* 1544–1550.

Costen, N.P., Parker, D.M., & Craw, I. (1994). Spatial content and spatial quantisation effects in face recognition. *Perception, 23,* 129–146.

Costen, N.P., Parker, D.M., & Craw, I. (1996). Effects of high-pass and low-pass spatial filtering on face identification. *Perception and Psychophysics, 58,* 602–612.

Coyne, A.C., Liss, L., & Geckler, C. (1984). The relationship between cognitive status and visual information processing. *Journal of Gerontology, 39,* 711–717.

Crawford, J.R., Hart, S., & Nelson, H.E. (1990). Improved detection of cognitive impairment with the NART: An investigation employing hierarchical discriminant function analysis. *British Journal of Clinical Psychology, 29,* 239–241.

Cronin-Golomb, A. (2001). Color vision, object recognition, and spatial localization in aging and Alzheimer's disease. In P.R. Hof and C.V. Mobbs (Eds.), *Functional Neurobiology of Aging* (pp. 517–529). San Diego: Academic Press.

Cronin-Golomb, A. (1995). Vision in Alzheimer's disease. *The Gerontologist, 35,* 370–376.

Cronin-Golomb, A. & Amick, M. (2001). Spatial abilities in aging. Alzheimer's disease, and Parkinson's disease. In F. Boller and S. Cappa (Eds.), *Handbook of Neuropsychology,* 2nd edition, Vol. 6 (pp. 119–143).

Cronin-Golomb, A., Corkin, S., & Growdon, J.H. (1995). Visual dysfunction predicts cognitive deficits in Alzheimer's disease. *Optometry and Vision Science, 72,* 168–176.

Cronin-Golomb, A., Corkin, S., Rizzo, J.F., Cohen, J., Growdon, J.H., & Banks, K.S. (1991). Visual dysfunction in Alzheimer's disease: Relation to normal aging. *Annals of Neurology, 29,* 41–52.

Cronin-Golomb, A., Corkin, S., & Rosen, T.J. (1993). Neuropsychological assessment of dementia. In P.J. Whitehouse (Ed.), *Dementia* (pp. 130–164). Philadelphia: F.A. Davis.

Cronin-Golomb, A., Cronin-Golomb, M., Dunne, T., Brown, A., Jain, K., Cipolloni, P.B., & Auerbach, S.H. (2000). Facial frequency manipulation normalizes face discrimination in Alzheimer's disease. *Neurology, 54,* 2316–2318.

Cronin-Golomb, A., Sugiura, R., Corkin, S., & Growdon, J.H. (1993). Incomplete achromatopsia in Alzheimer's disease. *Neurobiology of Aging, 14,* 471–477.

Cummings, J.L., Houlihan, J.P., & Hill, M.A. (1986). The pattern of reading deterioration in dementia of the Alzheimer type: Observations and implications. *Brain and Language, 29,* 315–323.

Curcio, C.A., & Drucker, D.N. (1993). Retinal ganglion cells in Alzheimer's disease and aging. *Annals of Neurology, 33,* 248–257.

Curry, R., Hung, G.K., Wilder, J., & Julesz, B. (1995). Context effect of common objects on visual processing. *Optometry and Vision Science, 72,* 452–460.

Davies, D.C., McCoubrie, P., McDonald, B., & Jobst, K.A. (1995). Myelinated axon number in the optic nerve is unaffected by Alzheimer's disease. *British Journal of Ophthalmology, 79,* 596–600.

Della Sala, S., Muggia, S., Spinnler, H., & Zuffi, M. (1995). Cognitive modeling of face processing: Evidence from Alzheimer's patients. *Neuropsychologia, 33,* 675–687.

den Heyer, K., & Benson, K. (1988). Constraints on the additive relationship between semantic priming and word recognition and on the interactive relationship between semantic priming and stimulus clarity. *Canadian Journal of Psychology, 42,* 399–413.

Dunne, T.E., & Neargarder, S. (1999). Applications of vision research to everyday activity. *Proceedings and Abstracts of the Annual Meeting of the Eastern Psychological Association, 70,* 55.

Dunne, I.E., Neargarder, S., & Cipolloni, P.B. (submitted). The effect of enhanced contrast on nutritional intake in individuals with advanced Alzheimer's disease.

Faber-Langendoen, K., Morris, J.C., Knesevich, J.W., LaBarge, E., Miller, J.P., Berg, L. (1988). Aphasia in senile dementia of the Alzheimer type. *Annals of Neurology, 23,* 365–370.

Filley, C.M., Kelly, J., & Heaton, R.K. (1986). Neuropsychologic features of early- and late-onset Alzheimer's disease. *Archives of Neurology, 43,* 574–576.

Folstein, M., Folstein, S., & McHugh, P.R. (1975). Mini-Mental State: A practical method for grading the cognitive state of patients for the clinician. *Journal of Psychiatric Research, 12,* 189–198.

Gilmore, G.C. (1995). Stimulus encoding in Alzheimer's disease: A multichannel view. In P. Allen & T.R. Bashore (Eds.), *Age differences in word and language processing* (pp. 199–219). Amsterdam: Elsevier.

Gilmore, G.C., Koss, E., Wenk, H.E., & Whitehouse, P.J. (1993). Visual dysfunction in Alzheimer's disease. *Neurology, 43,* A173.

Gilmore, G.C., & Levy, J.A. (1991). Spatial contrast sensitivity in Alzheimer's disease: A comparison of two methods. *Optometry and Vision Science, 68,* 790–794.

Gilmore, G. C., Seoane, M., Thomas, C. W., & Xue, P. (1995). Visual masking and aging: A sensory deficit. Presented at the annual meeting of the Psychonomics Society, Los Angeles, CA.

Gilmore, G.C., Thomas, C.W., Klitz, T., Persanyi, M.W., & Tomsak, R. (1996). Contrast enhancement eliminates letter identification speed deficits in Alzheimer's disease. *Journal of Clinical Geropsychology, 2,* 307–320.

Gilmore, G.C., Townsend, L., Lustig, C., & Morrison, S. (1998). Oral reading speed in Alzheimer's disease. Presented at the Cognitive Aging Conference, April 1998, Atlanta, GA.

Gilmore, G.C., Turner, J.B., Thomas, C.W., Koss, E., Ripich, D., Townsend, L.D., Greene, R., Stuve, T., Morrison, S.R., & Lustig, C. (1996). Picture recognition in Alzheimer's disease: Impact of contrast sensitivity deficit. Presented at the Cognitive Aging Conference, April 1996, Atlanta, GA.

Gilmore, G.C., Turner, J.B., Thomas, C.W., Koss, E., Townsend, L., Greene, R., Morrison, S. & Lustig, C.A. (1995). Impact of Alzheimer-type contrast filter on picture recognition and memory: Simulation and remediation of Alzheimer deficits. Presented at the conference on The Spectrum of Research on Aging and the Aged, November 1995, Case Western Reserve University, Cleveland, OH.

Gilmore, G.C., Wenk, H.E., Naylor, L.A., & Koss, E. (1994). Motion perception and Alzheimer's disease. *Journal of Gerontology, 49,* P52–57.

Gilmore, G.C., & Whitehouse, P. (1995). Contrast sensitivity in Alzheimer's disease: A one year longitudinal analysis. *Optometry and Vision Science, 72,* 83–91.

Grady, C.L., Haxby, J.V., Horwitz, B., Gillette, J., Salerno, J.A., Gonzalez-Aviles, A., Carson, R.E., Herscovitch, P., Schapiro, M.B., & Rapoport, S.I. (1993). Activation of cerebral blood flow during a visuoperceptual task in patients with Alzheimer-type dementia. *Neurobiology of Aging, 14,* 35–44.

Green, M.F., Nuechterlein, K.H., & Mintz, J. (1994). Backward masking in schizophrenia and mania: Specifying a mechanism. *Archives of General Psychiatry, 51,* 939–944.

Haxby, J.V., Grady, C.L., Horwitz, B., Ungerleider, L.G., Mishkin, M., Carson, R.E., Herscovitch, P., Schapiro, M.B., & Rapoport, S.I. (1991). Dissociation of object and spatial visual processing pathways in human extrastriate cortex. *Proceedings of the National Academy of Sciences USA, 88,* 1621–1625.

Haxby, J.V., Horwitz, B., Ungerleider, L.G., Maisog, J.M., Pietrini, P., Grady, C.L. (1994). The functional organization of human extrastriate cortex: a PET–rCBF study of selective attention to faces and locations. *Journal of Neuroscience, 14,* 6336–6353.

Hedges, T.R., Galves, R.P., Spiegelman, D., Barbas, N.R., Peli, E., & Yardley, C.J. (1996). Retinal nerve fiber layer abnormalities in Alzheimer's disease. *Acta Ophthalmologica Scandinavica, 74,* 271–275.

Henderson, A.S., Easteal, S., Jorm, A.F., Mackinnon, A.J., Korten, A.E., Christensen, H., Croft, L., & Jacomb, P.A. (1995). Apolipoprotein E allele epsilon 4, dementia, and cognitive decline in a population sample. *Lancet, 346,* 1387–1390.

Henderson, V.W., Mack, W., & Williams, B.W. (1989). Spatial disorientation in Alzheimer's disease. *Archives of Neurology, 46,* 391–394.

Higgins, K.E., Jaffe, M.J., Coletta, N.J., Caruso, R.C., & de Monasterio, F.M. (1984). Spatial contrast sensitivity: importance of controlling the patient's visibility criterion. *Archives of Ophthalmology, 102,* 1035–1041.

Hinton, D.R., Sadun, A.A., Blanks, J.C., & Miller, C.A. (1986). Optic nerve degeneration in Alzheimer's disease. *New England Journal of Medicine, 315,* 485–487.

Hodges, J., Salmon, D., & Butters, N. (1993). Recognition and naming of famous faces in Alzheimer's disease: A cognitive analysis. *Neuropsychologia, 31,* 775–788.

Hof, P.R., & Bouras, C. (1991). Object recognition deficit in Alzheimer's disease: Possible disconnection of the occipito-temporal component of the visual system. *Neuroscience Letters, 122,* 53–56.

Hof, P.R., Bouras, C., Constantinidis, J., & Morrison, J.H. (1989). Balint's syndrome in Alzheimer's disease: Specific disruption of the occipito-parietal visual pathway. *Brain Research, 493,* 368–375.

Hof, P.R., Bouras, C., Constantinidis, J., & Morrison, J.H. (1990). Selective disconnection of specific visual association pathways in cases of Alzheimer's disease presenting with Balint's syndrome. *Journal of Neuropathology and Experimental Neurology, 49,* 168–184.

Hof, P.R., Vogt, B.A., Bouras, C., & Morrison, J.H. (1997). Atypical form of Alzheimer's disease with prominent posterior cortical atrophy: A review of lesion distribution and circuit disconnection in cortical visual pathways. *Vision Research, 37,* 3609–3625.

Husain, M., Shapiro, K., Martin, J., & Kennard, C. (1997). Abnormal temporal dynamics of visual attention in spatial neglect patients. *Nature, 385,* 154–156.

Kaplan, E., Goodglass, H., & Weintraub, S. (1978). *The Boston Naming Test.* Boston: Kaplan & Goodglass.

Keane, M.M., Gabrieli, J.D., Fennema, A.C., Growdon, J.H., & Corkin, S. (1991). Evidence for a dissociation between perceptual and conceptual priming in Alzheimer's disease. *Behavioral Neuroscience, 105,* 326–342.

Kirshner, H.S., Webb, W.G., & Kelly, M.P. (1984). The naming disorder of dementia. *Neuropsychologia, 22,* 22–30.

Kiyosawa, M., Bosley, T.M., Chawluk, J., Jamieson, D., Schatz, N.J., Savino, P.J., Sergott, R.C., Reivich, M., & Alavi, A. (1989). Alzheimer's disease with prominent visual symptoms: Clinical and metabolic evaluation. *Ophthalmology, 96,* 1077–1086.

Kobayashi, R., Mikuno, K., Ishikawa, S., & Tasaki, Y. (1985). Hemisphere lateralization of spatial contrast sensitivity. *Annals of Neurology, 17,* 141–145.

Koss, E., Edland, S., Fillenbaum, G., Mohs, R., Clark, C., Galasko, D., & Morris, J.C. (1996). Clinical and neuropsychological differences between patients with earlier and later onset of Alzheimer's disease: A CERAD analysis, Part XII. *Neurology, 46,* 136–141.

Koss, E., & Gilmore, G.C. (1998). Environmental interventions and functional ability of AD patients. In B. Vellas, J. Fitten, & G. Frisoni (Eds.), *Research and Practice in Alzheimer's Disease* (pp. 185–191). Paris/New York: Serdi/Springer.

Kuljis, R.O. & Tikoo, R.K. (1997). Discontinuous distribution of senile plaques within striate cortex hypercolumns in Alzheimer's disease. *Vision Research, 37,* 3573–3591.

Kurylo, D.D., Corkin, S., Dolan, R.P., Rizzo, J.F., Parker, S.W., & Growdon, J.H. (1994). Broadband visual capacities are not selectively impaired in Alzheimer's disease. *Neurobiology of Aging, 15,* 305–311.

Kurylo, D.D., Corkin, S., & Growdon, J.H. (1992). Vision in Parkinson's disease: Several basic capacities are unimpaired. *Abstracts, Society for Neuroscience, 18,* 1082.

Kurylo, D.D., Corkin, S., & Growdon, J.H. (1994). Perceptual organization in Alzheimer's disease. *Psychology and Aging, 9,* 562–567.

Kurylo, D.D., Corkin, S., Rizzo, J.F., & Growdon, J.H. (1996). Greater relative impairment of object recognition than visuospatial abilities in Alzheimer's disease. *Neuropsychology, 10,* 74–81.

Lakshminarayanan, V., Lagrave, J., Kean, M.L., Dick, M., & Shankle, R. (1996). Vision in dementia: Contrast effects. *Neurological Research, 18,* 9–15.

Lanska, D.J. & Schoenberg, B.S. (1993). The epidemiology of dementia: Methodologic issues and approaches. In P.J. Whitehouse (Ed.), *Dementia* (pp. 3–33). Philadelphia: F.A. Davis.

Lerner, A.J., Koss, E., Patteson, M.B., Ownby, R.L., Hedera, P., Friedland, R.P., & Whitehouse, P.J. (1994). Concomitants of visual hallucinations in Alzheimer's disease. *Neurolgy, 44,* 523–527.

Leuba, G., & Saini, K. (1995). Pathology of subcortical visual centres in relation to cortical degeneration in Alzheimer's disease. *Neuropathology and Applied Neurobiology, 21,* 410–422.

Levine, D.N., Lee, J.M., & Fisher, C.M. (1993). The visual variant of Alzheimer's disease: A clinicopathologic case study. *Neurology, 43,* 305–313.

Lewis, D.A., Campbell, M.J., Terry, R.D., & Morrison, J.H. (1987). Laminar and regional distributions of neurofibrillary tangles and neuritic plaques in Alzheimer's disease, a quantitative study of visual and auditory cortices. *Journal of Neuroscience, 7,* 1799–1808.

Lindenberger, U. & Baltes, P.B. (1994). Sensory functioning and intelligence in old age: A strong connection. *Psychology and Aging, 9,* 339–335.

Loftus, G.R. (1974). Acquisition of information from rapidly presented verbal and nonverbal stimuli. *Memory & Cognition, 2,* 545–548.

Loftus, G.R., & Ruthruff, E. (1994). A theory of visual information acquisition and visual memory with special application to intensity-duration trade-offs. *Journal of Experimental Psychology, Human Perception and Performance, 20,* 33–49.

Lustig, C.A., Gilmore, G.C., Koss, E., Thomas, C.W., Ripich, D., Townsend, L., & Stuve, T. (1995). Boston Naming Test performance is affected by simulated Alzheimer-type visual deficit. Presented at the conference on The Spectrum of Research on Aging and the Aged, November 1995, Case Western Reserve University, Cleveland, OH.

Martinelli, V., Locatelli, T., Comi, G., Lia, C., Alberoni, M., Bressi, S., Rovaris, M., Franceschi, M., & Canal, N. (1996). Pattern visual evoked potential mapping in Alzheimer's disease: correlations with visuospatial impairment. *Dementia, 7,* 63–68.

McCormick, W.C., Kukull, W.A., Van Belle, G., Bowen, J.D., Teri, L., & Larson, E.B. (1994). Symptom patterns and comorbidity in the early stages of Alzheimer's disease. *Journal of the American Geriatrics Society, 42,* 517–521.

McFarland, R.A., Domey, R.G., Warren, A.B., & Ward, D.C. (1960). Dark adaptation as a function of age: I. A statistical analysis. *Journal of Gerontology, 15,* 149–154.

McFarland, R.A., & Fisher, B. (1955). Alterations in dark adaptation as a function of age. *Journal of Gerontology, 10,* 424–428.

McKhann, G., Drachman, D., Folstein, M., Katzman, R., Price, D., & Stadlan, E.M. (1984). Clinical diagnosis of Alzheimer's disease: Report of the NINCDS-ADRDA Work Group under the auspices of Department of Health and Human Services Task Force on Alzheimer's Disease. *Neurology, 34,* 939–944.

Mendez, M.F., Mendez, M.A., Martin, R., Smyth, K.A., & Whitehouse, P.J. (1990). Complex visual disturbances in Alzheimer's disease. *Neurology, 40,* 439–443.

Mendez, M., Turner, J., Gilmore, G.C., Remler, B., & Tomsak, R.L. (1990). Balint's syndrome in Alzheimer's disease: Visuospatial functions. *International Journal of Neuroscience, 54,* 339–346.

Mendola, J.D., Cronin-Golomb, A., Corkin, S., & Growdon, J.H. (1995). Prevalence of visual deficits in Alzheimer's disease. *Optometry and Vision Science, 72,* 155–167.

Mentis, M.J., Horwitz, B., Grady, C.L., Alexander, G.E., VanMeter, J.W., Maisog, J.M., Pietrini, P., Schapiro, M.B., & Rapoport, S.I. (1996). Visual cortical dysfunction in Alzheimer's disease evaluated with a temporally guided "stress test" during PET. *American Journal of Psychiatry, 153,* 32–40.

Miller, E. (1977). A note on visual information processing in presenile dementia: A preliminary report. *British Journal of Social and Clinical Psychology, 16,* 99–100.

Mirra, S.S., Hart, M.N., & Terry, R.D. (1993). Making the diagnosis of Alzheimer's disease. *Archives of Pathological and Laboratory Medicine, 117,* 132–144.

Morris, J.C., Mohs, R.C., Rogers, H., Fillenbaum, G., & Heyman, A. (1988). Consortium to Establish a Registry for Alzheimer's Disease (CERAD) clinical and neuropsychological assessment of Alzheimer's disease. *Psychopharmacology Bulletin, 24,* 641–644.

Murdock, B.B. (1989). Learning in a distributed memory model. In C. Izawa (Ed.), *Current issues in cognitive processes: The Tulane Floweree symposium on cognition* (pp. 69–106). Hillsdale: Erlbaum.

Neargarder, S.A., Stone, E., Cronin-Golomb, A., & Oross, S. (in press). The impact of acuity on contrast sensitivity performance in Alzheimer's disease. *Journal of Gerontology.*

Nebes, R.D., Martin, D.C., & Horn, L.C. (1984). Sparing of semantic memory in Alzheimer's disease. *Journal of Abnormal Psychology, 93,* 321–330.

Nissen, M.J., Corkin, S., Buonanno, F.S., Growdon, J.H., Wray, S.H., & Bauer, J. (1985). Spatial vision in Alzheimer's disease: general findings and a case report. *Archives of Neurology, 42,* 667–671.

Norris, D. (1984). The effects of frequency, repetition and stimulus quality in visual word recognition. *Quarterly Journal of Experimental Psychology, 36A,* 507–518.

Ogden, J.A. (1990). Spatial abilities and deficits in aging and age-related disorders. In F. Boller, & J. Grafman (Eds.), *Handbook of Neuropsychology: Vol. 4* (pp. 265–278). Amsterdam: Elsevier.

Owsley, C., Sekuler, R., & Boldt, C. (1981). Aging and low-contrast vision: Face perception. *Investigative Ophthalmology and Vision Science, 21,* 362–365.

Owsley, C., Sekuler, R., & Siemsen, D. (1983). Contrast sensitivity throughout adulthood. *Vision Research, 23,* 689–699.

Owsley, C., & Sloane, M.E. (1987). Contrast sensitivity, acuity, and the perception of "real-world" targets. *British Journal of Ophthalmology, 71,* 791–796.

Pearson, R.C.A., Esiri, M.M., Hiorns, R.W., Wilcock, G.K., & Powell, T.P.S. (1985). Anatomical correlates of the distribution of the pathological changes in the neocortex in Alzheimer's disease. *Proceedings of the National Academy of Sciences USA, 82,* 4531–4534.

Peli, E., Lee, E., Trempe, C., & Buzney, S. (1994). Image enhancement for the visually impaired: The effects of enhancement on face recognition. *Journal of the Optical Society of America A, 11,* 1929–1939.

Petersen, S.E., Fox, P.T., Snyder, A.Z., & Raichle, M.E. (1990). Activation of extrastriate and frontal cortical areas by visual words and word-like stimuli. *Science, 249,* 1041–1044.

Pietrini, P., Furey, M.L., Graff-Radford, N., Freo, U., Alexander, G.E., Grady, C.L., Dani, A., Mentis, M.J., & Schapiro, M.B. (1996). Preferential metabolic involvement of visual cortical areas in a subtype of Alzheimer's disease: Clinical implications. *American Journal of Psychiatry, 153,* 1261–1268.

Ramachandran, V.S., & Cobb, S. (1995). Visual attention modulates metacontrast masking. *Nature, 373,* 66–68.

Raz, N., Millman, D., & Moberg, P.J. (1990). Mechanism of age-related differences in frequency discrimination with backward masking: Speed of processing or stimulus persistence? *Psychology and Aging, 5,* 475–481.

Rebeck, G.W., & Hyman, B.T. (1993). Neuroanatomical connections and specific regional vulnerability in Alzheimer's disease. *Neurobiology of Aging, 14,* 45–47.

Regan, D., Raymond, J., Ginsburg, A.P., & Murray, T.J. (1981). Contrast sensitivity, visual acuity, and the discrimination of Snellen letters in multiple sclerosis. *Brain, 104,* 333–350.

Reid, W., Broe, G., Creasey, H., Grayson, D., McCusker, E., Bennett, H., Longley, W., Sulway, M.R. (1996). Age at onset and pattern of neuropsychological impairment in mild early-onset Alzheimer disease. A study of a community-based population. *Archives of Neurology, 53,* 1056–1061.

Ricker, J., Keenan, P., & Jacobson, M. (1994). Visuoperceptual-spatial ability and visual memory in vascular dementia and dementia of the Alzheimer type. *Neuropsychologia, 32,* 1287–1296.

Rizzo, J.F., Cronin-Golomb, A., Growdon, J.H., Corkin, S., Rosen, T.J., Sandberg, M.A., Chiappa, K.H., & Lessell, S. (1992). Retino-calcarine function in Alzheimer's disease: A clinical and electrophysiological study. *Archives of Neurology, 49,* 93–101.

Rocco, F.J., Cronin-Golomb, A., & Lai, F. (1997). Alzheimer-like visual profile in Down syndrome. *Alzheimer Disease and Associated Disorders, 11,* 88–98.

Rolls, E.T., & Tovee, M.J. (1994). Processing speed in the cerebral cortex and the neurophysiology of visual masking. *Proceedings of the Royal Society of London B, 257,* 9–15.

Rolls, E.T., Tovee, M.J., Purcell, D.G., Stewart, A.L., & Azzopardi, P. (1994). The responses of neurons in the temporal cortex of primates, and face identification and detection. *Experimental Brain Research, 101,* 473–484.

Roses, A.D. (1997). Alzheimer's disease: The genetics of risk. *Hospital Practice (Off Ed), 32,* 51–69.

Saccuzzo, D.S., Cadenhead, K.S., & Braff, D.L. (1996). Backward versus forward visual masking deficits in schizophrenic patients: Centrally, not peripherally, mediated? *American Journal of Psychiatry, 153,* 1564–1570.

Sadun, A.A., & Bassi, C.J. (1990). Optic nerve damage in Alzheimer's disease. *Ophthalmology, 97,* 9–17.

Sadun, A.A., Borchert, M., DeVita E., Hinton, D.R., & Bassi, C.J. (1987). Assessment of visual impairment in patients with Alzheimer's disease. *American Journal of Ophthalmology, 104,* 113–120.

Salthouse, T.A., Hancock, H.E., Meinz, E.J., & Hambrick, D.Z. (1996). Interrelations of age, visual acuity, and cognitive functioning. *Journal of Gerontology: Psychological Sciences, 51B,* P317–330.

Schlotterer, G., Moscovitch, M., & Crapper-McLachlan, D. (1984). Visual processing deficits as assessed by spatial frequency contrast sensitivity and backward masking in normal aging and Alzheimer's disease. *Brain, 107,* 309–325.

Selzer, B., Sherwin, I. (1983). A comparison of clinical features in early- and late-onset primary degenerative dementia. *Archives of Neurology, 40,* 143–146.

Silverman, S.E., Tran, D.B., Zimmerman, K.M., & Feldon, S.E. (1994). Dissociation between the detection and perception of motion in Alzheimer's disease. *Neurology, 44,* 1814–1818.

Spinks, R., Gilmore, G.C., & Thomas, C. (1996). Age simulation of a sensory deficit does impair cognitive test performance. Presented at the Cognitive Aging Conference, April 1996, Atlanta, GA.

Spinks, R., Gilmore, G., Thomas, C., & Hinman, J. (1998). What's the difference? Two sensory age simulations compared. Presented at the Cognitive Aging Conference, April 1998, Atlanta, GA.

Stevens, J.C., Cruz, L.A., Marks, L.E., & Lakatos, S. (1998). A multimodal assessment of sensory thresholds in aging. *Journal of Gerontology: Psychological Sciences, 53B,* P263–272.

Stone, E.R., Neargarder, S.A., Cronin-Golomb, A., & Oross, S. (1998). Exploring contrast sensitivity in Alzheimer's disease: Examination and comparison of four clinical tests. Presented at the Cognitive Aging Conference, April 1998, Atlanta, GA.

Theios, J., & Amrhein, P.C. (1989a). The role of spatial frequency and visual detail in the recognition of patterns and words. In C. Izawa (Ed.), *Current issues in cognitive processes: The Tulane Floweree symposium on cognition* (pp. 389–409). Hillsdale: L. Erlbaum.

Theios, J., & Amrhein, P.C. (1989b). Theoretical analysis of the cognitive processing of lexical and pictorial stimuli: Reading, naming, and visual and conceptual comparisons. *Psychological Review, 96,* 5–24.

Trick, G.L., Barris, M.C., & Bickler-Bluth, M. (1989). Abnormal pattern electroretinograms in patients with senile dementia of the Alzheimer type. *Annals of Neurology, 26,* 226–231.

Trick, G.L., & Silverman, S.E. (1991). Visual sensitivity to motion: Age-related changes and deficits in senile dementia of the Alzheimer type. *Neurology, 41,* 1437–1440.

Tsai, C.S., Ritch, R., Schwartz, B., Lee, S.S., Miller, N.R., Chi, T., & Hsieh, F.Y. (1991). Optic-nerve head and nerve fiber layer in Alzheimer's disease. *Archives of Ophthalmology, 109,* 199–204.

Uhlmann, R.F., Larson, E.B., Koepsell, T.D., Rees, T.S., & Duckert, L.G. (1991). Visual impairment and cognitive dysfunction in Alzheimer's disease. *Journal of General Internal Medicine, 6,* 126–132.

Ungerleider, L.G., & Mishkin, M. (1982). Two cortical visual systems. In D.J. Ingle, M.A. Goodale, & R.J.W. Mansfield (Eds.), *Analysis of visual behavior* (pp. 549–586). Cambridge, MA: MIT Press.

Walsh, D.A., Till, R.E., & Williams, M.V. (1978). Age differences in peripheral perceptual processing: A monoptic backward masking investigation. *Journal of Experimental Psychology: Human Perception & Performance, 4,* 232–243.

Wong-Riley, M., Antuono, P., Ho, K.-C., Egan, R., Hevner, R., Liebl, W., Huang, Z., Rachel, R., & Jones, J. (1997). Cytochrome oxidase in Alzheimer's disease: Biochemical, histochemical, and immunohistochemical analyses of the visual and other systems. *Vision Research, 37,* 3593–3608.

Wright, C.E., Drasdo, N., & Harding, G.F.A. (1987). Pathology of the optic nerve and visual association areas: Information given by the flash and pattern visual evoked potential and the temporal and spatial contrast sensitivity function. *Brain, 110,* 107–120.

Wright, C.E., Harding, G.F.A., & Orwin, A. (1984). Presenile dementia: The use of flash and pattern visual evoked potential in diagnosis. *Electroencephalography and Clinical Neurophysiology, 57,* 405–415.

Yonelinas, A.P., Hockley, W.E., & Murdock, B.B. (1992). Tests of the list-strength effect in recognition memory. *Journal of Experimental Psychology: Learning, Memory, and Cognition, 18,* 345–355.

Chapter 2

# Exploring Visual Perception Abilities in Individuals with Intellectual Disabilities: Assessment and Implications

*Stephen Oross III and Charles B. Woods*

> Marvelous as the eyes are, they are only tools. Seeing is a complexity consisting of the external world of physical objects and factors and the intricate world of psycho-physiological effects. The eyes are tools and doorways between these two worlds. The end products of seeing are human efficiency, behavior, and welfare
>
> —Luckiesh, 1934, pp. 2–3.

The sights, sounds, touches, tastes and other forms of stimulation received by sensory systems define the sensory/perceptual world, or *umwelt*, inhabited by individuals. Not surprisingly, variations across species in the structure and functioning of sensory systems define different sensory and perceptual worlds. Many species of birds are able to resolve higher rates of flicker than humans, thus enabling them to perceive the flicker of fluorescent lights (Emmerton, 1983). Fish are able to perceive ultraviolet light, enabling them to resolve patterns of polarized light that may serve to guide their navigation (Hawryshyn, 1992). Within a given species it is also well established that individuals at different points in development inhabit different *umwelts*. Human infants, for instance, are unable to perceive the depth depicted by monocular, or "pictorial," cues such as linear perspective and shading until approximately seven months of age, thus limiting the amount of three-dimensionality perceived in drawings or pictures (see e.g., Granrud, Yonas, & Opland, 1985; Oross, Francis, Mauk, & Fox, 1987; Yonas & Granrud, 1985).

A fact often overlooked is that even when two individuals of a given species are the same age, their *umwelts* will differ. Most often the difference is minor, as when slight variations in the optical power of the cornea and lens contribute to differences in visual acuity. A more pronounced difference exists when individuals lack the full complement of cone photopigments needed to enable normal trichromatic color vision. Severe differences often exist when damage has occurred to specific regions of the cerebral cortex. In one of the more fascinating—and tragic—cases, a woman who incurred bilateral damage to posterior brain regions as a result of a stroke exhibited "motion-blindness" that compromised her ability to perform many everyday actions and behaviors (Zihl, von Cramon, & Mai, 1983). Pouring liquid into a cup was challenging because the liquid appeared frozen in time and the level was not seen to rise. Crossing streets was difficult because a car would be seen in the distance and before the patient realized it, the car would be upon her. She also reported difficulty interacting with people because people would suddenly appear or disappear from her view, and she could not follow a conversation because she was unable to see the movements of faces and mouths. This case, along with similar cases, has stimulated an explosion of research devoted to specifying the *umwelts* of different clinical populations.

The charge of this chapter is to explore the possibility that the *umwelt* of individuals with intellectual disabilities differs from that of individuals without intellectual disabilities. It is important to realize that the statement of this possibility does not represent a novel or recent insight. Indeed, as nicely summarized by Berkson (1993), the implications of this possibility have been systematically explored, in one form or another, at least since the writings of John Locke. It served as the cornerstone of Sir Francis Galton's interest in accounting for intelligence-related differences. Jean Piaget sought to account for the cognitive development of children, at least in part, by considering how sensory and perceptual experience would reveal different types of information, thus prompting the accommodation of existing cognitive structures needed to assimilate this new information. Almost four decades ago, the publication of the *Handbook of Mental Retardation* (Ellis, 1963) contained two chapters that explicitly explored how the variations in sensory and perceptual processes often detected in individuals with mental retardation may account for many of the difficulties they experience (Kodman, 1963; Spivak, 1963).

Nevertheless, research into sensory and perceptual processing impairments associated with mental retardation has been relatively neglected since the mid-1970s. This may have been due to the fact that cognitive psychology and behaviorism began to dominate psychology at this time. Given that mental retardation is commonly associated with impairments in cognitive skills such as reasoning, forming and recalling memories, and language, it was natural that many investigators turned their energies to exploring cognitive abilities in individuals with mental retardation. Moreover, behavior analytic studies had begun to show that proper training procedures could lead to the acquisition of visual discrimination skills previously believed to be unattainable (e.g., Sidman & Stoddard, 1966).

The reappearance of research concerned with sensory and perceptual processing in individuals with intellectual disabilities summarized in this chapter derives from advances on several fronts. Paramount among these are advances in understanding the normal functioning of sensory systems, particularly vision, and the development of technologies that enable both the controlled presentation of stimuli and the imaging of neural activity to a greater degree than previously possible. These advances have enabled research devoted to investigating anomalies in sensory and perceptual processes to specify explicit relationships between neural functioning and behavior.

Scientists have long speculated on the links between central nervous system function and normal and impaired behavior. In the past there was a reliance on studies of gross brain abnormality (e.g., the case of Phineas Gage) in order to gain insight into this relationship. Studies of specific visual sensory processes now contribute considerably to our understanding of diseases and disorders that often involve more subtle forms of central nervous system pathology. A partial list of these disorders includes dyslexia, mental retardation, schizophrenia, and autism, as well as progressive neurodegenerative disorders such as Alzheimer's disease, Parkinson's disease, and multiple sclerosis. One challenge confronting researchers in these areas is that these disorders have a profile of spared and impaired behavioral abilities: some behaviors are relatively unaffected while others may be profoundly affected. For many diseases and disorders, therefore, scientists seek to identify the range of behaviors that show impairments, the relative magnitude of these impairments, and the cortical and subcortical areas of the brain that may be affected.

The topic addressed in this chapter encompasses extremely diverse fields and all aspects of the fields cannot possibly be addressed equally. For instance, the substantial variability evident among individuals with mental retardation obviously limits the ability to address them as if they belonged to a homogeneous population. The majority of research concerned with sensory and perceptual processing, however, has not yet been adequately concerned with exploring how variability among individuals with intellectual disabilities may either limit the conclusions that can be drawn or facilitate our understanding of why some, but not all, individuals with intellectual disabilities exhibit selective sensory and perceptual deficits. Additionally, knowledge about sensory systems is rapidly being gathered at many levels of analysis ranging from molecular interactions to synapses to systems, and it is impossible to discuss every level in great detail. Thus, the primary intent of this chapter is to present important empirical and theoretical findings at the systems level with a focus on exploring the implications of these findings.

In the following part, knowledge about visual processing in individuals with intellectual disabilities is summarized. A number of summaries devoted to sensory and perceptual processing were published prior to the 1970s (Kodman, 1963; Spitz, 1963; Spivak, 1963; Winters, 1969); therefore, the primary focus is on studies and issues currently receiving attention. The goal of the last part is to use vision as the model sensory system for addressing the origin and implications of these deficits.

## VISUAL PERCEPTION ABILITIES

In this part, we explore a variety of visual abilities in individuals with intellectual disabilities. This approach results, in part, from the growing awareness that measures of visual acuity alone do not provide a good prediction of behavioral difficulties in most individuals (see, e.g., National Advisory Eye Council report to the National Institutes of Health, 1995). Additionally, as we will summarize later, knowledge about the neural foundations of visual abilities has rapidly advanced over the last couple of decades. One result for this advancement is that a number of visual abilities have been identified as critical and we wish to explore what is known about these abilities in individuals with intellectual disabilities. Despite the identification of a number of visual abilities as critical, however, there is no clear consensus on which abilities must be examined to provide a comprehensive assessment of visual function.

We begin with a brief overview of our current understanding of the neural systems underlying visual information processing. We follow this with a consideration of ophthalmological investigations, including attempts to assess visual acuity in individuals with intellectual disabilities. Next, we move to a consideration of low-level followed by high-level visual perception abilities. Researchers in the area of vision and visual perception often distinguish, as a matter of descriptive convenience, between the earliest, low-level stages of sensory processing, and later high-level perceptual mechanisms. Visual tasks are sometimes described as low-level if our perceptual experiences can be described quite well by knowing the physical characteristics of the visual stimulus. Visual tasks are described as high-level, however, if prior context, experience, learning, and memory may play a role in determining our perceptual experiences. Because this is a common and useful way of categorizing our perceptual abilities, we will use it here. Where recent research is lacking on an important topic, we supplement our material with earlier work in the field. In some cases, research on an important topic is scarce, nonexistent, or in progress, and we feel it is important to draw the reader's attention to these topics as well.

### Neural foundations

Initially, the visual system, as with other sensory systems, can be viewed as a series of largely independent, parallel information-processing channels (see, e.g., Desimone & Ungerleider, 1989; Kaas & Garraghty, 1991; Livingstone & Hubel, 1987, 1988). Although there is presently a great deal of uncertainty about the exact number of pathways that exist and the degree of their independence, the literature continues to focus on two major pathways commonly referred to as the "transient" and "sustained" pathways. (For a review detailing the distinction between these pathways, see Merigan & Maunsell, 1993.) The transient pathway originates in the retina and is identified with axons that terminate in magnocellular layers of the lateral geniculate nucleus (LGN) before proceeding dorsally to

**Figure 2.1**
Schematic Representation of the Two Major Pathways in the Visual System often Referred to as the "What" and "Where" Pathways. These pathways differ in both their physiology and function as summarized here.

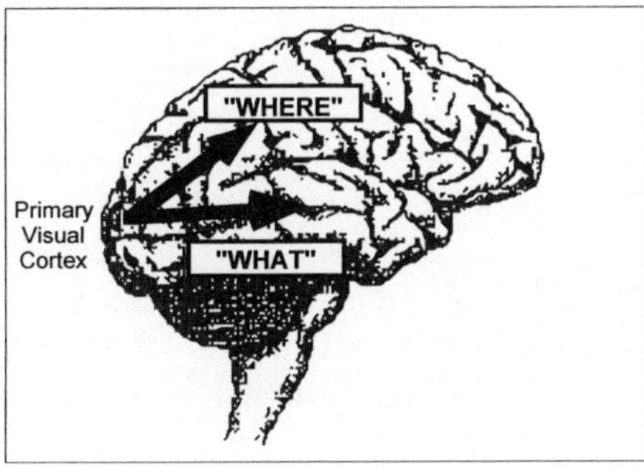


| VISUAL PATHWAYS | Magnocellular "Where" | Parvocellular "What" |
|---|---|---|
| FUNCTIONS | motion | color |
| | depth | texture |
| | high contrast sensitivity | low contrast sensitivity |
| | low spatial resolution | high spatial resolution |
| | high temporal resolution | low temporal resolution |
| PHYSIOLOGY | large neurons | small neurons |
| | large receptive fields | small receptive fields |
| | moderate myelination | light myelination |
| | fast axon conduction | slow axon conduction |
| | transient light response | sustained light response |
| | broad wavelength tuning | narrow wavelength tuning |

striate and extrastriate regions of the cortex. Cortical areas clearly associated with this pathway involve primary visual cortex (V1), V2, V3, middle temporal (MT) and those located in posterior parietal cortex. This pathway, roughly corresponding to the "where" pathway outlined by Ungerleider and Mishkin (1982), can be distinguished physiologically and behaviorally by its increased sensitivity to stimulus onsets and offsets, higher temporal frequencies, lower contrasts, depth, and motion. The sustained pathway is identified with axons that project to the parvocellular layers of the LGN. This pathway—the "what" pathway—then projects ventrally and involves areas V1, V2, V4, and inferior temporal (IT) and can be distinguished by its increased sensitivity to higher spatial frequencies, color, and form information. Figure 2.1 illustrates these pathways.

Evidence supporting the proposal that the visual system is organized as a series of parallel pathways appears relatively unassailable (although the exact differences among the pathways remains a matter of debate). Some of the evidence derives from human clinical observations of specific visual losses following specific neural damage such as with achromatopsia (selective loss of color vision). Other evidence originates from animal lesion research demonstrating specific cortical areas devoted to specific visual abilities (see, e.g., the careful linkage of area MT and the perception of coherent motion provided by Newsome and Pare (1988) and Newsome, Britten and Movshon (1989). Still other evidence is provided by anatomical patterns of connections between neural structures (e.g., DeYoe & Van Essen, 1988; Hubel & Livingstone, 1987; Ungerleider & Mishkin, 1982).

## Low-level visual perception abilities: Spatial vision

Spatial vision refers to our ability to detect and discriminate between different visual patterns or forms, such as size and shape, and our ability to accurately perceive the spatial arrangement or layout of the visual scene. Tests of spatial vision represent the most informative and most commonly employed assessments of visual function. In this part we present what is currently known about the spatial vision of individuals with intellectual disabilities.

### *Ophthalmological findings*

Generally, individuals with mental retardation are reported to have significantly more eye-related abnormalities and vision problems than nonclinical individuals. Mental retardation is associated with a high incidence of mild and severe refractive errors of all types: hypermetropia, myopia, and astigmatism (Haugen, Aasved, & Bertelsen, 1995; Levy, 1984). Mental retardation is also associated with a higher incidence of oculomotor anomalies (da Cunha & Moreira, 1996; Levy, 1984; Sacks, Goren, Burke, and White, 1991) and a large number of other ocular anatomical and pathological defects (see Levy, 1984; Maino, Maino, & Maino, 1990). For instance, Sacks, Goren, Burke and White (1991) conducted an ophthalmological screening of 113 adults with mental retardation varying in degree of retardation and etiology. Although the group exhibited few serious instances of ocular disease, there was high incidence of strabismus (32% in individuals with Down syndrome; 27% in individuals without Down syndrome) and 37 percent of the individuals had at least one eye that lacked appropriate refractive correction. Maino, Wesson, Schlange, Cibis, & Maino (1991) in their examination of individuals with Fragile X syndrome also report high incidences of strabismus (30%), refractive error of eyes (59%), and astigmatism (22%). As a point of comparison, the rate of strabismus in nonclinical individuals is 5 percent.

Of the studies investigating visual function in individuals with mental retardation that have been performed over the last two decades, a principal finding has been that individuals with mental retardation, whether institutionalized or in home care, do not receive adequate or timely ophthalmologic care (Evenhuis, 1995). This problem is pervasive. In fact, a recent study of Special Olympians at the 1995 Special Olympics World Summer Games found that 65 percent of the participants ($n = 905$) had not received any eye care in over three years (Block, Beckerman, & Berman, 1997). Because of the high incidence of ocular problems in this group, it has been recommended that individuals with mental retardation have *more* frequent eye examinations than individuals without mental retardation, and that these exams cover the entire lifespan (Evenhuis, 1995; Prokesova, Kriz, Berg, & Halvorsen, 1990).

### *Visual acuity*

The perceptual ability that exemplifies spatial vision, and, in fact, is tested most frequently, is visual acuity. Tests of visual acuity measure the minimum size (or visual angle) of detail that can just be resolved by an observer. Tests of visual acuity can rely on *recognition acuity* (e.g., Snellen letter chart), *detection/discrimination acuity* (e.g., Teller acuity cards), or *resolution acuity.* In many cases, the assessment of visual acuity in an individual with mental retardation requires a modified ophthalmic examination with the extent of this modification varying according to the intellectual capabilities of the individual. With individuals who have more severe intellectual difficulties, large allowances must be made. It indicates the importance of obtaining a satisfactory measure of visual acuity that so many alternative measures of visual acuity have been formulated for special needs populations. One commonly reported method is the use of acuity card or other forced-choice preferential looking (FCPL) techniques, where direction of gaze is observed for pairs of visual stimuli (Adoh, Woodhouse, & Oduwaiye, 1992; Birch, Hale, Stager, Fuller, & Birch, 1987; Chandna, Karki, Davis, & Doran, 1989; Courage, Adams, Reyno, & Kwa, 1994; Hall, Orel-Bixler, & Haegerstrom-Portnoy, 1991; Hertz, 1987; Lennerstrand, Axelson, & Andersson, 1983; O'Dell, Harshaw, & Boothe, 1993). However, other far less formal methods are sometimes used in practice (e.g., counting fingers; see Sacks, Goren, Burke, & White, 1991).

Possible limitations with using FCPL methods to examine visual abilities in individuals with mental retardation must be considered before they are used. These methods were developed to examine infant vision because they take advantage of the fact that infants show inherent preferences in their viewing: Infants generally prefer to view something over nothing, something new over something old, and something complex over something simple (see, e.g., Banks & Dannemiller, 1987). When these methods have been applied to other populations, there has been little attempt to determine whether similar viewing preferences hold. Thus, there is a potential problem in interpreting findings when no fixation differences are evident. Does no fixation mean that the stimulus was not perceived, or merely that the stimulus differences did not support preferential viewing?

Nevertheless, this procedure may have potential for assessing visual acuity because it has also been used successfully with children with mental retardation (e.g., Hertz, 1987; Hertz & Rosenberg, 1992; Lennerstrand, Axelson, & Andersson, 1983). Moreover, acuity estimates have been demonstrated to correlate well with visual-evoked potential (VEP) estimates of visual acuity in a population of multiply handicapped patients varying in age and handicapping condition (Orel-Bixler, Haegerstrom-Portnoy, & Hall, 1989). Other evidence supporting the use of this procedure was reported by Jacobsen, Magnussen, & Smith, (1997), who demonstrated that eight of eleven individuals with severe or profound mental retardation, previously diagnosed as deaf and blind, actually had visual acuities that were above the criterion for legal blindness. Indeed, several of these individuals had visual acuities that were normal or near normal. The importance of obtaining accurate and reliable measures of visual acuity cannot be overstated because it has been documented that improvements in motor skills, reading and writing abilities, and even social behavior, accompany supplying correct refractive lenses to individuals with mental retardation (Bader & Woodruff, 1980).

There are a number of potential problems, however, that must be addressed in a systematic manner before measures of visual acuity, or other indices of visual impairment, are used to prescribe corrective lenses or other interventions for individuals with mental retardation. For example, most estimates of visual acuity are obtained with little formal attempt to assess the impact of behavioral problems such as lack of diligence. This is a particularly glaring problem given that almost all researchers note that lack of diligence along with other behavior problems will contribute to inflated estimates of visual acuity. Consider, for instance, the performance in a visual acuity task of an individual who is concerned with making a mistake and, therefore, either stops responding or reverts to some idiosyncratic response pattern whenever the discriminations become more difficult (i.e., approach threshold). Such circumstances may lead to an underestimation of the threshold. The individual could exhibit, across sessions, reliable performance that is invalid because the threshold does not represent the true state of their acuity. Perhaps the most important problem that may arise from this situation is that the prescription of corrective lenses on the basis of an inaccurate assessment does more than simply blur vision; it may actually contribute to a worsening of vision.

Our standard practice for visual acuity assessments has evolved to include testing with the illiterate E visual acuity chart, providing individuals with a response board, and including a formal practice session that provides both familiarity with the test setting and establishes a formal criterion to demonstrate that instructions are understood. The response board places replicas of the response items immediately in front of the individuals. Participants, even those with good verbal skills, are instructed to point to the replica on the response board that matches the test item. It is our belief that this practice minimizes errors due to spatial confusions between the orientations of the test items.

With these methods in place, we routinely find that the acuity of clinical groups is significantly impaired relative to that of nonclinical groups. However, the visual acuity of the clinical groups is still fairly good. Consider, for example, the visual acuities obtained by Neargarder (1997) in a study that examined age-related declines in visual functioning of individuals with mental retardation. The visual acuities of the groups are as follows: young nonclinical adults (20/12.5); middle-aged nonclinical adults (20/14); young adults with mental retardation of unspecified etiology (20/24); middle-aged adults with mental retardation of unspecified etiology (20/22); young adults with Down syndrome (20/43); and middle-aged adults with Down syndrome (20/49). These values are similar to those that have been reported with other individuals with mental retardation (Catalano, 1990; Courage, Adams, & Hall, 1997; Courage, Adams, Reyno, & Kwa, 1994; Kim, 1995; Shimp & Oross, 1995). Overall, this data suggests that assessments of visual acuity that include methods to minimize the impact of nonvisual factors that are routinely present in the behavior repertoires of individuals with mental retardation may provide more accurate assessments of visual acuity.

### *Contrast sensitivity*

Our ability to see fine detail depends not only on the angular size of the stimulus but also on the brightness contrast, or difference in luminance, be-

**Figure 2.2**
A Stylized Representation of a Contrast Sensitivity Function in Humans. The area under the curve indicates the combinations of luminance contrast and spatial frequency that can be perceived.

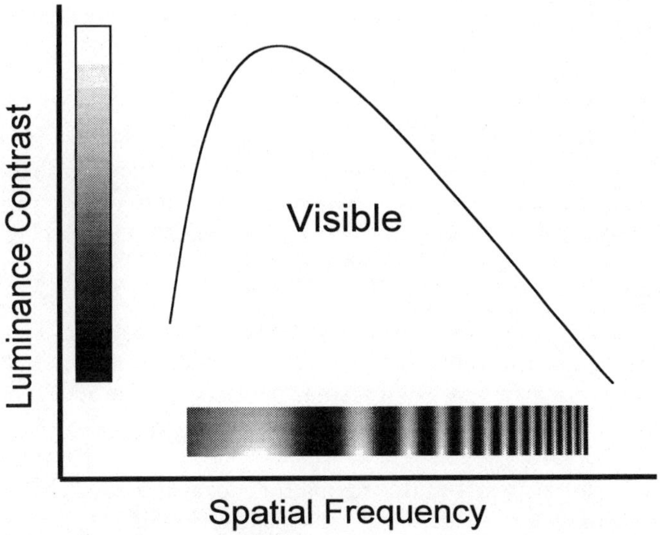


tween an object and its surroundings. Contrast sensitivity refers to the ability to discriminate small luminance differences for objects of different sizes. Contrast sensitivity is traditionally measured with spatially periodic patterns called "gratings," which differ in spatial frequency (the angular width of a pair of bars) and luminance contrast (the difference in luminance between adjacent bars). Using these stimuli, a function representing threshold contrast sensitivities for different spatial frequencies is determined (see Figure 2.2). The area underneath the curve labeled "visible" defines the typical human *umwelt*, or perceptual space, for size and contrast. Contrast sensitivity to high spatial frequencies helps facilitate our ability to see fine detail (like recognizing printed letters on this page), while this same measure at low spatial frequencies helps facilitate our ability to recognize shapes of larger objects such as signs, toys, or faces. It is also a good measure of our ability to see under sub-optimal conditions (e.g., low illumination, or retinal image blur due to uncorrected or miscorrected refractive error).

Despite the theoretical importance and practical significance of testing contrast sensitivity, there are few reports in the literature that have examined this ability in individuals with mental retardation. Courage, Adams, & Hall (1997), in a study of children and adults with Down syndrome, reported small but significant losses at all spatial frequencies, with the relative losses becoming greater with increasing spatial frequency. Rocco, Cronin-Golomb, & Lai (1997) found that adults with Down syndrome exhibited losses in contrast sensitivity relative to adults with mental retardation of unspecified etiologies, a finding that we have replicated in our lab. Moreover, we have found that the contrast sensitivity of adults with mental retardation (including those with Down syndrome and those with unspecified etiologies) is qualitatively similar to that of nonclinical adults, but shows significant losses at all spatial frequencies (Shimp & Oross, 1994; Neargarder, 1997).

### *Depth perception*

Depth perception describes our ability to determine the distance between objects in the world (relative depth) and the ability to determine the distance between an object and ourselves (absolute depth). A variety of cues contribute to these abilities including *monocular*, or pictorial, cues, *binocular* cues, and *motion* cues. As stated earlier, sensitivity to these individual classes of cues is evidenced at different points in development; however, by seven months of age, typically developing individuals possess sensitivity to all three classes of cues (see, e.g., Oross, Francis, Mauk, & Fox, 1987). Whether or not these cues interact to provide more precise depth information that can be provided by any single cue is an issue that has received little attention (however, see Cutting & Bruno, 1988; Massaro, 1988). We hope this will become a priority for future research because it may provide information for addressing issues of remediation among individuals who exhibit deficits in perceiving the depth specified by these classes of cues.

***Monocular cues.*** Monocular cues are often referred to as "pictorial" cues because they are used by artists to depict depth in paintings and drawings. Some of the types of monocular cues that have been identified are texture gradients, linear perspective, size, interposition, and relative height in a scene. To our knowledge, there have been only a few published studies that have examined the ability of individuals with mental retardation to use monocular depth cues to judge relative or absolute depth. An early review by Spivak (1963) mentions only one published report (Barnett & Pryer, 1958) but this study did not compare performance to individuals without mental retardation. There are, however, early studies that examined related phenomena of the perceptual abilities of individuals with mental retardation. Winters (1969) and Spitz (1967) both examined size–constancy under varied cue environments and reported very small and no differences, respectively, between individuals with mental retardation and comparison participants. A similar conclusion was drawn by Leibowitz (1961), who found that individuals with mental retardation related the apparent size of objects to distance in a similar manner to individuals without mental retardation, thus suggesting similar depth-perception abilities. Leibowitz & Heisel (1958) studied a geometric illusion, the Ponzo illusion, that is thought to be due to misapplied depth cues. They report that individuals with mental retardation show a larger illusion, suggesting enhanced sensitivity to monocular depth cues. Although not thorough investigations, these findings are consistent with the hypothesis that individuals with mental retardation accurately perceive monocular depth.

This hypothesis, however, does not receive unequivocal support because there are also a few reports that suggest that individuals with mental retardation do not accurately perceive monocular depth. In one of these reports, individuals with profound mental retardation were tested using a variant of the well-known "visual cliff" task (Gibson & Walk, 1960). This task presents individuals with a checkerboard pattern that depicts an apparently shallow surface on one side of an apparatus and an apparently deep surface on the other side. Individuals are then coaxed into crawling or walking across both the shallow and the deep surface to reach a desired goal (usually a mother or other caregiver). The premise of the task is that individuals who possess accurate depth perception will refuse to cross the "deep" side. Garcia, Cleland, Rago, Wayne, & Swartz (1974) found that approximately 39 percent of the individuals with mental retardation refused to cross the "deep" side, indicating that they might have perceived the depicted depth. However, a majority of the group (61%) exhibited behaviors that could have indicated a failure to accurately perceive the depicted depth (e.g., crossing the deep side).

To examine sensitivity to monocular depth cues in our laboratory, we have assessed the ability to perceive depth specified by texture and linear perspective information in individuals with and without mental retardation. Two circles were placed against either a blank background (control condition) or against a background containing information that simulates depth. This condition, with monocular viewing, induces a variant of the well-known moon illusion. Specifically, the lower circle appears smaller than the upper circle even when they are of the same physical size (see Figure 2.3).

By varying the sizes of the circles, it was possible to index the strength of the illusion. In the texture–perspective condition, individuals without mental retardation exhibited the classic illusion of stating that the bottom circle is physically smaller when, in fact, it was physically larger. The majority of the individuals with mental retardation did not succumb to this illusion. Furthermore, control condition data indicated that both groups were equally able to discriminate between circles of different sizes.

***Binocular cues.*** Binocular vision and stereopsis, the depth information provided by retinal disparity, has been studied in individuals with mental retardation. Stereopsis is best studied with random dot stereograms, first described by Julesz (1960). These stimuli consist of two patterns of seemingly random dots, one pattern presented to each eye. These patterns are subtly different in that some of the dots in one pattern have been slightly displaced horizontally relative to the other pattern, producing retinal disparities that simulate the situation of viewing real depth. To an observer who possesses stereopsis, the resulting image will be a form (corresponding to the shape of the displaced subset of dots) appearing to float either in front of or behind a randomly textured

**Figure 2.3**
Upper Panel Presenting One Variant of an Illusion Caused by Monocular Depth Cues, Both Texture Gradients and Linear Perspective. When this display is viewed under monocular conditions (with one eye closed) it is commonly reported that the upper circle appears larger than the lower circle despite the fact that they are physically identical. The lower left panel quantifies this illusion for individuals without mental retardation. The open circles refer to the control condition wherein the size judgments are made in the absence of the inducing background; the filled circles refer to the texture-perspective condition.

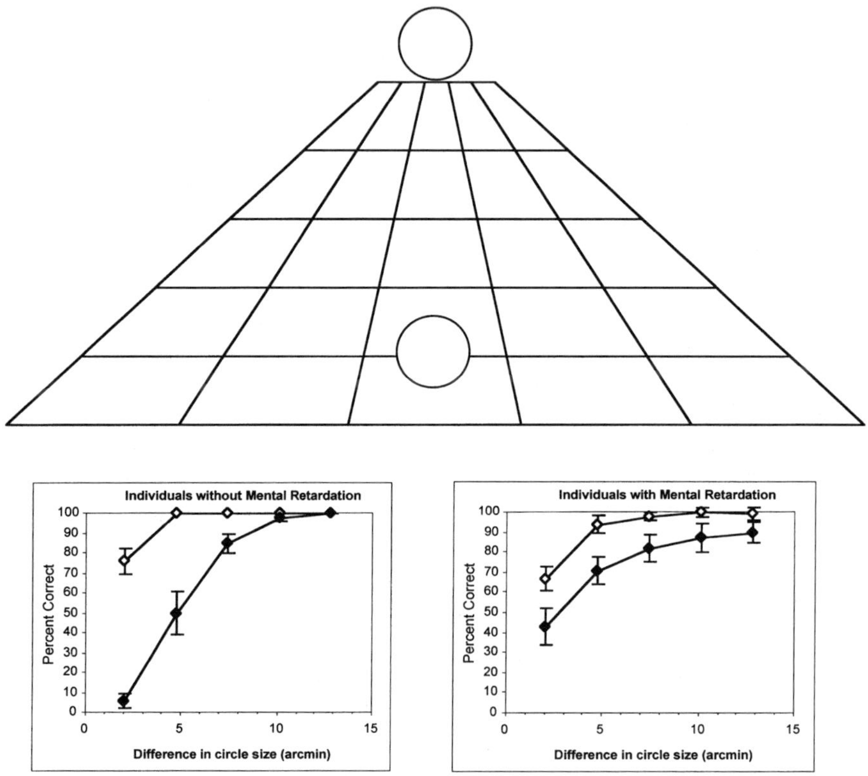


background in accord with the geometry of vision (see Cormack & Fox, 1985). One of the advantages of these stimuli is that they do not contain monocular cues to depth and the target stimulus is only visible when the brain fuses the left and right eye images. Absent stereopsis, all that is seen is a flat, two-dimensional surface of dots.

Webb (1972) used these stimuli to test stereopsis in individuals with mental retardation by presenting two tasks that asked participants to locate the two-dimensional position of stereoscopic forms and to discriminate between different shapes (square, triangle, and cross). The data revealed no significant differences between individuals with or without mental retardation, supporting the hypothesis that individuals with mental retardation possess accurate binocular stereopsis. It should be noted that Webb did not examine depth perception associated with stereograms and, therefore, "whether the discrimination of depth itself yields similar group equivalencies remains to be seen" (Webb, 1972, p. 701).

Using similar stimuli, Fox & Oross (1988) also discovered that adults with mild mental retardation were able to detect the left–right spatial position of stereograms when the density of dots used to create the stereoscopic form was high; however, as dot density was reduced, this ability was lost. Moreover, even when stereograms were detected (i.e., when dot density was high), shape discrimination and depth-localization abilities were impaired. The ability to discriminate between shapes was not completely absent. Rather, similar shapes, such as squares and rectangles or triangles and diamonds, were confused. The localization of the stereograms in depth, however, was never accurate. These results suggested fundamental deficits in depth perception associated with mental retardation. Moreover, the impact of dot-density reductions gave rise to a hypothesis that individuals with mental retardation possess impaired interpolative processes that are needed for compensating for missing information.

A study by Block, Beckerman, & Berman (1997) studied stereopsis in a sample of individuals with mental retardation who had gathered for the 1995 Special Olympics World Summer Games. These authors successfully screened 93 percent of the athletes for stereopsis and approximately 20 percent failed this test. This was about the same proportion that was found to have strabismus. Of those demonstrating stereopsis, however, sensitivity was found to be an order of magnitude larger (poorer) than population norms. In addition, Sadler, Olitsky, & Reynolds (1996) and Olitsky, Sadler, & Reynolds (1997) have found reduced stereoacuity in Williams syndrome. These findings, coupled with those of Fox & Oross (1988b), suggest that individuals with mental retardation inefficiently process retinal disparity information.

***Motion cues.*** Although motion will be discussed in more detail later, it is important to note that motion does provide important information regarding depth. The term "optic flow" was coined by Gibson (1966) to describe the information provided by the global flow of surfaces and textures that occurs with motion. Empirical investigations of human sensitivity to optic flow conducted over the last twenty years have revealed a number of intriguing results as well as raise a number of questions. One interesting result found is that human infants are sensitive to optic flow patterns very early in life, perhaps as young as five months of age (Bertenthal & Bai, 1989; Butterworth & Cicchetti, 1978; Delorme, Frigon, & Lagace, 1989). Such findings are by no means surprising when the importance of maintaining postural stability for both safety and performing almost any other task is considered. Other empirical results suggest that there are a number of aspects in the flow pattern, including location of flow (central vs. peripheral vision) and geometrical structure of flow, which influence postural responses to optic flow.

Wade (1990) explored the degree to which individuals with mental retardation may differ in their ability to control postural stability in the presence of optic flow patterns. Participants stood on a force plate that recorded postural sway both forward and backward and side to side while they viewed a centrally located, patterned disk that rotated. Quite interestingly, individuals with mental retardation were found to be less influenced by the optic flow pattern than were the individuals without mental retardation. The conclusion drawn from this finding was that the motor systems of individuals with mental retardation are less sensitive to visual information than are the motor systems of individuals without mental retardation. The possibility that the differences were due to reduced perceptual, rather than motor, sensitivities was acknowledged but not explored.

### *Texture*

Texture, in addition to providing monocular depth information, is considered to be an elementary feature that is used to segregate figure from ground in a visual scene (Beverley & Regan, 1983; Nothdurft, 1991). Oross, Fryer, & Woods (1997) examined the texture-perception abilities of individuals with specific reading disability (SRD), individuals with mental retardation, and typically developing comparison individuals who provided chronological and mental age matches, using two tasks nearly identical to ones previously used to examine perception of depth- and motion-defined forms (Fox and Oross, 1988b; 1990). Stimuli were composed of arrays of oriented (vertical and horizontal) line segments. Texture-defined forms were created by presenting a background containing vertical line segments and a foreground, shaped as various letters, containing horizontal line segments. In one experiment—detection task—participants had to indicate which of two successively presented arrays contained the form. In the second experiment—discrimination task—participants had to indicate which of ten texture-defined forms was presented by selecting a matching form from a response board. Within each experiment, forms were degraded by either decreasing the density of the line segments (bar density) or intermixing horizontal and vertical line segments in the background and foreground areas (horizontal–vertical segmentation).

Figure 2.4 depicts the results for both experiments and indicates (a) no deficits in either detecting or discriminating texture-defined forms within the SRD group, (b) small impairments in both detecting and discriminating texture-defined forms that varied in horizontal–vertical segmentation among individuals with mental retardation, and (c) large impairments in both detecting and discriminating among texture-defined forms that varied in bar density among individuals with mental retardation. In brief, the small deficits in perceiving texture-defined forms, evident across variations in horizontal–vertical

**Figure 2.4**
Discrimination Performance of Individuals with SRD (Open Squares), Individuals with Mental Retardation (Filled Diamonds), and Typically Developing Comparison Individuals. Typically developing individuals provided chronological (open triangles) and mental age matches (filled circles) for texture-defined forms. The left panel quantifies the influence of variations in bar density; the right panel quantifies the influence of variations in degree of horizontal–vertical segmentation.

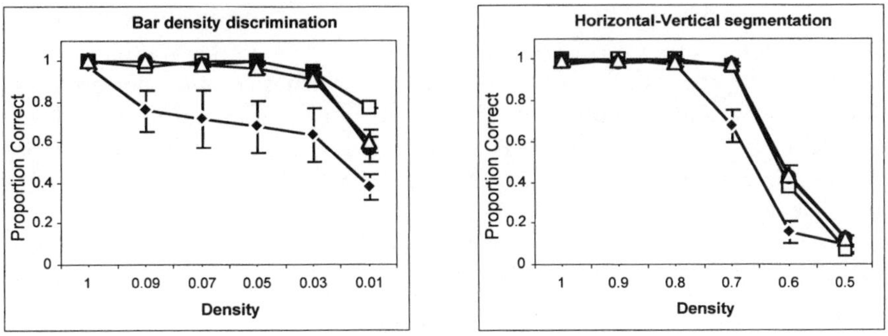


segmentation, suggested that texture perception abilities were relatively intact within individuals with mental retardation. Interestingly, the impairment that became evident with variation in bar density mirrored the impairment found in binocular stereopsis and may argue for a general deficit in interpolative abilities.

### *Color vision*

Color vision is tested with specialized stimuli that require the discrimination of lights or the recognition of forms that are similar in brightness but differ in wavelength composition. Color vision deficiency can be caused by congenital factors such as failing to inherit the normal complement of color vision genes, by optical factors such as the yellowing of the lens that often accompanies presbyopia, by localized brain injury and, as discussed below, possibly by neurodegenerative processes such as Alzheimer's disease. Although there is a substantial body of literature describing the outcomes of routine visual exams on individuals with intellectual disabilities, few of these reports describe assessments of color vision (but see Erickson & Block, 1999).

A few recent investigations of color discrimination by individuals with mental retardation have been prompted by three observations. First, middle-age adults with Down syndrome show a high prevalence of the behavioral and neuropathological changes that also accompany Alzheimer's disease (Lai & Williams, 1989; Wisniewski, Wisniewski, & Wen, 1985; Wisniewski, Silverman, & Wegiel, 1994). Second, there is accumulating evidence that many of the visual deficits that are observed in Alzheimer's disease are also found in individuals with Down syndrome (e.g., Neargarder, 1997; Rocco, Cronin-Golomb, & Lai, 1997). Finally, tritan vision impairments (blue–yellow discrimination losses) may be disproportionately prevalent in individuals with Alzheimer's disease and individuals with Down syndrome (Cronin-Golomb, Corkin, Rizzo, Cohen, Growdon, & Banks, 1991; Cronin-Golomb, Suguira, Corkin, & Growdon, 1993; Neargarder, 1997; Perez-Carpinell, de Fez, & Climent, 1994; Rocco, Cronin-Golomb, & Lai, 1997). In contrast, this pattern of deficits has not been observed in individuals with mental retardation of unspecified etiologies (Shimp & Oross, 1994).

It should be noted that most assessments of color vision and color discrimination have been made with tests designed to offer a quick pass/fail assessment that is used primarily to diagnose the most common congenital color vision deficiencies (e.g., Erickson & Block, 1999). Many of these tests, such as the Ishihara plates, can reliably identify and categorize the more severe color-vision defects, but they are not designed to quantify color discrimination. Other tests, such as the Farnsworth–Munsell 100, or laboratory tests such as the color matching of spectral lights, would offer insight into this ability.

### *Areas for future research*

Obviously, a broad topic such as spatial vision, combined with the relative lack of research involving individuals with intellectual disabilities, leaves quite a number of areas ripe for future research. To limit this part, we restrict our discussion to building on some of the topics that we have presented in this spatial vision section to highlight approaches and questions that we believe warrant investigation.

An individual's visual acuity is constrained by a combination of optical, neural, and behavioral factors (for extended discussions of this point, see, e.g., Bailey, 1998; Banks & Salapatek, 1983; Spear, 1993). Indeed, the recognition of this point is critical because different methods for assessing visual acuity often yield different estimates. Autorefraction methods provide a good estimate of the optical quality of the eye, but they provide no indication of how the neural processing proceeds (see, e.g., Bennett & Rabbetts, 1989). Visual-evoked potential (VEP) estimates of visual acuity, including sweep VEP methods, provide some indication of the early neural processing of visual stimuli; however, they usually do not indicate how visual information is processed by the higher cortical areas that underlie behavioral responses (see, e.g., Orel-Bixler, Haegerstrom-Portnoy, & Hall, 1989). Measures of eye movements such as optokinetic nystagmus (OKN) must be interpreted carefully because OKN and psychophysical estimates of visual acuity are differential-affected by a number of stimulus parameters (Schor & Narayan, 1981).

Research that attempts to isolate the factors contributing to visual acuity would contribute greatly to the field by identifying the relative contributions of optical, neural, and behavioral factors. This information would assist in determining the type of intervention or correction to apply and ascertaining the potential impact of visual deficiencies on other skills. Some recent findings, indicating that visual acuity limitations may, in fact, be determined by factors other than the optics of the eye in individuals with intellectual impairments, have confirmed the potential importance of adopting this approach. For instance, Woodhouse et al. (1996) report that poor visual acuity observed in a large sample of infants and children with Down syndrome could not be explained by refractive errors or by poor accommodation. They comment that their data suggests, instead, developmental differences in the maturation of visual cortex. Similarly, Courage et al. (1994), found that lower visual acuity estimates in children with Down syndrome remained even after excluding individuals who had ocular disorders or uncorrected refractive errors. Their conclusion also implicates neural deficits as a possible explanation.

There are other spatial abilities that, while offering little in the way of practical significance for everyday living, are theoretically important and would offer great insight into fundamental visual mechanisms. One of these is a type of visual acuity not mentioned above: positional acuity. Positional acuity describes the resolution of spatial localization—where objects are in the world relative to each other. Vernier acuity is a good example. A common variant of a vernier acuity task asks observers to judge the degree to which a pair of fine lines are collinear or aligned to each other.

Acuity measured with a vernier task is about ten times greater than that measured with a traditional visual acuity task. In fact, in a vernier acuity task, it is possible to distinguish the offset of a pair of lines even when the offset is much smaller than the diameter of a single photoreceptor in the retina. For this reason, vernier acuity is one of a class of very special abilities called "hyperacuities." (Other examples of hyperacuities include stereoacuity and wavelength discrimination; see Morgan, 1991.)

The high resolution of vernier acuity is possible because a line, for example, stimulates many photoreceptors in the retina, and neural processing of the input from these photoreceptors that occurs at later visual stages (including cortex) fine tunes object location. It is of special theoretical interest because (a) it has been studied extensively in nonclinical populations, (b) it has a unique time course of early development, (c) it is one of few visual abilities to show little decline with age, and (d) it offers an assessment of cortical function (for reviews, see Morgan, 1991; Enoch, Werner, Haegerstrom-Portnoy, Lakshminarayanan, & Rynders, 1999).

Interestingly, it has been documented that children with cortical visual impairment (CVI) exhibit greater losses in vernier acuity than in grating acuity. Cortical visual impairment refers to a developmental disorder with bilateral loss of visual acuity that results from damage to optic radiations or cortex that is usually due to perinatal trauma such as hypoxia and ischemia (Good, Jan, Burden, Skoczenski, & Candy, 2001; Huo, Burden, Hoyt, & Good, 1999). Identification of this disorder is also dependent upon eye exams revealing normal pupillary response and normal outcomes on ophthalmological examinations. In some cases, however, CVI has been noted to occur concurrently with other ocular abnormalities (Huo, et al., 1999). Another developmental disorder, cortical visual dysfunction (CVD), shows mpairments in other aspects of vision, such as problems with perception that include many of the difficulties (e.g., difficulty with perceiving motion and depth) seen in individuals with intellectual disabilities (Dutton, 1994). It would be important for future efforts to examine more closely the possibility of a link of CVI and CVD with mental retardation.

## Low-level: Temporal vision

For spatial vision, the important stimulus information involves how the stimulus appears or changes across space. In temporal vision, however, the important stimulus information involves how the stimulus changes across time. Recent work on the visual perceptual abilities of individuals with mental retardation suggests that tasks that require the accurate encoding of the temporal characteristics of, and relationships between, sensory stimuli show some of the largest impairments (Croce, Horvat, & Roswal, 1995; Fox & Oross, 1990; Nettelbeck & Brewer, 1981; Wade, 1990; Woods, DeFord, & Oross, 1996).

### *Motion*

Early studies of motion perception commonly examined various illusions and aftereffects created by the movement of real objects (see, e.g., Spitz & Lipman, 1959; Spivack & Levine, 1959; Werner & Thuna, 1942). More recent studies of motion have benefited from the development of computers that can generate and display specialized stimuli. The random-dot kinematogram provides a unique stimulus for studying motion processing by the visual system (see, e.g., Anstis, 1978; Nakayama, 1985). Fox & Oross (1990; 1992) tested the ability of adults with and without mental retardation to perceive forms generated by kinematograms to explore whether deficits noted in Fox & Oross (1988b) were restricted to the depth domain, or whether they reflected a more general perceptual impairment. Kinematograms are the temporal analogue of stereograms; thus, a series of seemingly random dot patterns, presented in rapid succession, contain an identical subset of dots that cannot be detected in any single pattern due to the randomly arranged surrounding dots. A form is perceived only when this subset of dots in the display is displaced in a correlated manner—across temporal frames—and the visual system detects this displacement.

Fox & Oross (1990) instructed participants to simply indicate the spatial position, left or right, of a rectangular form whose dots were vertically displaced as a correlated unit across frames. It was superimposed on backgrounds whose dots were uncorrelated across frames and possessed the same average displacement as that of the form. Upon the successful completion of a practice condition, sensitivity to the kinetic forms was assessed. Dot density and temporal correlation (the percentage of elements that move in a correlated manner across frames) were factorially combined to determine their effect on detection performance.

The results revealed that when either dot density or temporal correlation was high, both groups of subjects performed well. As the value of either of these variables was reduced, however, the performance of the adults with mental retardation quickly fell to chance. Their difficulty with reductions in temporal correlations suggested basic problems with motion perception. The difficulty with reductions in element density suggested the possibility of a general deficit in interpolation. Consistent with this hypothesis was the phenomenal report offered by adults without mental retardation that an apparent "filling-in" occurred that permitted the perception of the boundaries of the forms.

Two recent reports that also utilize random-dot kinematograms have revealed intriguing deficits among specific subtypes of individuals with intellectual disabilities. Atkinson, King, Braddick, Nokes, Anker, & Braddick (1997) demonstrated that individuals with Williams syndrome exhibited a pattern of specific deficits in visual functioning with deficits in detecting coherent motion and a relative sparing of detecting a form defined by texture created by oriented line segments. A more striking dissociation as a function of subtype was demonstrated in a task that examined the discrimination of forms defined by kinematograms (Fox, Yang, Feurer, Butler, & Thompson, in press). They compared the performance of four groups of individuals: a subtype of Prader-Willi syndrome with chromosome 15q deletion; a subtype of Prader-Willi syndrome with uniparental maternal disomy; individuals without Prader-Willi syndrome matched on age, intelligence, and fat mass; and nonclinical control individuals. In brief, all of the clinical groups performed worse than the nonclinical group. Interestingly, the subtype of Prader-Willi syndrome with the uniparental disomy performed significantly worse than the other two clinical groups. This finding may point the way for exploring how variability among individuals with mental retardation may either limit the conclusions that can be drawn or facilitate our understanding of why some, but not all, individuals with mental retardation exhibit selective sensory and perceptual deficits.

Paradigms that do not utilize random-dot kinematograms have also revealed evidence consistent with the hypothesis that individuals with mental retardation are impaired in perceiving motion. Shinkfield, Sparrow, & Day (1997) presented computer simulation of arm movements and asked individuals with mental retardation to make visual discriminations of the extent and duration of the arm movements. Their findings revealed that the visual discrimination of both extent and duration of arm movement were significantly impaired in individuals with mental retardation.

Using a very different experimental paradigm, Conners, Wyatt, and Dulaney (1998) studied motion processing in adults with mental retardation. These researchers studied the "representational momentum effect," an error in the mental representation of moving objects. Observers frequently err in determining the stopping point of an object that has been in motion. This error manifests itself as a selected stopping point that is farther along the path of motion, and it is thought to be due to a lag in the mental processing of this information: a "cognitive inertia." In this experiment, adults with and those without mental retardation both showed the "representational momentum effect," but the magnitude was significantly smaller for the adults with mental retardation. The authors concluded that these findings were consistent with the idea that adults with mental retardation cognitively process motion similarly to adults without mental retardation, but that they do so less efficiently.

### *Visual duration discrimination*

Duration discrimination describes the ability to (a) discriminate the passage of time between two visual events or (b) discriminate visual stimuli based only on their length of presentation. In children and adults without intellectual disabilities, the ability to make these discriminations has been studied extensively and there are now several theoretical treatments of this perceptual process (Allan, 1979; Allan & Kristofferson, 1974; Ivry & Hazeltine, 1995). Temporal discrimination tasks have been said to tap an internal clock, or "central timekeeping mechanism" (Grondin & Rousseau, 1991), or rely on an internal representation of the stimulus event (Ivry & Hazeltine, 1995). The interest in describing the functional properties of the internal clock has been great, partly because a single central timing mechanism is thought to be shared by the different perceptual, cognitive, and motor behaviors (Keele, Pokorny, Corcos, & Ivry, 1985; Rousseau & Rousseau, 1996). Except for preliminary work from our laboratories, as described later, this is an important area of research that has not been examined.

We asked individuals with and without mental retardation to perform a task in which they had to discriminate between two brief flashes of light. One flash was always 50 milliseconds in duration and the second was always 50 milliseconds plus one of five added time increments ranging from 20 to 640 milliseconds. We measured the percent correct discrimination as a function of this added increment and the data is presented in Figure 2.5. Individuals with mental retardation required substantially larger differences in the duration of the two visual flashes to reach the performance level seen in individuals without mental retardation.

Duration discrimination (the ability to discriminate short-duration stimuli) is usually distinguished from time estimation (the ability to discriminate long-duration stimuli). When the stimulus durations are brief, the task is thought to reflect perceptual processes, but when the durations are on the order of seconds, the task has been theorized to instead reflect cognitive processes (Macar, Grondin, & Casini 1994). Although there is a large literature on this topic in typically developing individuals, we do not know of any studies directed at the

**Figure 2.5**
Visual Duration Discrimination Obtained from Individuals with (Filled Squares) and Without Mental Retardation (Open Diamonds) with a Base Stimulus Value of 50 Msec.

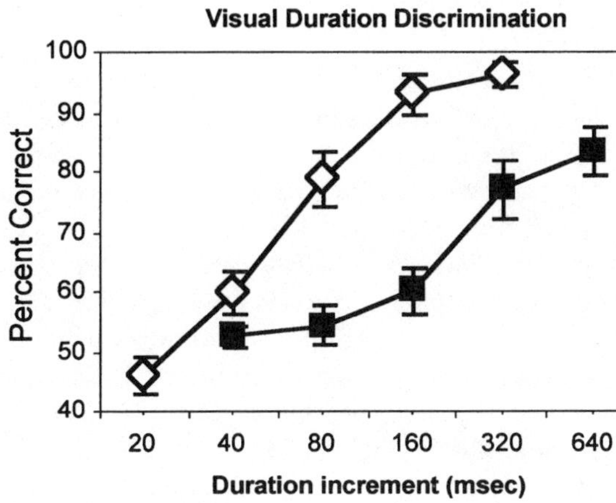


visual time-estimation abilities of individuals with intellectual disabilities. Interesting findings have been reported, however, in two previous studies of auditory time estimation abilities of individuals with mental retardation (McNutt & Melvin, 1968; Mulhern, Warm, and Clark, 1974). Both studies report that individuals with mental retardation produced underestimates of the auditory stimulus durations.

### *Areas for future research*

One of the most fundamental measures of the temporal resolution of the visual system is the Critical Flicker Frequency (CFF) task. The CFF measures individuals' abilities to detect flicker in a light stimulus whose luminance is intermittent, that is, turning on and off at some periodic rate. The highest frequency (flicker rate) that can be seen as flickering by an observer is used to infer the temporal resolution of the visual system. The perceptual quality being judged in this task is flicker or no-flicker, it does not directly reflect our perception of the passage of time or of temporal duration. The CFF task is thought to be mediated little by cognitive processes and to incur little cognitive demand (Macar, Grondin, & Casini, 1994; Rammsayer, 1993; Rammsayer & Lima, 1991).

Despite its importance, however, only a few studies have used the CFF to characterize the temporal processing of individuals with intellectual disabilities. Jensen (1983) summarized much of the early work attempting to correlate the CFF with measures of intelligence. Most of these studies did not include individuals with mental retardation, and it is difficult to adequately characterize the nature of the retardation among those studies that did. Recently, Ali, Khaleque, Khanam, al-Shatti, & Ahmed (1994) found significant impairments in individuals with mental retardation compared to individuals without mental retardation.

Other topics related to temporal processing that would seem to warrant investigation include systematic analyses of various aspects of motion perception. As one example, research in nonclinical adults and children has shown that adding color to a global motion stimulus (i.e., noise dots one color; signal dots a different color) greatly facilitates image segmentation and consequently improves motion coherence thresholds (Croner & Albright, 1997). This robust finding is intriguing, as it has been typically assumed that the visual pathway that processes motion information is insensitive to color. Thus, these results may point to a greater interconnection between pathways than has previously been acknowledged. It seems important, therefore, that we examine one question regarding the cause of past motion-perception deficits observed in individuals with mental retardation. Specifically, could image-segmentation deficits be one underlying cause of motion-perception deficits? We are currently exploring the role of image segmentation in motion perception by manipulating stimulus features in the global motion display.

## High-level visual abilities

Visual tasks are considered low-level if our perceptual experiences can be described quite well by knowing the physical characteristics of the visual stimulus. In contrast, we stated that visual tasks are described as high-level if prior context, experience, learning, and memory play a role in determining our perceptual experiences. Although this serves as a common and useful way of categorizing our perceptual abilities, it is becoming increasingly difficult to segregate visual tasks, due to an increasing number of reports outlining various feedforward and feedback connections between distinct cortical areas. Our goal for this part, therefore, is to address three new areas of research we are involved with that have been, to at least a number of investigators, thought to be higher-level. Elements within each of these areas, however, can be identified as also involving low-level abilities.

### *Visual search*

As nicely described in Soraci, Carlin, and Wiltse (1998), one use of visual-search tasks is to index the ability of individuals to identify relevant visual information from a scene that can be utilized in more complex problem-solving settings. In a typical visual-search task, a variable number of items is presented in an array and individuals attempt to identify a particular target item (or items) from among the distractors. It has been well documented that visual search can, depending on the characteristics of the items, occur in a serial manner in which individuals appear to search the display item by item to find the target. In this type of search, the amount of time needed to find the target increases as the number of distractors is increased. Alternatively, individuals may also search the display in a parallel manner, indicated by a rapid detection of the target, the speed of which is unaffected by the number of distractors (Treisman & Gelade, 1980).

Carlin, Soraci, Goldman, and McIlvane (1995) very clearly established that individuals with mental retardation are less efficient in their visual-search abilities than are nonclinical individuals. Specifically, individuals with mental retardation exhibited search patterns that were slower and that appeared serial in nature when presented with items (varying in color, form, and size) known to induce parallel search in nonclinical individuals. Moreover, their data revealed the existence of subgroups among the individuals with mental retardation that indicated varying levels of deficits in visual search.

Inspired by these results, we sought to explore the visual-search abilities of individuals with mental retardation when presented with visual-search asymmetries. Visual-search asymmetries occur when items yield both serial and parallel search patterns, depending on whether a given item is presented as the target or as distractor items. For example, searching for a *Q* from among a set of *O*s occurs in a parallel manner, whereas searching for an *O* from among a set of *Q*s occurs in a serial manner (Treisman & Souther, 1985). Utilizing these types of items permitted us to characterize how individuals with mental retardation perform when presented with items known to induce both parallel and serial search.

In the first of two experiments, participants were instructed to search for either a *Q* among *O*s or an *O* among *Q*s. Both groups of individuals demonstrated search patterns that could be characterized as parallel or serial. Individuals with mental retardation, however, were significantly slower in both serial and parallel searches than were the nonclinical individuals. This slowing was not expressed as a simple elevation in response times that remained constant across a varying number of distractor items. Instead, the difference in response times increased as the number of distractors increased. Moreover, similar to the results obtained by Carlin et al. (1995), we found evidence for subgroups. Even when engaged in a condition that should yield parallel search (*Q* target, *O* distractors), approximately half of the individuals with mental retardation exhibited search strategies that could be characterized as serial.

The second experiment also presented items known to generate asymmetrical search strategies (*C* target with *O* distractors yields parallel search; *O* target with *C* distractors yields serial search). In this experiment, we manipulated target salience by varying the size of the gap in the *C*s. Again, a subgroup of individuals with mental retardation exhibited serial search strategies when presented with the task that should yield parallel search. We also found systematic variations in the serial-search condition related to both the number of distractor items and the salience of the gap sizes.

Thus, across these two experiments, some of our results point to simple quantitative differences in performance. Specifically, individuals with mental retardation require more time than nonclinical individuals to find targets from among distractors. Other results, such as the existence of subgroups and the increases in slowing that are coupled to increases in number of distractors, however, appeared to be qualitative in form and may reflect limits in processing capacity. The observed differences in performance between subgroups were not related to differences in IQ, visual acuity, age, or gender. The basis for subgroups is a topic that we believe warrants increased attention in future studies.

### *Boundaries and objects*

The goal of this part is not to exhaustively explore research examining object perception in individuals with mental retardation. Other researchers more intimately involved in these studies have reported many of these findings (see, e.g., Carlin & Soraci, 1993; Soraci, Carlin, & Wiltse, 1998). Rather, our goal is to briefly discuss a topic that was suggested by our investigations of low-level visual abilities in individuals with mental retardation. Specifically, we consider one of the processes that underlie the perception of object boundaries—interpolation. We would like to note that many investigators, including ourselves, believe that the process of interpolation begins very early in the analysis of visual information and, therefore, may be considered a low-level visual ability. Nevertheless, we include this part under the topic of high-level visual abilities in this chapter due to the fact that we are concentrating on the perception of forms defined by interpolation.

The process of interpolation, although somewhat incomplete in definition, is presumed to govern the perceptual "filling-in" that occurs under many conditions (Kellman & Shipley, 1991; Ramachandran, 1992). Several of these conditions include the filling-in of our blind spot, the perception of objects as intact when portions of the objects are occluded (e.g., a person standing behind a tree), the perception of subjective contours, and the phenomenological report of filling-in as the element density of stereograms, kinematograms, and texture-defined forms is reduced. The importance of interpolation is obvious for maintaining stable percepts across changing visual conditions.

One difficulty, however, in assessing interpolation abilities is in determining objective indicators. This difficulty is responsible, at least in part, for the reliance on phenomenological reports in investigations of interpolation abilities. Consider, for example, the display labeled "subjective form condition" below in Figure 2.6. Nonclinical observers commonly report "seeing" faint edges that define the square, apparent brightness differences between the square and background, and that the square appears to be placed on top of the four black disks. In examining interpolation abilities in animals, very young infants and children, and individuals with intellectual disabilities, a reliance on phenomenolog-

**Figure 2.6**
Schematic Representation of the Four Test Conditions Used in this Investigation of Subjective Contour Perception. In the actual stimulus, the 4 inducing elements shown here were placed into a 4 x 4 grid of inducing elements. The pattern of rotations applied to these elements generated a form that was moving either up and down or back and forth in the middle of the grid.

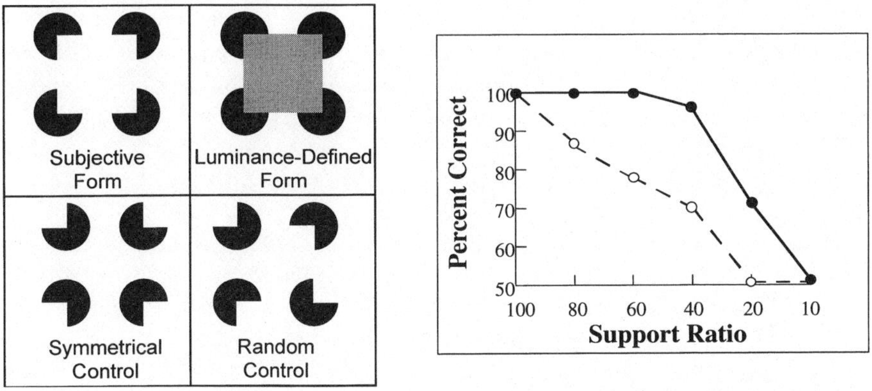


ical reports is not possible because of the limitations in verbal skills along with other behavioral limitations. For these reasons, it is necessary to involve objective indicators that can illustrate the presence or absence of interpolation.

Following a series of pilot tests, we conducted formal testing of individuals with and without mental retardation in the following manner. Participants were shown a grid (4 x 4) of "inducing elements." These elements changed across a sequence of frames so that proper alignment of the missing segments in the inducing elements would yield the percept of a form moving either up and down or left and right. Four conditions were created (see Figure 2.6): a subjective-form condition, a luminance-defined form condition, a symmetrical control condition, and a random-configuration control condition. For each of these conditions, we presented trials in which participants had to make a forced-choice judgment regarding the direction of movement of the form. We also varied the magnitude of interpolation required by making the inducing elements small, thus requiring interpolation to bridge a large amount of visual space, or making the inducing elements large.

The condition of interest was the subjective-form condition. Indeed, the performance of the two groups was identical for the three control conditions. The data for the subjective-form condition revealed that individuals with mental retardation were able to detect the subjective form. Note, however, they could do so only for visual distances of about one third the distance spanned by the individuals without mental retardation. These data, along with deficits previously noted for dot density reductions in stereograms, kinematograms, and texture-defined forms, is consistent with a fundamental interpolation deficit.

### *Face perception*

Across a number of studies, individuals with mild mental retardation are reported to exhibit difficulties in making judgments about emotional expression. Hobson, Ouston, and Lee (1989) presented individuals with and without mental retardation a task that required matching a photograph of an emotional expression to some emotional sounds (e.g., sad sighs and groans), or a photograph of common objects to their appropriate sounds. Their results indicated that the individuals with mental retardation were impaired, relative to the typically developing individuals, on matching the emotional expressions with their sounds but not with respect to matching the common objects to their sounds. What is unclear, and acknowledged by the authors, is the basis for the impairments. Among the possibilities suggested are mistakes in identifying the emotional vocalizations, the photographs of the emotional expressions, or both.

Other investigators have argued that the use of static photographs may lead to an underestimation of face-perception abilities. Evidence supporting this argument was provided by Harwood, Hall, and Shinkfield (1999), who presented both static and dynamic displays of emotional expressions to individuals with and without mental retardation. They found that individuals with mental retardation were relatively deficient for both display types; however, both groups did significantly better when presented with dynamic displays.

In one of the few studies to contrast the performance of individuals with varying levels of mental retardation, McAlpine, Kendall, and Singh (1991) found that individuals with mental retardation were impaired in their ability to match the facial expression depicted in a photograph to a short story. In addition, their data indicated that the magnitude of the impairment was correlated with the severity of the mental retardation. Using line drawings of facial expressions and short stories, Xeromeritou (1992) replicated the findings of McAlpine et al. (1991). As summarized in the review of Rojahn, Lederer, and Tassé (1995), the overall conclusion is that individuals with mental retardation possess difficulties in understanding emotional expressions depicted in faces. What also emerges from this review is the unmistakable conclusion that the basis for many of these difficulties is unknown.

Our new and ongoing research program involves presenting individuals with various developmental disabilities (e.g., autism, pervasive developmental disorder, and mental retardation) with a battery of tests to assess their face-perception abilities. The tests include probe trials embedded in baseline trials. The baseline trials measure the ability of the participants to discriminate between everyday objects; the probe trials are of several types. One type of trial measures visual–visual matching of identity with different individuals displaying a neutral emotional expression. A second type measures visual–visual matching of identity with different individuals displaying the same emotional expression (across trials, happy, sad, angry, fearful, disgusted or surprised). A third type measures visual–visual matching of emotional expression with the same individual displaying different emotions. A fourth type measures visual–visual matching of emotional expression with different individuals displaying different emotions. A fifth type measures auditory–visual matching of emotional expression when presented with auditory samples (e.g., spoken word "happy") and different individuals displaying different emotions. A sixth type measures expressive language abilities by asking participants to label emotional expressions.

The results that we have been obtaining reveal that profiles of abilities and disabilities vary across individuals. Some individuals have exhibited accurate performance only on trials that involve matching on the basis of identity. Still other individuals exhibit accurate performance on emotional expression trials only when the visual–visual matching involves the same individual displaying the emotional expressions. In still other individuals, the performance profiles are more complex. In sum, combined with similar clinical neuropsychological findings (see, e.g., Rapcsak, Comer, & Rubens, 1993), our findings point to the importance of examining a wide range of abilities designed to more fully assess understanding of faces and the identity and emotion information that they convey.

### *Areas for future research*

Determining the interaction between low-level and high-level abilities will no doubt be one of the most fascinating lines of research to be conducted in the coming years. As is shown repeatedly, the most common approach taken when examining visual abilities (low and high level) is to isolate the ability of interest. According to Geisler and Chou (1995) this approach has " . . . done little to reveal the relationships between the low-level and high-level mechanisms, nor have they provided much insight into which mechanisms are more important for predicting performance in complex real-world tasks" (p. 356).

Over the last several years, there have been a number of reports that begin to correct this shortcoming in the field of vision science. Adelson (1993) described a new set of visual illusions that examined how brightness perception (a low-level ability) was influenced by aspects of perceptual organization (a high-level ability). Combining patch areas of varying shades of gray, white, and black created geometrical visual patterns that generated these illusions, complex offspring of the displays used to create the effect known as "simultaneous contrast." The phenomena and the data obtained when participants are asked to adjust the luminance of various patch areas to match the perceived brightness of other selected areas are striking. Changes in the geometrical organization of the displays that should play little, if any, role in brightness judgments are found instead to play a dramatic role.

Similar findings across a variety of domains have also pointed to a greater degree of interaction between low-level and high-level visual abilities. Hubner (1997) demonstrated that the global precedence effect that is evident when observers have to identify larger forms or letters that are made from smaller forms or letters is influenced by both sensory and attentional mechanisms. Geisler and Chou (1995) demonstrated that many of the effects in visual search tasks that have been related to high-level abilities, such as attention and decision processes, can be accounted for by a careful consideration of low-level abilities. Across a series of reports, motion perception has been shown to be more reliant on high-level attentional mechanisms than previously believed (Cavanagh, 1992, 1993; Culham & Cavanagh, 1994).

In a related vein, attention has been shown to modulate the response of neural structures and, in turn, to improve behavioral performance (see, e.g., Spitzer, Desimone, & Moran, 1988). Corbetta, Miezin, Dobmeyer, Shulman, and Petersen (1990), using positron emission tomography (PET) methodologies, found that the degree of attention given to a particular stimulus dimension (e.g., color, shape, or movement velocity) increased both behavioral sensitivity and the activation of neural areas that are presumed to process the stimulus dimension. In more recent investigations, the prefrontal cortex, an area involved in higher-level skills such as decision making, has been shown to modulate the neuronal activity of extrastriate cortex during tasks that require visual discrimination (Barcelo, Suwazono, & Knight, 2000).

Taken together, the studies cited in this section clearly indicate that the processing of visual information cannot occur in a simple bottom-up manner. Rather, the end result of processing visual information is generated by a complex set of feedback and feedforward connections that involve multiple areas of cortex and multiple domains of behavior.

## IMPLICATIONS AND DIRECTIONS

In this concluding part, our goal is to introduce, in a brief manner, a number of topics, questions, and concerns to stimulate more detailed considerations in the future. These topics range from non-sensory influences on assessments of sensory and perceptual abilities, to an exploration of the link between neurological structures and function, to impact of sensory/perceptual deficits on other skills, and, finally, to issues of prevention and remediation.

### Non-sensory influences on assessments of sensory and perceptual abilities

A factor that must be carefully evaluated when assessing the sensory or perceptual abilities of individuals with mental retardation is the extent to which non-sensory factors may be influencing task performance. Some of the potential confounds that must be addressed include prescription medication use, cognitive load (task difficulty, comprehension, attention, and memory demands), willingness to participate (diligence, distractibility), and psychophysical responses and response bias.

It is important to identify individuals using psychoactive medications. These can compromise the data in many ways, and individuals with mental retardation are more likely to have been prescribed medication. Consider, as one example, an experiment in which human subjects were administered either haloperidol or a placebo and duration discrimination was measured for briefly presented (50 msec) stimuli. Discrimination performance was severely impaired in the group administered haloperidol: discrimination thresholds were larger by approximately a factor of two (Rammsayer, 1993).

Concerns regarding the cognitive demands of perceptual tasks are ordinarily the most problematic. Task difficulty and comprehension can be addressed with initial practice trials that demonstrate criterion (as defined) performance. However, the attentional and memory resources required by a perceptual task must be addressed with compelling argument or clever experimental design. The nature of the psychophysical task, and the required perceptual decision and response, will also determine the decision-making process and the cognitive demand. The same–different task, for example, is a desirable choice because the perceptual decision is not dependent on the subject understanding the stimulus dimension upon which they are making their judgment (Macmillan & Creelman, 1991). This task is especially appropriate when " ... the physical dimension itself is difficult to characterize for subjects, or the experimental design precludes training, or subjects are unsophisticated, and forced-choice instructions are difficult to convey" (Macmillan & Creelman, p. 141).

Psychophysical experiments require the observer to make a response (usually spoken) concerning the stimuli that have been presented and this response is the result of a decision-making process. It is well known that in many settings the experimental subject may show a response bias, that is, an inclination to respond in a certain way (e.g., automatically responding "no" when unsure). Response bias can greatly affect performance and, in fact, percent correct responses must be interpreted with caution because they may reflect true sensitivity differences, criterion (bias) differences, or both. Individuals with mental retardation, just like nonclinical subjects, may be prone to response biases in experimental settings (Heal & Sigelman, 1995). Experimental designs that permit signal-detection analyses allow the subjects' sensitivity to the stimulus dimension and any criterion differences to be measured independently. This is a methodological concern that has infrequently been addressed in research on mental retardation (but see Ellis, McCartney, Feretti, & Cavalier, 1977; Laine & Baumeister, 1985; Lally & Nettelbeck, 1977).

We believe that many of the concerns above can be addressed by screening for prescription drug use, providing preliminary practice trials, carefully selecting a psychophysical method, and paying close attention to the shape of the measured psychometric functions. Research methods that permit the entire psychophysical function to be described offer several advantages. If differences exist in these plotted functions for clinical and nonclinical populations, for example, the exact nature of these differences offers information about the likely

**Figure 2.7**
Hypothetical Data Representing Impact of Differences in Sensitivity (A), Response Bias (B), and Non-sensory Factors such as Lack of Diligence (C) on Performance in a Psychophysical Task.

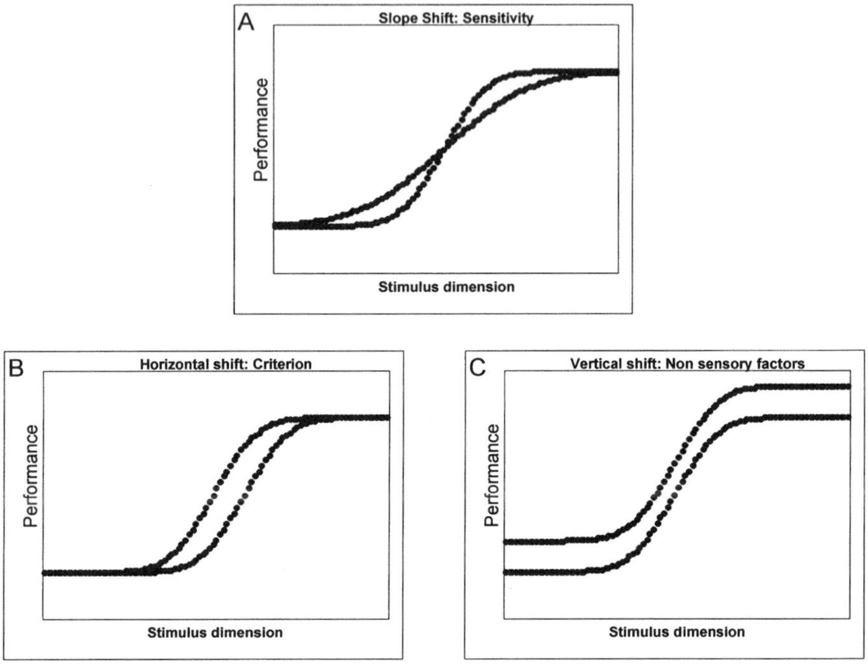


cause. The psychophysical function, or ogive, generally has a shape similar to that of the cumulative normal distribution. For any two groups of observers, as shown in Figure 2.7, the psychophysical functions can be different in slope (A), different in horizontal placement (B), or different in vertical placement (C). Slope differences (A) would indicate differences in their sensitivity to the stimulus dimension. A horizontal shift (B) would be indicative of a difference in criterion or response bias; and a vertical shift (C) would represent differences in performance that are most likely due to non-sensory factors such as attention or motivation. This "curve-shift paradigm" can be used to identify how different experimental manipulations affect performance and behavior (see Meck, 1995). This approach might also be ideally suited for specifying how task performance is influenced by training.

## Linkage of neurological structures and sensory and perceptual abilities

Low-level visual impairments in individuals with mental retardation appear to be selective: Aspects of motion and depth perception are impaired, whereas visual acuity, static contrast sensitivity, color vision, and the perception of texture appear relatively unimpaired. Within a given etiology, subgroups of adults with mental retardation appear to be differentially impaired. Even within a group of adults with mental retardation of unspecified etiology, approximately 60 to 75 percent of adults appear to possess moderate levels of impairments whereas the other 25 to 40 percent of adults are distributed, roughly evenly, into two groups—one that appears to have typical levels of sensitivity and the other with either severe impairments or an apparent inability to comprehend task instructions.

Anomalies in visual abilities, especially when selective, have provided researchers with information about both the structure and function of mechanisms in the visual system that are otherwise difficult to study. Even a casual inspection of the research on visual anomalies reveals how susceptible the brain is to insult. Although it is beyond the scope of this chapter to consider all of this research, it is helpful to consider specific anomalies associated with some diseases and injuries in order to compare them with the deficits found in individuals with mental retardation. As described in the introduction, perhaps the most famous case in recent history involves the motion-blindness of a patient who incurred bilateral posterior brain damage (Zihl, von Cramon, & Mai, 1983). However, this is by no means the only interesting case.

Degenerative diseases such as Alzheimer's disease and multiple sclerosis also selectively impair perceptual abilities. Alzheimer's disease is known to differentially attack the magnocellular pathway of the visual system, which is presumed to be primarily responsible for motion and depth perception (see, e.g., Livingstone & Hubel, 1988). Thus, Trick and Silverman (1991) examined whether the degeneration seen in retinocortical pathways with Alzheimer's disease would result in impaired motion perception. Threshold estimates for perceiving the direction of global motion in kinematograms indicated that motion perception was indeed impaired, and, furthermore, severity of Alzheimer's was positively correlated with the degree of loss. In their discussion it was speculated that the spatial disorientation often experienced by patients suffering from Alzheimer's may be due, at least in part, to their perceptual impairment because the ability to detect the motion of objects is crucial for spatial localization ability.

This brief survey suggests many research areas that are worthy of inquiry. One area is dedicated to determining the underlying neural substrate responsible for various perceptual abilities. For instance, determining what visual areas (either in terms of retinocortical pathways or cortical structures) play what role in enabling specific abilities such as perceiving the direction of motion (e.g., Livingstone & Hubel, 1987; 1988; Newsome, Britten, & Movshon, 1989; Ungerleider & Desimone, 1986; Ungerleider & Mishkin, 1982). Despite the potential promise that this area holds, however, it is difficult to speculate about neurological impairments and perceptual impairments in individuals with mental retardation at this time. As mentioned earlier, vastly different types of diseases or injuries may result in similar perceptual deficits. For example, the perception of motion is impaired as temporal correlation is reduced for individuals with mental retardation of unspecified etiology, stroke patients with localized damage to cortical areas, and cats that have lost directionally selective neurons throughout the brain. Thus, it is unlikely that research conducted in this area will provide useful information concerning the role that damaged structural features play in the perceptual impairments of individuals with mental retardation until a battery of sensitive tasks is designed that can differentiate between different diseases or injuries.

Attempts to differentiate between different diseases or injuries are reliant on evidence for a profile of selective impairments. The profiles of sensory and perceptual impairments that have been observed in adults with mental retardation and, in fact, in other clinical conditions, have been generally derived from different participants tested in different experiments. Moreover, almost all of these experiments find that only a subset of individuals can complete the testing and there is little, if any, attempt to determine what variables differentiate between these subsets of individuals. When a descriptive profile of a clinical condition is derived from the results of different participants tested in different experiments, the possibility exists that differential inclusion of subgroups may cause the overall group profile to be distorted. For this reason, it is incumbent upon future investigations to test a variety of abilities within the same individuals to establish visual ability profiles. These efforts will no doubt be aided by the advances that have been made in marrying neural imaging methodologies to behavioral testing (see, e.g., Deutsch, Oross, DiFiore, & McIlvane, 2000; DiFiore, Dube, Oross, Wilkinson, Deutsch, & McIlvane, 2000).

## Impact of sensory/perceptual difficulties on other abilities

That sensory and perceptual difficulties influence other skills is well documented in the literature. Individuals with various kinds of mental retardation exhibit a variety of deficits in navigating their environment that may originate, at least in part, due to their visual impairment in processing spatial information. Some specific examples include spatial confusions and stumbling (see, e.g., Henderson, Mack, & Williams, 1989; Liu, Gauthier, & Gauthier, 1990; Rider, Mahler, & Ishee, 1983; Rieser, Guth, & Weatherford, 1987; Uecker, Mangan, Obrzut, & Nadel, 1993; Wade, Newell, & Hoover, 1982).

Temporal processing efficiency and the perception and estimation of time are central and influential qualities of perceptual and cognitive processes. A visual temporal-processing deficit would impact many behaviors. For example, impairments in visual temporal information may contribute to difficulties observed in perception-action tasks among individuals with mental retardation. It has been suggested that manual tracking impairments derive from difficulties in establishing the temporal coordination of motor commands with sensory information (Henderson, Morris, & Frith, 1981). Impairments have been reported for individuals with mental retardation on simple and complex motor tasks

(Blais & Kerr, 1990; Croce, Horvat, & Roswal, 1995; Kerr & Blais, 1987) and in simple and choice–reaction-time tasks (Lally & Nettlebeck, 1977; Nettelbeck & Brewer, 1981). These impairments may also be driven by impaired temporal information processing.

Revealing the specific nature of temporal processing impairments may provide insight into other perceptual, cognitive, and motor abilities. For example, speech perception relies on the accurate coding of the duration of speech sounds and the gaps between sounds. It has been shown that the same sounds are perceived differently with variations in duration (Repp, Liberman, Eccardt, & Pesetsky, 1978). Impairments in auditory duration discrimination, therefore, may contribute to some of the speech comprehension difficulties noted in individuals with mental retardation. Also, a large body of literature links auditory temporal-processing impairments with language deficits in children with dyslexia (Tallal, Miller, & Fitch, 1993).

A number of research lines converge on the conclusion that impairments in producing and recognizing emotional expressions may negatively impact on social skills. Research in social referencing, for example, has demonstrated that parents and caregivers who interact with individuals with mental retardation often report that they have little confidence in their interpretation of facial expression in these individuals (see Walden, Blackford, & Carpenter, 1997). Cicchetti and Sroufe (1978) have reported that individuals with Down syndrome exhibit facial expressions that are atypical relative to the expressions exhibited by typically developing individuals. Children with pervasive developmental disorder have been found to exhibit deficits in facial emotion recognition that correlated with the socialization domain of the Vineland Adaptive Behavior Scales (Braverman, Fein, Lucci, & Waterhouse, 1989).

One obvious question that is prompted by these observations concerns how improvements in visual functioning would influence performance in other domains. In addition to the research cited earlier in this chapter, this question has been examined in some detail in individuals with Alzheimer's disease. Changing lighting conditions to enhance the amount of visual contrast has been shown to improve performance on a variety of tasks. These tasks have included picture recognition (Gilmore, 1995; Gilmore et al. 1995; Lustig et al. 1995), reading speed (Gilmore, Thomas, Klitz, Persanyi, & Tomsak, 1996), and face discrimination (Brown, Dunne, Jain, Cronin-Golomb, & Cronin-Golomb, 1996). At times, the improvements have been so large that the original behavioral deficits were functionally eliminated. Similarly, Salthouse, Hancock, Meinz, and Hambrick (1996) have demonstrated how variations in visual acuity impact working memory and associative learning. It remains to be established whether or not similar benefits can be realized in individuals with intellectual disabilities.

## Issues of prevention and remediation

A particularly helpful take on prevention and remediation was offered by Warren (1992), who defined primary, secondary, and tertiary prevention. Primary prevention was defined as the approach designed to reduce the number of new instances of disabilities or problems through action that eliminates risk factors. The goal of secondary prevention is to reduce the number of existing instances of disabilities or problems. Tertiary prevention attempts to reduce the impact of a disability or problem.

The vast majority of research conducted to explore the origins of sensory and perceptual impairments does not concern itself with issues of primary prevention. However, as outlined below, methods that touch on secondary and tertiary prevention are now being applied to instances of sensory and perceptual impairments in individuals with mental retardation. Often these methods involve more frequent vision testing and prescribing corrective lenses to counteract losses in visual acuity. More importantly, a rapidly growing area that has come to be called "perceptual learning" may hold particular promise for addressing other forms of sensory and perceptual impairments.

One of the most overlooked aspects of perceptual abilities is that many of them are plastic. That is, these abilities can be enhanced (or sometimes degraded) by the provision of experience (see, e.g., Ball & Sekuler, 1982, 1987; Carlin, Soraci, Goldman, & McIlvane, 1995; Gibson, 1969; Ramachandran & Braddick, 1973). Karni and Sagi (1993) suggest that experience-dependent perceptual learning may consist of two stages, fast and slow. Fast learning (defined as a rapid improvement evident over the first few blocks of trials that saturates quickly) is theorized to reflect the learning of a task-specific routine, whereas slow learning (defined as the small improvements in psychophysical thresholds across daily sessions) is theorized to reflect structural changes in neural mechanisms.

With respect to individuals with intellectual disabilities, Sal Soraci, Mike Carlin, and colleagues have obtained an intriguing set of findings. These findings, summarized in Soraci, Carlin, and Wiltse (1998), document a number of domains of visual functioning including visual search, perception-of-object symmetry, and motion perception, in which specialized training procedures have proven to significantly improve the performance of individuals with intellectual disabilities. Consider, for example, a recent study by Carlin, Hobbs, Bud, and Soraci (1999) that used random-dot kinematograms similar to those employed by Fox and Oross (1990) to explore motion perception and perceptual learning in individuals with mental retardation and autism. Individuals were initially presented with a kinematogram containing high levels of both element density and temporal correlation. When an initial loss of motion-perception ability (defined as a failure to exhibit 85% or better discrimination) was found in a given individual, that individual was presented with a training procedure. This training included two components: response-based training and surround fading. In brief, the response-based training taught a matching procedure that may have facilitated understanding of how to indicate the presence or absence of a target. The surround-fading training initially removed the background of moving dots and, over time, reintroduced the background. The outcome of the training was both successful and impressive. Following training, individuals with mental retardation and autism exhibited sensitivity levels comparable to a group of nonclinical individuals.

The magnitude of the training effect on performance by individuals with mental retardation and autism is an exciting finding. However, because the nonclinical adults and children were not similarly trained, it is difficult to draw conclusions about differences in performance between the two groups and how they relate to sensory and non-sensory factors that may limit performance. As the authors note, the exact cause of the observed improvements remains open to speculation.

It also remains open to speculation the degree to which various perceptual learning tasks may improve the performance of individuals who exhibit reductions, rather than a complete loss, in sensitivity. In an initial attempt to examine this issue, Oross, Carlin, and Fox (1990) presented individuals with, and without, mental retardation random-dot kinematograms. For each individual, pilot testing was conducted to identify two classes of stimuli, one that initially yielded approximately 75 percent correct detections and one that yielded chance (50%) detections. Two training procedures were used. Fade training involved initially making the test stimulus quite discriminable by introducing large differences in density between it and the background and, over trials, gradually reducing the difference. Repeated-exposure training involved simply presenting the same stimuli over trials. For each procedure, subjects were asked to indicate the left–right spatial position of the kinetic form. Neither training procedure resulted in substantial improvement. For stimuli, initially 75 percent discriminable there was some small improvement (approx. 10%) for each of the two groups. No improvements were found for the stimuli that initially yielded chance levels of discrimination. This outcome was interpreted as being consistent with the hypothesis that the deficits in motion perception originate early within the visual system rather than being due to factors such as task comprehension, motivation, or attention deficits. Obviously, perceptual learning is one topic that may hold great promise for issues of secondary and tertiary prevention of sensory/perceptual impairments.

One limitation of perceptual learning research conducted to date, however, is that little effort has been directed toward establishing that the improvements in sensitivity seen in the laboratory have a direct impact on real-world skills. Moreover, there has been precious little research directed toward determining compensatory strategies that may be applied to various problems when efforts to eliminate or reduce sensory and perceptual impairments are unsuccessful. Fortunately, the application of perceptual abilities and limitations to real-world settings comprises a substantial portion of the field known as "human factors." The field of human factors tries to insure that our environment and we are a good fit, or to put it another way, that our physical surroundings are not at odds with our *umwelt*. The goal of human factors is to match the person to the environment to reduce error, increase productivity, and enhance safety and comfort (Wickens & Hollands, 2000). To that end, the field of human factors relies on five different approaches to achieve this goal: equipment design, task design, environmental design, training, and selection.

The application of human factors, therefore, can take many different forms. As one example, consider that many employment opportunities for individuals with intellectual disabilities involve perceptual-motor tasks. Based on research findings, it may be possible to predict the kinds of tasks that would be problematic. It may also be possible to predict the extent to which training or practice may eliminate or reduce the impact of a specific deficit (secondary or tertiary prevention) on specific tasks.

A desire to apply our understanding of sensory and perceptual abilities outside the laboratory would ideally lead to different kinds of research methods. Methods that could accommodate ecologically valid stimuli and settings would need to be studied in order to expect some degree of predictive power in real-world settings. These studies should require individuals with intellectual disabilities to categorically synthesize different kinds of sensory information in real-time, as is required in the real world. It is possible that advancing technologies that permit the presentation of virtual environments may provide the means for manipulating multiple visual inputs or cues (as well as auditory or tactile ones) simultaneously. In individuals with intellectual impairments, even more so than in nonclinical adults, it may be important to investigate how the presence, absence, or degradation of visual cues affects performance in multi-cue displays or virtual environments that offer situations more analogous to the real world.

## NOTE

This work was supported by research grants HD29556, HD35975, HD25995, and HD35349 from the National Institute of Child Health and Human Development. We would especially like to thank Dr. Sal Soraci for exhibiting patience far beyond the call of duty. Our appreciation is also sent to the many assistants, participants, parents and guardians, testing site staff members, and administrators who made our efforts possible.

## REFERENCES

Adelson, E.H. (1993). Perceptual organization and the judgment of brightness. *Science, 262,* 2042–2044.

Adoh, T., Woodhouse, J., & Oduwaiye, K. (1992). The Cardiff test: A new visual acuity test for toddlers and children with intellectual impairment. Preliminary report. *Optometry and Vision Science, 69,* 427–432.

Ali, M.R., Khaleque, A., Khanam, M., al-Shatti, A., Ahmed, R.U. (1994). Critical flicker frequency of mentally retarded and normal persons. *Perceptual &Motor Skills, 79,* 1235–1238.

Allan, L.G. (1979). The perception of time. *Perception and Psychophysics, 26,* 340–354.

Allan, L.G., & Kristofferson, A.B. (1974). Psychophysical theories of duration discrimination. *Perception and Psychophysics, 16,* 26–34.

Anstis, S. (1978). Apparent motion. In R. Held, H. Leibowitz, & H.L. Teuber (Eds.), *Handbook of sensory physiology: Vol. 8.* New York: Springer–Verlag.

Atkinson, J., King, J., Braddick, O., Nokes, L., Anker, S., & Braddick, F. (1997). A specific deficit of dorsal stream function in Williams syndrome. *Neuroreports, 8,* 1919–1922.

Bader, D., & Woodruff, M.E. (1980). The effects of corrective lenses on various behaviors of mentally retarded persons. *American Journal of Optometry & Physiological Optics, 57,* 447–459.

Bailey, I. (1998). Visual acuity. In Benjamin, W.J., & Borish, I.M., Eds. *Borish's clinical refraction.* Philadelphia: W.B. Saunders.

Ball, K., & Sekuler, R. (1982). A specific and enduring improvement in visual motion detection. *Science, 218,* 697–698.

Ball, K., & Sekuler, R. (1987). Direction-specific improvement in motion discrimination. *Vision Research, 27,* 953–965.

Banks, M.S., & Dannemiller, J.L. (1987). Infant pattern vision: A new approach based on the contrast sensitivity function. *Journal of Experimental Child Psychology, 31,* 1–45.

Banks, M.S., & Salapatek, P. (1983). Contrast sensitivity function of the infant visual system. *Vision Research, 16,* 867–869.

Barcelo, F., Suwazono, S., & Knight, R.T. (2000). Prefrontal modulation of visual processing in humans. *Nature Neuroscience, 3,* 399–403.

Barnett & Pryer. (1958). Note on depth perception in defectives. *Perceptual and Motor Skills, 8,* 130.

Bennett A., & Rabbetts, R. (1989). Proposals for new reduced and schematic eyes. *Ophthalmic and Physiological Optics, 9,* 228–230.

Berkson, G. (1993). *Children with handicaps: A review of behavioral research.* Hillsdale: Erlbaum.

Bertenthal, B.I., & Bai, D.L. (1989). Infants' sensitivity to optic flow for controlling posture. *Developmental Psychology, 25,* 936–945.

Beverley, K.I., & Regan, D. (1983). Texture changes versus size changes as stimuli for motion in depth. *Vision Research, 23,* 1387–1400.

Birch, E., Hale, L., Stager, D., Fuller, D., & Birch, D. (1987). Operant acuity of toddlers and developmentally delayed children with low vision. *Journal of Pediatric Ophthalmology and Strabismus, 24,* 64–69.

Blais, C., & Kerr, R. (1990). Nonballistic coincident timing. *Perceptual and Motor Skills, 71,* 161–162.

Block, S.S., Beckerman, S.A., & Berman, P.E. (1997). Vision profile of the athletes of the 1995 Special Olympics World Summer Games. *Journal of the American Optometric Association, 68,* 699–708.

Braverman, M., Fein, D., Lucci, D., & Waterhouse, L. (1989). Affect comprehension in children with pervasive developmental disorders. *Journal of Autism and Developmental Disorders, 19,* 301–315.

Brown, A., Dunne, T., Jain, K., Cronin-Golomb, M., & Cronin-Golomb, A. (1996). Contrast sensitivity and face discrimination in Alzheimer's disease. *Abstracts, Society for Neuroscience, 22,* 433.

Butterworth, G., & Cicchetti, D. (1978). Visual calibration of posture in normal and motor retarded Down's syndrome infants. *Perception, 7,* 513–525.

Carlin, M., & Soraci, S. (1993). Similarities in the detection of stimulus symmetry by individuals with and without mental retardation. *American Journal on Mental Retardation, 98,* 336–348.

Carlin, M., Hobbs, K.L., Bud, M.J., & Soraci, S. (1999). Detection of motion-defined forms by individuals with mental retardation and autism: Evidence of modifiability. *Intelligence, 27,* 141–156.

Carlin, M., Soraci, S., Goldman, A., & McIlvane, W. (1995). Visual search in unidimensional arrays: A comparison between subjects with and without mental retardation. *Intelligence, 21,* 175–196.

Catalano, R.A. (1990). Down syndrome. *Survey of Ophthalmology, 34,* 385–398.

Cavanagh, P. (1992). Attention-based motion perception. *Science, 257,* 1563–1565.

Cavanagh, P. (1993). The perception of form and motion. *Current Opinions in Neurobiology, 3,* 177–182.

Chandna, A., Karki, C., Davis, J., & Doran, R. (1989). Preferential looking in the mentally handicapped. *Eye, 3,* 833–839.

Cicchetti, D., & Sroufe, L.A. (1978). An organizational view of affect: Illustration from the study of Down's syndrome infants. In M. Lewis & L. Rosenblum (Eds.), *The development of affect* (pp. 309–350). New York: Plenum Press.

Conners, F., Wyatt, B., & Dulaney, C. (1998). Cognitive representation of motion in individuals with mental retardation. *American Journal of Mental Retardation, 102,* 438–450.

Corbetta, M., Miezin, F.M., Dobmeyer, S., Shulman, G.L., & Petersen, S.E. (1990). Attentional modulation of neural processing of shape, color and velocity in humans. *Science, 248,* 1556–1559.

Cormack, R., & Fox, R. (1985). The computation of disparity and depth in stereograms. *Perception & Psychophysics, 38,* 375–380.

Courage M., Adams R., Hall E. (1997). Contrast sensitivity in infants and children with Down syndrome. *Vision Research, 37,* 1545–1555.

Courage, M., Adams, R., Reyno, S., & Kwa, P. (1994). Visual acuity in infants and children with Down syndrome. *Developmental Medicine and Child Neurology, 36,* 586–593.

Croce, R., Horvat, M., & Roswal, G. (1995). Coincident timing by nondisabled, mentally retarded, and traumatic brain-injured individuals under varying target-exposure conditions. *Perceptual and Motor Skills, 80,* 487–496.

Croner, L. & Albright, T. (1997). Image segmentation enhances discrimination of motion in visual noise. *Vision Research, 37,* 1415–1427.

Cronin-Golomb, A., Corkin, S., Rizzo, J., Cohen, J., Growdon, J., & Banks, K. (1991). Visual dysfunction in Alzheimer's disease: Relation to normal aging. *Annals of Neurology, 29,* 41–52.

Cronin-Golomb, A., Suguira, R., Corkin, S., & Growdon, J. (1993). Incomplete achromatopsia in Alzheimer's disease. *Neurobiology of Aging, 14,* 471–477.

Culham, J.C., & Cavanagh, P. (1994). Motion capture of luminance stimuli by equiluminance color gratings and by attentive tracking. *Vision Research, 34,* 2701–2706.

Cutting, J.E., & Bruno, N. (1988). Additivity, subadditivity, and the use of visual information: A reply to Massaro. *Journal of Experimental Psychology: General, 117,* 422–424.

da Cunha, R., & Moreira, J. (1996). Ocular findings in Down's syndrome. *American Journal of Ophthalmology, 122,* 236–244.

Delorme, A., Frigon, J., & Lagace, C. (1989). Infants' reaction to visual movement of the environment. *Perception, 18,* 667–673.

Desimone, R. and Ungerleider, L.G. (1989). Neural mechanisms of visual processing in monkeys. In F. Boller & J. Grafman (Eds.), *Handbook of Neuropsychology* (pp. 267–299). Amsterdam: Elsevier Science Publishers B.V.

Deutsch, C.K., Oross, S., DiFiore, A., & McIlvane, W.J. (2000). Measuring brain activity correlates of behavior: A methodological overview. *Experimental Analysis of Human Behavior Bulletin, 18,* 36–42.

DeYoe, E.G., & Van Essen, D.C. (1988). Concurrent processing streams in monkey visual cortex. *Trends in Neuroscience, 11,* 219–226.

DiFiore, A., Dube, W.V., Oross, S., Wilkinson, K., Deutsch, C.K., & McIlvane, W.J. (2000). Studies of brain activity correlates of behavior in individuals with and without developmental disabilities. *Experimental Analysis of Human Behavior Bulletin, 18,* 33–35.

Dutton, G. (1994). Cognitive visual dysfunction. *British Journal of Ophthalmology, 78,* 723–726

Ellis, N.R. (1963). *Handbook of mental deficiency.* New York: McGraw-Hill.

Ellis, N.R., McCartney, J., Feretti, R., & Cavalier, A. (1977). Recognition memory in mentally retarded persons. *Intelligence, 1,* 310–317.

Emmerton, J. (1983). Vision. In M. Abs (Ed.), *Physiology and behaviour of the pigeon.* San Diego: Academic Press.

Enoch, J., Werner, J., Haegerstrom-Portnoy, G., Lakshminarayanan, V., & Rynders, M. (1999). Forever young: Visual functions not affected or minimally affected by age: a review. *Journal of Gerontology Series A: Biological Sciences and Medical Sciences, 54,* B336–351.

Erickson, G., & Block, S. (1999). Testability of a color vision screening test in population with mental retardation. *Journal of the American Optometry Association, 70,* 758–763.

Evenhuis, H. (1995). Medical aspects of aging in a population with intellectual disability: I. Visual impairment. *Journal of Intellectual Disability Research, 39,* 19–25.

Fox, R., & Oross, S. III. (1988a). Complex biological motion perception by human infants. *Investigative Ophthalmology & Visual Science Supplement, 29,* 26.

Fox, R., & Oross, S. III. (1988b). Deficits in stereoscopic depth perception by mildly mentally retarded adults. *American Journal on Mental Retardation, 93,* 232–244.

Fox, R., & Oross, S. III. (1990). Mental retardation and perception of global motion. *Perception and Psychophysics, 48* , 252–258.

Fox, R., & Oross, S. III. (1992). Perceptual deficits in mildly mentally retarded adults. *International Review of Research in Mental Retardation, 18,* 1–27.

Fox, R., Yang, G.S., Feurer, I.D., Butler, M.G., & Thompson, T. (in press). Kinetic form discrimination in Prader-Willi syndrome. *Journal of Intellectual Disabilities Research.*

Garcia, R., Cleland, C.C., Rago, W., Wayne, P., & Swartz, J.D. (1974). Perception of depth in the profoundly retarded. *Bulletin of the Psychonomic Society, 4,* 185–187.

Geisler, W.S., & Chou, K. (1995). Seperation of low-level and high-level factors in complex tasks: Visual search. *Psychological Review, 102,* 356–378.

Gibson, E.J. (1969). *Principles of perceptual learning and development.* New York: Appleton-Century-Crofts.

Gibson, E.J., & Walk, R.D. (1960). The visual cliff. *Scientific American, 202,* 64–71.

Gibson, J.J. (1966). *The senses considered as perceptual systems.* Boston: Houghton Mifflin.

Gilmore, G.C. (1995). Stimulus encoding in Alzheimer's disease: A multichannel view. In P. Allen & T. R. Bashore (Eds.), *Age differences in word and language processing* (pp. 199–219). Amsterdam: Elsevier.

Gilmore, G.C., Thomas, C.W., Klitz, T., Persanyi, M.W., and Tomsak, R. (1996). Contrast enhancement eliminates letter identification speed deficits in Alzheimer's disease. *Journal of Clinical Geropsychology, 2,* 307–320.

Gilmore, G.C., Turner, J.B., Thomas, C.W., Koss, E., Townsend, L., Greene, R., Morrison, S., & Lustig, C.A. (1995, November). Impact of Alzheimer-type contrast filter on picture recognition and memory: Simulation and remediation of Alzheimer deficits. Presented at the conference on The Spectrum of Research on Aging and the Aged, Case Western Reserve University, Cleveland, OH.

Good, W.V., Jan, J.E., Burden, S.K., Skoczenski, A., & Candy, R. (2001). Recent advances in cortical visual impairment. *Developmental Medicine & Child Neurology, 43,* 56–60.

Granrud, C.E., Yonas, A., & Opland, E.A. (1985). Infants' sensitivity to the depth cue of shading. *Perception & Psychophysics, 37,* 415–419.

Grondin, S., & Rousseau, R. (1991). Judging the relative duration of multimodal short empty time intervals. *Perception & Psychophysics, 49,* 245–256.

Hall, A., Orel-Bixler, D. & Hagerstrom-Portnoy, G. (1991). Special visual assessment techniques for multiply handicapped persons. *Journal of Visual Impairment and Blindness, 85,* 23–29.

Harwood, N., Hall, L., & Shinkfield, A. (1999). Recognition of facial emotional expressions from moving and static displays by individuals with mental retardation. *American Journal on Mental Retardation, 104,* 270–278.

Haugen, O., Aasved, H, & Bertelsen, T., (1995). Refractive state and correction of refractive errors among mentally retarded adults in a central institution. *Acta Ophthalmologica Scandanavia, 73,* 129–132.

Hawryshyn, C.W. (1992). Polarization vision in fish. *American Scientist, 80,* 164–175.

Heal, L., & Sigelman, C. (1995). Response biases in interviews of individuals with limited mental ability. *Journal of Intellectual Disability Research, 39,* 331–340.

Henderson, S.E., Morris, J., & Frith, U. (1981). The motor deficit in Down's syndrome children: A problem of timing? *Journal of Child Psychology Psychiatry, 22*(3), 233–245.

Henderson, V.W., Mack, W., and Williams, B.W. (1989). Spatial disorientation in Alzheimer's disease. *Archives of Neurology, 46,* 391–394.

Hertz, B. (1987). Acuity card testing of retarded children. *Behavioral and Brain Research, 24,* 85–92.

Hobson, R.J., Ouston, J., & Lee, A. (1989). Recognition of emotion by mentally retarded adolescents and young adults. *American Journal on Mental Retardation, 93,* 434–443.

Hubel, D.H., & Livingstone, M.S. (1987). Segregation of form, color, and stereopsis in primate area 18. *Journal of Neuroscience, 7,* 3378–3415.

Hubner, R. (1997). The effect of spatial frequency on global precedence and hemispheric differences. *Perception & Psychophysics, 59,* 187–201.

Huo, R., Burden, S., Hoyt, C., & Good, W. (1999). Chronic cortical visual impairment in children: aetiology, prognosis, and associated neurological deficits. *British Journal of Ophthalmology, 83,* 670–675.

Ivry, R.B., & Hazeltine, R.E. (1995). Perception and production of temporal intervals across a range of durations: Evidence for a common timing mechanism. *Journal of Experimental Psychology: Human Perception and Performance, 21,* 3–18.

Jacobson, K., Magnussen, S., & Smith, L. (1997). Hidden visual capabilities in mentally retarded subjects diagnosed as deaf-blind. *Vision Research, 37,* 2931–2935.

Jensen, A. (1983). Critical flicker frequency and intelligence. *Intelligence, 7,* 217–225.

Julesz, B. (1960). Binocular depth perception of computer-generated patterns. *Bell System Technical Journal, 30,* 1125–1162.

Kaas, J.H., & Garraghty, P.E. (1991). Hierarchical, parallel, and serial arrangements of sensory cortical areas: Connection patterns and functional aspects. *Current Opinions in Biology, 1,* 248–251.

Karni, A. & Sagi, D. (1993). The time course of learning a visual skill. *Nature, 365,* 250–252.

Keele, S., Pokorny, R., Corcos, D., & Ivry, R. (1985). Do perception and motor production share common timing mechanisms: A correlational analysis. *Acta Psychologica, 60,* 173–191.

Kellman, P.J., & Shipley, T.F. (1991). A theory of visual interpolation in object perception. *Cognitive Psychology, 23,* 141–221.

Kerr, R., & Blais, C. (1987). Down syndrome and extended practice of a complex motor task. *American Journal of Mental Deficiency, 91,* 591–597.

Kim, L.Y. (1995). *Contrast sensitivity in adults with Down syndrome and its effects on cognitive abilities and age.* Unpublished doctoral dissertation, Case Western Reserve University, Cleveland, OH.

Kodman, F. (1963). Sensory processes and mental deficiency. In N.R. Ellis, (Ed.) *Handbook of mental deficiency,* p p. 463–479. New York: McGraw-Hill.

Lai, F, & Williams, R. (1989). A prospective study of Alzheimer disease in Down syndrome. *Archives of Neurology, 46,* 849–853.

Laine, R., & Baumeister, A. (1985). Short-term memory for a pure tone stimulus among mentally retarded and nonretarded persons. *Intelligence, 9,* 237–257.

Lally, M., & Nettelbeck, T. (1977). Intelligence, reaction time, and inspection time. *American Journal of Mental Deficiency, 82,* 273–281.

Leibowitz, H. (1961). Apparent visual size as a function of distance for mentally deficient subjects. *American Journal of Psychology, 74,* 98–100.

Leibowitz, H., & Heisel, M. (1958). The development of the Ponzo illusion as a function of age. *Archives de Psychologie, 36,* 328–331.

Lennerstrand, G., Axelson, A., & Andersson, G. (1983). Visual acuity testing with preferential looking in mental retardation. *Acta Ophthalmologica, 61,* 624–633.

Levy, B. (1984). Incidence of oculo-visual anomalies in an adult population of mentally retarded persons. *American Journal of Optometry and Physiological Optics, 61,* 324–326.

Liu, L., Gauthier, L., and Gauthier, S. (1990). Spatial disorientation in persons with early senile dementia of the Alzheimer type. *American Journal of Occupational Therapy, 45,* 67–74.

Livingstone, M., & Hubel, D. (1987). Psychophysical evidence for separate channels for the perception of form, color, movement, and depth. *Journal of Neuroscience, 7,* 3416–3468.

Livingstone, M., & Hubel, D. (1988). Segregation of form, color, movement, and depth: Anatomy, physiology, and perception. *Science, 240,* 740–750.

Luckiesh, M. (1934). *Seeing and human welfare.* Baltimore: Williams & Wilkins.

Lustig, C.A., Gilmore, G.C., Koss, E., Thomas, C.W., Ripich, D., Townsend, L., and Stuve, T. (1995, November). Boston Naming Test performance is affected by simulated Alzheimer-type visual deficit. Presented at the conference on The Spectrum of Research on Aging and the Aged, Case Western Reserve University, Cleveland, OH.

Macar, F., Grondin, S., & Casini, L. (1994). Controlled attention sharing influences time estimation. *Memory and Cognition, 22,* 673–686.

Macmillan, N., & Creelman, C. (1991). *Detection theory: A users guide.* Cambridge: Cambridge University Press.

Maino, D., Maino, J., & Maino, S. (1990). Mental retardation syndromes with associated ocular defects. *Journal of the American Optometry Association, 61,* 707–716.

Maino, D., Wesson, M., Schlange, D., Cibis, G., & Maino, J. (1991). Optometric findings in the fragile X syndrome. *Optometry and Visual Science, 68,* 634–640.

Massaro, D. (1988). Ambiguity in perception and experimentation. *Journal of Experimental Psychology: General, 117,* 417–421.

McAlpine, C., Kendall, K.A., & Singh, N.N. (1991). Recognition of facial emotion by persons with mental retardation. *American Journal on Mental Retardation, 96,* 29–36.

McNutt, T., & Melvin, K. (1968). Time estimation in normal and retarded subjects. *American Journal of Mental Deficiency, 72,* 584–589.

Meck, W. (1995). Neuropharmocology of timing and time perception. *Cognitive Brain Research, 3,* 227–242.

Merigan, W., & Maunsell, J. (1993). How parallel are the primate visual pathways? *Annual Review of Neuroscience, 16,* 369–402.

Morgan, M. (1991). Hyperacuity. In Cronly-Dillion, (Ed.) *Vision and visual dysfunction: Vol. 10. Spatial vision.* Boca Raton: CRC Press.

Mulhern, T., Warm, J., & Clark, D. (1974). Intensity modulation and the reproduction of auditory duration by nonretarded and retarded children. *American Journal of Mental Deficiency, 78,* 721–726.

Nakayama, K. (1985). Biological image motion processing: A review. *Vision Research, 25,* 625–660.

National Advisory Eye Council. (1995). *Vision research: A national plan 1994–1998.* (NIH Publication No. 95–3186). Washington, DC: National Eye Institute.

Neargarder, S.A. (1997). *Exploring dementia of the Alzheimer's type in Down syndrome: An examination of visual perception and cognitive abilities.* Unpublished dissertation, Vanderbilt University, Nashville, TN.

Nettelbeck, T., & Brewer, N. (1981). Studies of mild mental retardation and timed performance. *International Review of Research in Mental Retardation, 10,* 61–106.

Newsome, W.T., Britten, K.H., & Movshon, J.A. (1989). Neuronal correlates of a perceptual decision. *Nature, 341,* 52–54.

Newsome, W.T., & Pare, E.B. (1988). A selective impairment of motion perception following lesions of the middle temporal visual area (MT). *Journal of Neuroscience, 8,* 2201–2211.

Nothdurft, H.C. (1991). Texture segmentation and pop-out from orientation contrast. *Vision Research, 31,* 1073–1078.

O'Dell, C., Harshaw, K., & Boothe, R. (1993). Vision screening of individuals with severe or profound mental retardation. *Mental Retardation, 31,* 154–160.

Olitsky, S., Sadler, L., & Reynolds, J. (1997). Subnormal binocular vision in the Williams syndrome. *Journal of Pediatric Ophthalmology and Strabismus, 34,* 58–60.

Orel-Bixler D., Haegerstrom-Portnoy G., Hall A. (1989). Visual assessment of the multiply handicapped patient. *Optometry and Visual Science, 66,* 530–536.

Oross, S. III, Carlin, M.T., & Fox, R. (1990). Motion perception deficits in mildly mentally retarded adults: An experimental attempt to ameliorate deficits via training. *Gatlinburg Conference on Research and Theory in Mental Retardation and Developmental Disabilities Abstracts, 23,* 66.

Oross, S. III, Francis, E., Mauk, D., & Fox, R. (1987). The Ames window illusion: Perception of illusory motion by human infants. *Journal of Experimental Psychology: Human Perception and Performance, 13,* 609–613.

Oross, S. III, Fryer, T.L., & Woods, C.B. (1997). Recognition of contrast and texture defined letters in individuals with developmental disabilities. *Society for Neuroscience Abstracts, 23,* 858.

Perez-Carpinell, J., de Fez, M.D., & Climent, V. (1994). Vision evaluation in people with Down's syndrome. *Ophthalmic and Physiological Optics, 14,* 115–121.

Prokesova, V., Kriz, S., Berg, L., & Halvorsen, I. (1990). Ophthalmic examination of the mentally retarded at a central institution. *Tidsskr Nor Laegeforen, 110,* 1659–1662. English abstract.

Ramachandran, V.S. (1992). Blind spots. *Scientific American, 266,* 86–91.

Ramachandran, V.S. & Braddick, O. (1973). Orientation-specific learning in stereopsis. *Perception, 2,* 371–376.

Rammsayer, T. (1993). On dopaminergic modulation of temporal information processing. *Biological Psychology, 36,* 209–222.

Rammsayer, T., & Lima, S. (1991). Duration discrimination of filled and empty auditory intervals: Cognitive and perceptual factors. *Perception and Psychophysics, 50,* 565–574.

Rapcsak, S.Z., Comer, J.F., & Rubens, A.B. (1993). Anomia for facial expressions: Neuropsychological mechanisms and anatomical correlates. *Brain and Language, 45,* 233–252.

Repp, B., Liberman, A., Eccardt, T. & Pesetsky, D. (1978). Perceptual integration of acoustic cues for stop, fricative, and affricate manner. *Journal of Experimental Psychology: Human Perception and Performance, 4,* 621–637.

Rider, R.A., Mahler, T.J., and Ishee, J. (1983). Comparison of static balance in trainable mentally handicapped and nonhandicapped children. *Perceptual and Motor Skills, 56,* 311–314.

Rieser, J., Guth, D., & Weatherford, D. (1987). Mentally retarded and nonretarded adults' sensitivity to spatial structure. *American Journal of Mental Deficiency, 91,* 379–391.

Rocco, F.J., Cronin-Golomb, A., & Lai, F. (1997). Alzheimer-like visual deficits in Down syndrome. *Alzheimer Disease and Associated Disorders, 11,* 88–98.

Rojahn, J., Lederer, M., Tassé, M.J. (1995). Facial emotion recognition by persons with mental retardation: A review of the experimental literature. *Research in Developmental Disabilities, 16,* 393–414.

Rousseau, L., & Rousseau, R. (1996). Stop-reaction time and the internal clock. *Perception and Psychophysics, 58,* 434–448.

Sacks, G., Goren, M, Burke, M., & White, S. (1991). Ophthalmologic screening of adults with mental retardation. *American Journal on Mental Retardation, 95,* 571–574.

Sadler, L., Olitsky, S., & Reynolds, J. (1996). Reduced stereoacuity in Williams syndrome. *American Journal of Medical Genetics, 66,* 287–288.

Salthouse, T.A., Hancock, H.E., Meinz, E.J., & Hambrick, D.Z. (1996). Interrelations of age, visual acuity, and cognitive functioning. *Journal of Gerontology: Psychological Sciences, 51,* P317–330.

Schor, C. & Narayan, V. (1981). The influence of field size upon the spatial frequency response of optokinetic nystagmus. *Vision Research, 21,* 985–994.

Shimp, S.A. & Oross, S. III. (1994). Assessment of visual abilities in Alzheimer's disease and mental retardation. *Society for Neuroscience Abstracts, 20,* 1579.

Shimp, S.A. & Oross, S. III. (1995). Assessment of visual abilities in individuals with mental retardation and Alzheimer's disease. *Gatlinburg Conference on Research and Theory in Mental Retardation and Developmental Disabilities Abstracts, 28,* 26.

Shinkfield, A.J., Sparrow, W.A., & Day, R.H. (1997). Visual discrimination and motor reproduction of movement by individuals with mental retardation. *American Journal on Mental Retardation, 102,* 172–181.

Sidman, M., & Stoddard, L.T. (1966). Programming perception and learning for retarded children. In N.R. Ellis (Ed.), *International review of research in mental retardation: Vol. 2.* (pp. 151–208). New York: Academic Press.

Siegel, R.M., & Anderson, R.A. (1988). Perception of three-dimensional structure from motion in monkey and man. *Nature, 331,* 259–261.

Soraci, S.A., Carlin, M.T., & Wiltse, L.J. (1998). Facilitating visual attention. In S.A. Soraci, & W.J. McIlvane (Eds.), *Perspectives on fundamental processes in intellectual functioning. Vol. 1: A survey of research approaches.* Greenwich, CT: Ablex.

Spear, P.D. (1993). Neural bases of visual deficits during aging. *Vision Research, 33,* 2589–2609.

Spitz, H.H. (1963). Field theory in mental deficiency. In N.R. Ellis (Ed.), *Handbook of mental deficiency: Psychological theory and research.* New York: McGraw-Hill.

Spitz, H.H. (1967). A comparison of mental retardates and normals on the distorted room illusion. *American Journal of Mental Deficiency, 72,* 34–49.

Spitz, H.H., & Lipman, R.S. (1959). Some parameters in the perception of the spiral aftereffect. *Perceptual and Motor Skills, 9,* 81.

Spitzer, H., Desimone, R., & Moran, J. (1988). Increased attention enhances both behavioral and neuronal performance. *Science, 240,* 338–340.

Spivack, G. (1963). Perceptual processes. In N. Ellis (Ed.) *Handbook of mental deficiency: Psychological theory and research* (pp. 480–511). New York: McGraw-Hill.

Spivack, G. & Levine, M. (1959). Spiral aftereffect and measures of satiation in brain-injured and normal subjects. *Journal of Personality, 27,* 211–227.

Tallal, P., Miller, S., & Fitch, R. (1993). Neurobiological basis of speech: A case for the preeminence of temporal processing. In P. Tallal, A.M. Galaburda, R. Linas, & C. von Euler (Eds.), *Temporal information processing in the nervous system. Annals of the New York Academy of Sciences,Vol. 682.*

Trick, G.L., & Silverman, S.E. (1991). Visual sensitivity to motion: Age-related changes and deficits in senile dementia of the Alzheimer type (SDAT). *Neurology, 41,* 1437–1440.

Treisman, A., & Gelade, G. (1980). A feature integration theory of attention. *Cognitive Psychology, 12,* 97–136.

Treisman, A., & Souther, J. (1985). Search asymmetry: A diagnostic for preattentive processing for separable features. *Journal of Experimental Psychology: General, 114,* 285–310.

Uecker, A., Mangan, P.A., Obrzut, J.E., and Nadel, L. (1993). Down syndrome in neurobiological perspective: An emphasis on spatial cognition. *Journal of Clinical Child Psychology, 22,* 266–276.

Ungerleider, L.G., & Desimone, R. (1986). Cortical connections of visual area MT in macaque. *Journal of Comparative Neurology, 248,* 190–222.

Ungerleider, L.G., & Mishkin, M. (1982). Two cortical visual systems. In D.J. Ingle, M.A. Goodale, & R.J.W. Mansfield (Eds.), *Analysis of visual behavior* (pp. 549–586). Cambridge, MA: MIT Press.

Wade, M.G. (1990). Impact of optical flow on postural control in normal and mentally handicapped persons. In A. Vermee (Ed.), *Motor development, adapted activity and mental retardation* (pp. 21–29). Karger: Med Sport Sci. Basel.

Wade, M.G., Newell, K.M., and Hoover, J.H. (1982). Coincident timing behavior of young mentally retarded workers under varying conditions of target velocity and exposure. *American Journal of Mental Deficiency, 86,* 643–649.

Walden, T.A., Blackford, J.U., & Carpenter, K.L. (1997). Differences in social signals produced by children with developmental delays of differing etiologies. *American Journal on Mental Retardation, 102,* 292–305.

Warren, S. (1992). Some thoughts about prevention. *Kennedy Center News, 19,* 1–2.

Webb, T.E. (1972). Stereoscopic contour perception in mental retardation. *American Journal of Mental Deficiency, 76,* 699–702.

Werner, H., & Thuna, B. (1942). Critical flicker-frequency in children with brain injury. *American Journal of Psychology, 55,* 394–399.

Wickens, C. & Hollands, J. (2000). *Engineering psychology and human performance (3rd Edition).* Upper Saddle River, NJ: Prentice-Hall.

Winters, J.J. (1969). A comparison of normals and retardates on physiological and experimental visual illusions. *American Journal of Mental Deficiency, 73,* 956–962.

Wisniewski, H., Silverman, W., & Wegiel, J. (1994). Aging, Alzheimer disease and mental retardation. *Journal of Intellectual Disability Research, 38,* 233–239.

Wisniewski, K.E., Wisniewski, H.M., and Wen, G.Y. (1985). Occurrence of neuropathological changes and dementia of Alzheimer's disease in Down Syndrome. *Annals of Neurology, 17,* 278–282.

Woodhouse, J., Pakeman, V., Saunders, K., Parker, M., Fraser, W., Lobo, S., & Sastry, P. (1996). Visual acuity and accommodation in infants and young children with Down's syndrome. *Journal of Intellectual Disability Research, 40,* 49–55.

Woods, C.B., DeFord, K.J., & Oross, S. III. (1996). Temporal information processing deficits associated with mental retardation: Investigations of vision and audition. *Gatlinburg Conference on Research and Theory in Mental Retardation and Developmental Disabilities Abstracts, 29,* 137.

Xeromeritou, A. (1992). The ability to encode facial and emotional expressions by educable mentally retarded and nonretarded children. *Journal of Psychology, 126,* 571–584.

Yonas, A., & Granrud, C.E. (1985). The development of sensitivity to kinetic, binocular, and pictorial depth information in human infants. In D.J. Ingle, M. Jeannerod, & D.N. Lee (Eds.), *Brain mechanisms and spatial vision* (pp 113–145). Boston: Matinus Nijhoff.

Zihl, J., von Cramon, D., & Mai, N. (1983). Selective disturbance of movement vision after bilateral brain damage. *Brain, 106,* 313–340.

Chapter 3

# Enhancing Performances of Individuals with Mental Retardation: Manipulations of Visual Structure

*Michael T. Carlin, Sal Soraci Jr., and Christina Strawbridge*

## INTRODUCTION

Developing methods for directing attention to task-relevant information in a visual array, and inhibiting attention to irrelevant stimuli, is one of the greatest challenges for the design of educational technologies for individuals with mental retardation. One of the most basic prerequisites for establishing new performances in individuals with mental retardation is to induce such individuals to focus on the relevant information for successful completion of a task. Once attention is oriented to the proper component(s) of the task, the detection of the critical target(s) is more likely to occur. This focus on understanding visual search processes is important for cognitive perspectives on attention (Yantis & Egeth, 1999; Soraci, Carlin & Chechile, 1998), as well as for behavior-analytic approaches to observing behavior (Dinsmoor, 1985; Serna & Carlin, in press). In behavioral parlance, the initial observing responses are critical for establishing the "first instances" of learning (Ray & Sidman, 1967; Skinner, 1935), such that the observing behavior is repeatedly reinforced and eventually established as a reliable event. Success in this endeavor is based on a sound understanding of attentional abilities and limitations at both an individual and group level. In effect, the scientist is placed in the position of being an "attention engineer" (cf. Brown, 1970), who is attempting to "guide" visual attention (Wolfe, Cave, & Franzel, 1989; Soraci, Carlin & Wiltse, 1998) to critical elements of complex visual arrays.

Effective guidance of visual attention is dependent upon knowledge of the structural properties of visual arrays that serve to direct attention within the visual field. This chapter presents a review of recent research that focuses on this and related issues. We will begin with a review of basic research on visual search processes in individuals with mental retardation. This program of research is concerned with whether the visual search behaviors of individuals with and without mental retardation are governed by the same principles. If so, then the extensive knowledge of these processes in individuals without mental retardation may be extended to the problem of designing computer-based learning environments for children with mental retardation. Alternatively, use of well-established methodologies may allow us to identify important differences in the observing behaviors of individuals with and without mental retardation. This would allow us to design specialized learning programs for individuals with mental retardation as well as provide some insight into the basis or nature of mental retardation; such findings have import for advancing the development of theory and intervention techniques. We also will discuss recent extensions of this focus on the analysis and manipulation of visual structure to the design of programs for augmenting higher-order skills such as matching-to-sample, a behavior important for language development programs, and free recall, an ecologically critical skill.

## RELATIONAL LEARNING THEORY

Soraci, Baumeister, and Carlin (1993) outlined a program of research that indicated (a) that individuals with mental retardation are less sensitive to relational aspects of visual structure (e.g., similarity–difference), and (b) that particular manipulations of the structure of visual displays can enhance relevant target–distractor relations and their detectability. When this is accomplished, the performances of individuals with mental retardation often can be improved dramatically. For example, in the context of an oddity task in which the participant must learn to pick the stimulus that is different from the other two, the individual must selectively attend to and respond on the basis of the oddity relation. However, young children often fail to do so when explicit verbal instructions are not provided. Whereas many authors previously attributed this failure to demonstrate oddity-based responding to a developmental limitation (e.g., Ellis & Sloan, 1959; Penn, Sindberg, & Wohlheuter, 1969; Schadler, 1973; Scott & House, 1978), Soraci and his colleagues (e.g., Soraci, Alpher, Deckner, & Blanton, 1983; Soraci, Deckner, Haenlein, Baumeister, Murata-Soraci, & Blanton, 1987; Soraci et al., 1991) demonstrated that young children without mental retardation and children with developmental disabilities with low mental ages (i.e., $< 4$ years) can perform the task when provided with perceptual guidance. One technique used by Soraci et al. (1983, 1987) was to simply increase the number of non-odd stimuli from two to eight, thus increasing the salience of the difference relation. The children who had previously failed the three-choice oddity task now succeeded on the nine-element task and a subsequent three-choice post-test, demonstrating rapid rule acquisition and transfer. Other techniques for enhancing the salience of the "odd" stimulus also proved to be effective for the majority of children tested.

In fact, Soraci et al. (1991) demonstrated that oddity performance could be enhanced through use of a bimodal training procedure in which visual forms were paired with tones. An odd tone was associated with the odd stimulus in each array, whereas non-odd stimuli were paired with identical tones. Using this procedure, Soraci et al. (1991) demonstrated the rapid induction of rule-based oddity learning in children as young as three years old.

The theory that individuals with mental retardation have lowered sensitivities to critical stimulus relations was expanded to explain intelligence-related differences in several other areas of research as well. Similar evidence of reduced sensitivity to stimulus relations was reported by Soraci, Carlin, Deckner, and Baumeister (1990) in the context of a rapid-presentation, two-choice matching-to-sample paradigm utilizing checkerboard patterns varying in symmetry. It was shown that individuals with mental retardation performed comparably to individuals without mental retardation when the target–distractor disparities were maximized or minimized. However, at moderate levels of distractor structure, the performances of the individuals with and without mental retardation diverged. This was seen as further support for the belief that individuals with mental retardation are less able to detect critical stimulus relations when the salience of relational information is reduced. As a final example, Soraci, Barlean, Haenlein, and Baumeister (1986), using patterned sequences of tones and cardiac deceleration as an attentional index, demonstrated that individuals with mental retardation have lowered sensitivities to auditory relational information as well. Thus, lowered sensitivities to stimulus relations seem to underlie performance differences between individuals with and without mental retardation on a wide range of experimental tasks involving both visual and auditory processing.

## VISUAL SEARCH EFFICIENCY

Given our interest in structural manipulations of visual arrays that can guide observing of task-relevant stimuli or relations (Soraci et al., 1998; Carlin, Soraci, Dennis, Chechile, & Loiselle, in press), we have pursued the study of visual search processes in individuals with mental retardation. We have been particularly interested in using methodologies that assess individuals' sensitivities to various stimulus dimensions or features, and their ability to restrict attention to task-relevant stimuli while inhibiting attention to task-irrelevant stimuli. These are the processes that are related directly to our goals of understanding the functioning of individuals with mental retardation and designing the most efficient remediation procedures for enhancing the performances of these individuals.

### Feature search

Our initial formal study of the visual search abilities of individuals with mental retardation employed a basic feature-search task. A feature-search task

**Figure 3.1**
Example of the Feature-Search Arrays. The target is the large triangle. Typically, size-based differences of this magnitude can be detected with little or no increase in search time as set size increases.

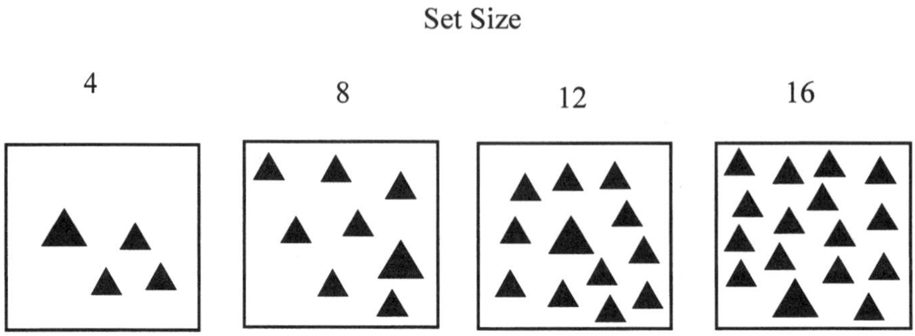


involves presentation of multiple stimuli in each array, with a particular predefined target that differs from a homogenous field of distractors on a single dimension. An example of a size-based feature search task is shown in Figure 3.1. In this example, the participant would be instructed to indicate as rapidly as possible whether the target (i.e., a large triangle) is present in the array. The target differs in size from the other stimuli in the array but is identical on all other dimensions (e.g., form and color). Treisman and her colleagues (e.g., Treisman, 1988; Treisman & Gelade, 1980; Treisman & Gormican, 1988) have used this task to identify features on particular dimensions that can be detected with little or no increase in reaction time as set size increases. These features are said to "pop out" of the visual context in which they are embedded; detection requires minimal attentional resources on the part of the individual.

Importantly, the presence of the popout effect is highly dependent upon the degree of difference between the target and distractor stimuli on the critical dimension of difference (Duncan & Humphreys, 1992; Theeuwes, 1992). Stimuli that differ along a particular dimension may or may not be immediately identifiable depending upon their disparity from the alternative feature (i.e., perceptual salience). Thus, a popout effect may be evident for a red circle among blue circles, but not for a light blue circle among dark blue circles. The popout effect, when demonstrated, is restricted to particular pairings of features used and not to the dimension of difference in a global sense.

Recognition of this feature–pair specificity is critical for properly interpreting performance differences between individuals with and without mental retardation. If feature-search efficiency varies across groups for a given pair of features, it may indicate that individuals with mental retardation are less sensitive to the difference(s) between the target and distractor stimuli (i.e., the difference is less salient). Often, differences in search efficiency are described dichotomously by use of the terms parallel and serial search. A parallel search is one in which search rates do not increase as the number of distractors is increased. Alternatively, a serial search is one in which search rates increase linearly with increases in set size. A finding that individuals with mental retardation perform a serial search of arrays for which individuals without mental retardation exhibit parallel search does not mean that individuals with mental retardation cannot perform parallel searches of visual arrays. Individuals with mental retardation may do so with a different pairing of features on the same dimension. Thus, if a particular individual or group is less sensitive to differences along a certain dimension, then that sensitivity difference would be manifested in less efficient search (i.e., an increased RT x Set Size slope), or a qualitatively different mode of search if one adheres to the parallel–serial dichotomization.

In our study of the feature search abilities of individuals with mental retardation (Carlin, Soraci, Goldman, & McIlvane, 1995), we identified several differences across the groups of individuals with and without mental retardation. First, overall search times for the individuals with mental retardation were about 275 milliseconds (ms) slower across the dimensions of color, form, and size. We subsequently have shown that the magnitude of this difference can be substantially reduced, though not eliminated entirely, if the participants with mental retardation are given extended practice with the task. Thus, much of the difference was due to the influence of factors associated with performing this type of task, but that do not involve visual search processes per se. Second, there were differences across groups in terms of the slopes of the RT x Set Size functions as well. For the pairs of features selected, the individuals without mental retardation showed consistent search rates within each dimension regardless of the number of distractors presented (i.e., parallel search). The individuals with mental retardation also showed such consistent search times across set sizes for the color dimension. However, the individuals with mental retardation showed increasing RT x Set Size slopes (i.e., serial search) specific to the dimensions of form and size. Focused analyses indicated that about half of the individuals with mental retardation were utilizing serial searches of these arrays and the other half were showing the consistent search rates that had been shown by the individuals without mental retardation. Thus, a subgroup of the participants with mental retardation were unable to immediately identify the specified target for the form and size search tasks. This may indicate that these individuals are less sensitive to differences along these dimensions than individuals without mental retardation, and the other individuals with mental retardation. Though the basis for this within-group variability has not been identified, we have found that it is not related to the age or IQ of the participants. As our testing continues with greater numbers of individuals, we will assess the correlation between this subgrouping and other likely sources of individual differences within this population, such as developmental history (e.g., institutionalization) and etiology of mental retardation.

### Role of inhibition

Though the feature-search task allows us to determine whether individuals and/or groups differ in their ability to respond to a highly salient stimulus in a visual array, efficient and adaptive search requires that the individual focus attention on task-relevant items and inhibit attention to irrelevant items (e.g., Dempster, 1991; Iarocci & Burack, 1998). Increased distractibility and an inability to inhibit attention to irrelevant items in a visual display have been forwarded as potential causes of performance differences across individuals who differ in intelligence (Aks & Coren, 1990; Cha & Merrill, 1994). Several studies using a negative priming task have indicated that individuals with mental retardation may be able to inhibit attention to irrelevant stimuli but do not do so in all circumstances. Cha and Merrill (1994) showed that their participants with mental retardation did not demonstrate a negative priming effect on an identity-based task, and therefore concluded that these individuals had an inhibitory deficit. However, Merrill, Cha, and Moore (1994) did find a negative priming effect for individuals with mental retardation on a location-based task. These authors concluded that individuals with mental retardation do have the ability to inhibit attention to irrelevant information in visual displays, but do not do so in all circumstances. The challenge, then, is to identify circumstances under which the use of inhibitory mechanisms is demonstrated by individuals with mental retardation and "the variables that determine when and how individuals with mental retardation will use them" (Merrill et al., 1994, p. 212). We have been investigating these inhibitory mechanisms in individuals with mental retardation by employing two types of visual search tasks: guided search and attentional capture.

#### *Guided search*

The guided search tasks that we have employed in our research have been designed to determine whether individuals with mental retardation can limit search to task-relevant items when provided with a guiding (i.e., goal-directed) verbal instruction. In our research, the relevant verbal instruction is the identification of the target's shape and color (e.g., "find the black triangle"). Given arrays in which elements vary with regard to shape and color, this instruction should limit attention to only those items that share the target's color. Then, a serial search should be conducted among the target-color items to identify whether the target shape is present or not (see Figure 3.2 for an example of guided search arrays). If visual attention can be limited to task-relevant items only, this would be indicative of top-down control of visual attention in individuals with mental retardation. The goals of the observer exert control over the allocation of attention. Thus, attention would be inhibited to task-irrelevant items (i.e., those that do not share the goal or target color; gray items in Figure 3.2), and limited to the task-relevant items (i.e., black forms in Figure 3.2).

**Figure 3.2**
Example of the Guided Search Arrays. The target is the black triangle. If the participant limits attention to the black items (i.e., uses color to guide search to likely targets), search times should be independent of total set size.

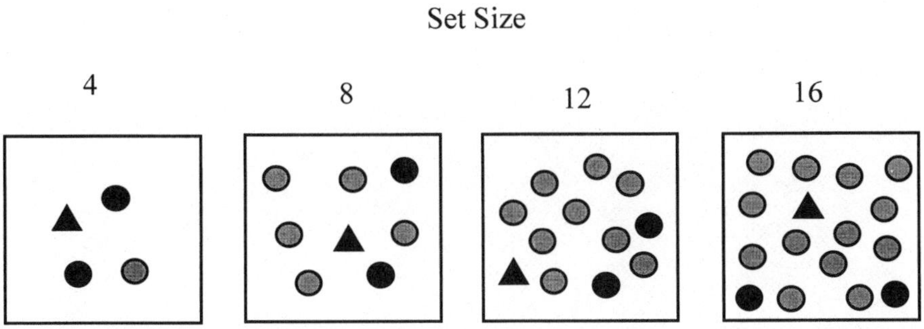


In our first study of guided search processing in individuals with mental retardation, we varied the total number of elements in each array (4, 8, 12, or 16) but held the number of target-color elements constant (i.e., 3), as shown in Figure 3.2. Thus, if the participants were utilizing the color dimension to limit search to target-color elements, search times would not be expected to vary as a function of the total number of elements. Rather, each array should become functionally a three-element search. Results, as shown in Table 3.1, indicated that the individuals with mental retardation did demonstrate the predicted patterns of visual-search times for the three tasks. Search rates were consistent across set sizes for the color-based feature search task, indicating that it was a candidate for serving as the guiding cue. The form-based feature-search task results showed that the participants' search times increased dramatically as set size increased (i.e., serial search). Finally, the guided-search task results indicated that target identification times were significantly faster than the form-based feature search task, and that detection times were faster than the four-element arrays in the form-based feature search task. The latter finding is consistent with the prediction that search would be limited to only three elements (e.g., the black elements in Figure 3.2) in each of the guided-search arrays. These data support the contention that individuals with mental retardation can utilize a verbal, or conceptually based instruction, to guide their observing behaviors on a visual search task.

Our second study of this phenomenon varied both the total number of elements in the array (4, 8, 12, or 16) and the number of target-color elements (2, 3, or 4). The predictions were that search times on the guided-search task would be independent of total elements in the array, but a direct function of the number of target-color elements in the array. Specifically, it was expected that as the number of target-color elements increased, search times would become longer. The supposition is that attention is limited to the target-relevant elements in

**Table 3.1**
Mean Search Times (in ms) for the First Guided-Search Study Involving Individuals with Mental Retardation

| | | Set Size | | |
|---|---|---|---|---|
| | 4 | 8 | 12 | 16 |
| Feature Search | | | | |
| Color | 695 | 750 | 682 | 664 |
| Form | 893 | 1015 | 1105 | 1188 |
| Guided Search | 836 | 837 | 839 | 870 |

the array, but these elements are searched serially. Again, the data supported the hypotheses (see Table 3.2). Search rates were positively correlated with set size for the form-based feature-search task. For the guided-search tasks, search times did not increase significantly as set size increased (3 ms/item), but search times did increase significantly as the number of target-color elements increased (30 ms/item). In conjunction, these guided-search studies indicate that individuals with mental retardation can control their allocation of attention across a visual array to meet task demands and maximize performance. These individuals with mental retardation were able to limit attention to search-relevant items while inhibiting attention to search-irrelevant items. Once the search-relevant items were selected for further processing, the data indicate that a serial search of the relevant items was conducted by the participants. Thus, it appears that pairs of features on a given dimension (e.g., color) that result in parallel search (i.e., no increase in RT [reaction time] as set size increases) may be used to guide attention to increase the efficiency of search for an element in a visual array defined by a difference on another dimension (e.g., form). Individuals with mental retardation demonstrate this increase in search efficiency, indicating that they can exert substantial top-down control over the allocation of visual selective attention on a spatial-search task (Carlin, Soraci, Dennis, Strawbridge, & Chechile, 2002).

### *Attentional capture*

The attentional-capture literature focuses on the issue of whether individuals can inhibit attention to highly salient stimuli that are irrelevant to the search goals of the task. Specifically, interest focuses on whether spatial attention is involuntarily drawn to particular elements in an array, even if they are unrelated to the search goals of the organism. This phenomenon is largely a

**Table 3.2**
Mean Search Times (in ms) by Set Size and the Number of Relevant Stimuli for the Second Guided-Search Study

| | Set Size | | | |
|---|---|---|---|---|
| | 4 | 8 | 12 | 16 |
| Task | | | | |
| Feature Search (Form) | 829 | 970 | 1128 | 1142 |
| Guided Search | | | | |
| Two Relevant | 731 | 766 | 774 | 810 |
| Three Relevant | 831 | 797 | 796 | 830 |
| Four Relevant | 805 | 852 | 866 | 825 |

bottom-up processing effect because it is the salience of the stimulus that draws attention and not the goal orientation of the observer. Whereas in the previous section we were concerned with the voluntary direction of attention to goal-relevant stimuli in a visual array, this section deals with the involuntary "capturing" of attention by goal-irrelevant stimuli. However, as was true of the guided-search studies, we are assessing the ability of individuals with mental retardation to inhibit attention to task-irrelevant elements of visual arrays.

The conditions under which attentional capture effects occur and do not occur have been studied extensively in the past decade. As with all visual search tasks, the phenomenon is governed by a complex interplay between the goals of the participant (top-down processing) and the relations between the elements in the visual array (bottom-up activation for individual stimuli). Though attentional capture is dependent to a great extent on the conspicuousness of elements in the visual array (e.g., Cole & Hughes, 1984), the attentional control "settings" of the observer do influence whether capture is likely to occur (Folk, Remington, & Johnston, 1992; Yantis, 1993; Yantis & Egeth, 1999). Typically, capture effects do not occur if the target differs from a homogeneous field of distractors along a single dimension (e.g., size) and the targets and distractors are consistent across trials. This is true regardless of the salience, or conspicuity, of the capture stimulus.

In the standard attentional capture task that we utilized in assessing the performances of individuals with mental retardation, we employed methods that typically do not result in significant capture effects with people without mental retardation. In these tasks, the single target and multiple distractors were consistent across trials, and the capture stimulus was a single stimulus that differed

**Figure 3.3**
Examples of the Three Attentional Capture Tasks Employed. The target in all three conditions is the large gray triangle. The size of the target is reduced in the Strong Capture condition. The salience of the capture stimulus is reduced in the Weak Capture condition.

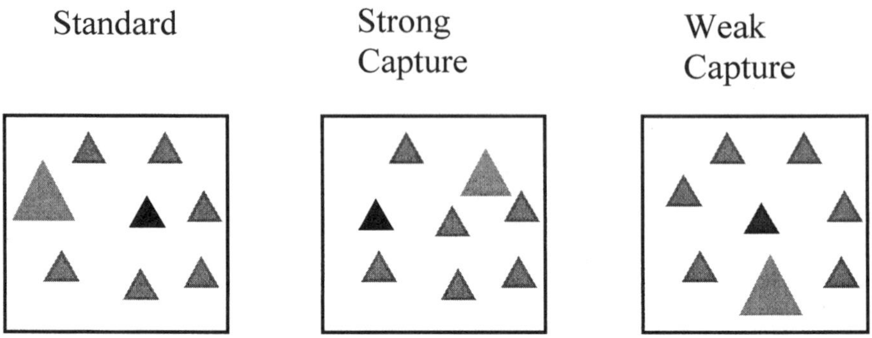


in color (e.g., shade of gray) from the other stimuli in the array. Example arrays for each of the three tasks are shown in Figure 3.3. The Standard task required the participant to find a large gray triangle among small gray triangles with the capture stimulus being a highly salient small black triangle. The other two capture tasks involved manipulations of the relations between either the target and primary distractors (Strong Capture) or the capture stimulus and the other elements of the array (Weak Capture). The former manipulation was designed to make the search task more difficult (e.g., result in increasing search times as set size increases) and to increase the relative activation strength of the capture stimulus, even though it was not relevant to the search goal of the observer. The Weak Capture task maintained the degree of difference between the target and primary distractors, but decreased the difference between the capture stimulus and the other elements, decreasing the activation strength of the capture stimulus.

Results for these tasks indicated that the individuals with and without mental retardation demonstrated a significant capture effect for the Strong Capture task only. Both groups also showed positive RT x Set Size search functions for the Strong Capture task. Thus, both groups of individuals showed the capture effect only when the search task was so difficult that serial search for the target was necessitated. If the target of search was highly salient (i.e., the Standard and Weak Capture tasks), then attention was directed immediately to the target stimulus and the capture stimulus did not promote interference. The individuals with mental retardation were able to inhibit attention to the salient, but goal-irrelevant, capture stimulus in certain circumstances. In summary, for the conditions tested, the individuals with mental retardation were not more deleteriously affected by a highly salient distractor stimulus than individuals without mental retardation.

In conjunction, the results of our work on guided search and attentional capture indicate that individuals with mental retardation (a) can utilize goal-directed instructions to control the allocation of attention and enhance visual search efficiency, and (b) are not more susceptible to interference from irrelevant stimuli than are individuals without mental retardation. These results indicate that the basic top-down and bottom-up processes controlling the allocation of attention in visual arrays operate similarly for individuals with and without mental retardation. It should be noted, however, that the search tasks utilized to date have targeted the extremes of performance (i.e., highly salient targets and capture stimuli). More subtle differences between groups may emerge as the relative saliences of array elements are manipulated or as more ecologically valid search tasks (e.g., conjunction search; heterogeneous arrays of distractors) are used.

### *Scene perception and observing behavior*

Typically, when individuals view a visual scene, they have the impression that their representation of the scene is nearly complete and quite detailed. However, a series of recent studies of a phenomenon known as "change blindness" have called this impression into question. Several labs have demonstrated that individuals fail to detect changes that take place in their everyday surroundings. Change blindness has been induced in a number of settings by changing scene details across saccades (e.g., Henderson & Hollingworth, 1999), blank periods (e.g., Rensink, O'Regan, & Clark, 1997), and other forms of visual interruption (e.g., Simons & Levin, 1998). For example, Rensink et al. (1997) used the flicker paradigm in which two versions of a scene are presented to participants with a brief (80 ms) blank interval between the presentations: the participant then has the perception of viewing a blinking scene. Across the two versions of each scene a single object was changed either in color, location, or presence versus absence. Participants had difficulty identifying many of the changes in the flicker condition, especially if the changes were in the periphery of the scene. Simons and Levin (1998) demonstrated that change blindness could occur in a real-world situation. In their study, an experimenter carrying a campus map approached a pedestrian and asked for directions to a building on campus. While the pedestrian was giving the directions to the first experimenter, two other confederates carrying a door passed between them obstructing their views of one another. While the door was passing, the experimenter that requested directions switched positions with one of the door-carrying confederates. Only seven of the fifteen pedestrians noticed that the individual's identity changed when the door interrupted the conversation. These results seem to indicate that only the small portion of a complex scene that receives focal attention is encoded in detail. Peripheral changes and changes that occur across visual interruptions are more difficult to detect because they require a focused, serial search of the scene and/or longer dwell times on objects.

**Figure 3.4**
Two Versions of a Scene Used in the Flicker Task. The two pictures were flashed alternately until the participant detected the object that was changing. In this scene, the leg of the table is changing from rectangular to triangular.

We have begun recently to employ the flicker methodology to study the ability of individuals with mental retardation to detect changes to complex scenes (Carlin, Soraci, Strawbridge, Dennis, & Coiselle, in press). In our study, individuals with and without mental retardation were presented with thirty-six real-world scenes. Each scene was presented in a flicker sequence comprising the original picture, a blank screen, an altered version of the picture, and another blank screen. This sequence was shown repeatedly until the change was identified. The three types of changes studied were color changes, form changes, and presence or absence of an object. In addition, changes occurred either in the focal area of the scene or in the periphery. Figure 3.4 shows an original and altered version of one of the scenes used. This example involves a centrally located form change (i.e., the legs of the table).

Our results were consistent with the change blindness literature in that changes were not identifiable immediately, color changes were detected more quickly than form or presence or absence changes, and centrally located changes were detected more quickly than peripheral changes. In addition, we found that individuals without mental retardation detected changes more quickly (10 seconds) than individuals with mental retardation (23 s), and a significant group × location interaction, indicating that peripheral changes were more difficult for individuals with mental retardation than for individuals without mental retardation. We currently are investigating the basis for the latter effect using eye-tracking technology. Preliminary analyses from videotapes of three individuals with mental retardation indicate that the individuals with mental retardation scan the displays inefficiently. That is, the individuals with mental retardation observe individual objects for prolonged periods of time without shifting gaze to other objects in the display. Thus, their identification times are inflated due to longer dwell times on non-target objects in the scenes. However, once the target object is observed, they seem to rapidly detect the change and respond quickly. We believe this is a fertile area for additional research into intelligence-related differences in ecologically relevant visual-search behavior (particularly differential top-down control) and for application of attentional guidance programs similar to those discussed in the next part.

## MANIPULATING VISUAL STRUCTURE: APPLICATIONS

A challenge to those concerned with educating individuals with mental retardation is that such individuals often do not understand or consistently follow verbal instructions. This fact places the burden on the scientist or teacher to structure tasks in which attention is guided to the target stimulus or stimulus relation, without the assistance or goal-directness that verbal instructions typically provide. The basic research discussed has provided substantial knowledge about the effects of particular forms of structural manipulation on the observing responses of individuals with mental retardation. In the following segments, we describe two "front-end" procedures (Brown, Collins, & Duguid,

1989; Soraci, Carlin, Sharp, Franks, Vye, & Bransford, 1993) based on these basic principles, that were designed to guide observing responses to critical components of visual arrays to enhance subsequent performance. Both of the procedures described are designed to work without the need for detailed verbal instruction; the techniques should be useful for a wide range of individuals with and without mental retardation, even those with limited verbal skills.

### Inducing matching-to-sample performance

As mentioned briefly earlier in this chapter, we have previously induced oddity rule learning in young children with and without mental retardation (Soraci et al., 1983, 1987) by increasing the number and contiguity of non-odd stimuli. These manipulations were designed to make the oddity relation (same–different) more salient, and thus more likely to be the basis for responding. Similarly, we have utilized multi-element arrays to induce matching-to-sample performances in young children with and without mental retardation (Mackay, Soraci, Carlin, Dennis, & Strawbridge, 2002).

The stimuli employed in these studies were five geometric forms: circle, square, triangle, rectangle, and hexagon. The sample and comparison stimuli were randomly selected on each trial from the set of five figures. The sample was presented alone until the participant touched it to activate the comparison array. For children who did not demonstrate consistent matching on a two-choice pretest, we introduced a progressive training procedure in which the number of comparison stimuli was decreased incrementally across trials. The training procedure is depicted in Figure 3.5. Initial trials during the training phase involved nine-element arrays. These arrays were designed to make the correct choice (i.e., the matching comparison, or S+) "pop out" of the array, making the correct choice more likely to be selected and reinforced. If the participant made three consecutive correct responses, then the number of comparison stimuli was reduced by one. This progression continued until the participant either failed to advance within twenty trials or until the individual completed the eight-stage training program.

During the initial stages of training, selection is likely to be perceptually based. That is, the participant is expected to select the most salient comparison stimulus, without regard to its relation to the sample. Evidence of effective perceptual guidance would be a lack of incorrect responses, even on initial trials. If guidance is effective, then the correct comparison should be selected even on the first trial, and the number of errors that occur throughout the training sequence should be minimal. Ideally, the procedure should induce matching errorlessly. The purpose of utilizing the progressive procedure was not only to reduce the number of errors, but also to slowly reduce the perceptual guidance afforded by the arrays and increase the likelihood that responding would be governed by the identity relation between the sample and correct comparison. If this transition from perceptually based responding to conceptually based responding did not occur, failure to respond correctly on the final stage of training (i.e., two-choice arrays) would be expected.

**Figure 3.5**
Depiction of the Match-to-Sample Training Procedure Employed. The number of comparison stimuli was reduced progressively until the participant returned to the previously failed two-choice match-to-sample task.

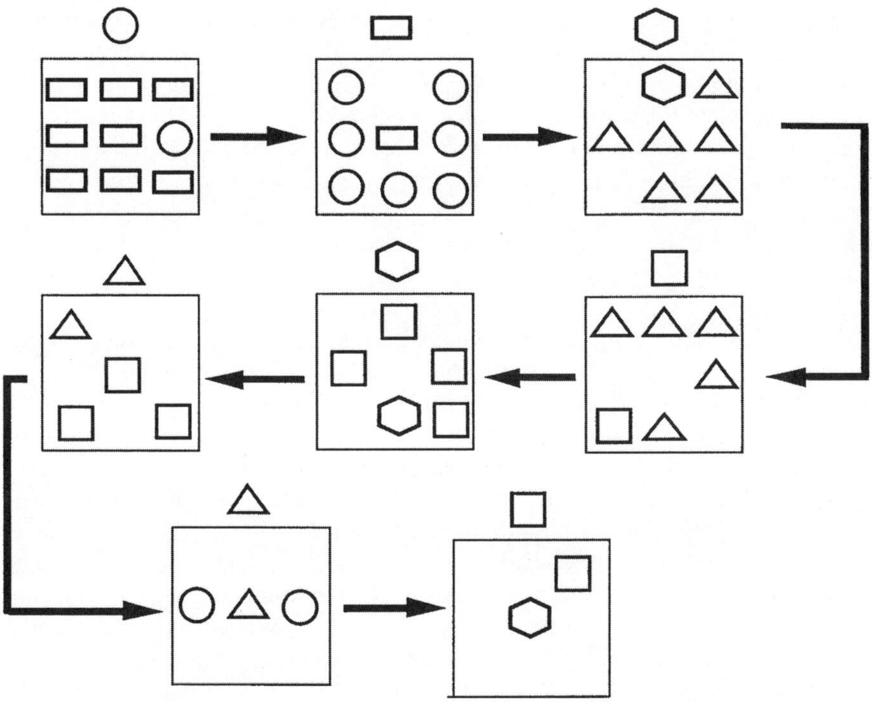

From a behavioral perspective, stimulus control must transfer at some point during the training sequence to the identity relation between the sample and correct comparison. The training procedure includes an apparent risk of the participant not attending to the critical relation between the sample and correct comparison. Correct responding, and reinforcement, may occur throughout the training sequence without establishment of the identity relation. It is only in the final stage (i.e., the two-choice arrays) that identity-based responding must occur. At all other stages of the training sequence, the participant may respond correctly by selecting the "odd" comparison, without attending to the sample or its relation to the comparison stimuli. If transfer of control does not occur at some point in the training sequence, then participants should fail to respond correctly when presented with two-choice matching arrays.

Our initial test of the efficacy of this procedure included seventeen three- and four-year-old children without mental retardation who had failed an initial two-choice match-to-sample test. Thirteen of the seventeen children completed the training sequence, and eleven of these thirteen performed perfectly on the two-choice matching post-test. Of the other two who completed the training sequence, one was unavailable for the post-test and the other responded correctly on fourteen of the twenty (70%) post-test trials. Attesting further to the power of this training technique was the finding that two individuals completed the training sequence without errors, and five others committed only a few errors initially and then performed errorlessly throughout the progressive reduction procedure. Thus, the perceptual enhancement technique was effective in guiding initial responses to the correct comparison stimulus and resulted in rapid learning of the matching-to-sample response. Of the four children who did not complete the training sequence, three failed to progress beyond the initial nine-element stage and the fourth failed to pass the seven-element stage. Thus, for these latter four individuals, the perceptual enhancement was not effective, or at minimum, they did not respond to guiding. There were no individuals who progressed to the two-choice training stage and failed. Thus, the concern with the possibility of failing to establish the critical relation or teaching the incorrect relation (i.e., oddity) was alleviated.

Given these promising findings, we next assessed the efficacy of this training procedure with individuals with mental retardation who failed an initial matching pretest. Twelve individuals with mild mental retardation participated. Of these twelve individuals, nine successfully completed the training sequence. Six of the nine committed five or fewer errors throughout the entire training program, again attesting to the power of the perceptual enhancement manipulation. Of the three participants who did not complete the training sequence, all three failed to pass the initial nine-element stage. Thus, these data are quite similar to the performance patterns and rates observed in the study involving young children without mental retardation. Of the nine participants with mental retardation who advanced to the post-test, seven performed at rates of 92 percent or higher, one responded correctly on 70 percent of the post-test trials and the final individual was successful on only 35 percent of the post-test trials. Finally, the seven top performers were assessed a month or more later for maintenance of the matching behavior, and all seven performed errorlessly on the maintenance test.

This training procedure has been shown to be quite effective in establishing matching-to-sample behaviors in individuals with and without mental retardation, and does so in most cases with a minimal number of errors. However, many questions remain regarding the nature of the training effects observed. Are all of the training stages necessary, or can the procedure be designed more efficiently? At what point in the training sequence does the transfer of stimulus control, or change from perceptually based to conceptually based responding, occur? Will the procedure be as effective for establishing arbitrary relations? In addition, analyses of eye-movement patterns of individuals performing the training would indicate whether the correct comparison truly pops out of the initial arrays and whether participants scan the entire array or respond rapidly to the most salient, and engaging, comparison. Answering these types of questions will be critical for determining the nature of these types of perceptually based interventions, and designing the most efficient training programs possible.

The experimental basis for the design of this applied training technique was our extensive research concerning particular manipulations of visual structure that increase the salience of critical relations among stimuli and that induce observation of the critical element (i.e., popout) in the visual array. Further improvements in the development of successful training techniques for all, or most, individuals with mental retardation will likely come about through expanding our repertoire of effective structural manipulations and individualization of training. The latter will likely be critical for increasing the success rates of the procedures utilized. For example, those individuals who did not progress beyond the nine-element stage in the current training studies may respond better to other types of manipulations or to more salient differences (e.g., color) among training stimuli. The latter would be consistent with our visual-search studies indicating that color differences result in popout effects in a much higher percentage of individuals with mental retardation than do form- or size-based variations. Thus, individual differences on the basic visual-search tasks may be manifested in poorer performances on educationally relevant training tasks. Recognition of these within-group differences, and development of methods for identifying children in need of augmented training procedures, is a necessity for increasing the generality of training effects of the type described above.

### Enhancing free recall rates

Much research on the memory abilities of individuals with mental retardation has focused on identifying various types of memory deficiencies (e.g., Belmont & Butterfield, 1971; Borkowski, 1985; Ellis, 1970; Turnure, 1985) and the development of programs for teaching various memory-enhancement strategies. More recently, however, greater attention has been given to the importance of the learning and testing contexts on memory performance (Bray, Fletcher, & Turner, 1997; Bray et al., 1998). Bray and his colleagues have demonstrated that individuals with mental retardation may be capable of employing memory enhancement techniques, but do so only if contextual factors, or "situational supports," are optimal. This places the burden on the teacher or experimenter to identify and implement optimal learning environments for these individuals.

Toward this end, we have been studying generative learning processes in memory with undergraduates for some time. Generative learning refers to advantages that accrue as a result of generating items rather than simply being provided with items to remember. The common finding is a recall advantage for generated responses (e.g., A drinking vessel: C__P) over responses that simply are read (e.g., A drinking vessel: CUP). This generation effect has been instantiated across a range of encoding conditions and experimental stimuli, including words (Slamecka & Graf, 1978; Soraci et al., 1994), nonwords (Johns & Swanson, 1988), mathematical problems (McNamara & Healy, 1995), homographs (Soraci, Carlin, Chechile, Franks, Wills, & Watanabe, 1999), and pictures (Kinjo & Snodgrass, 2000; Peynircioglu, 1989; Wills, Soraci, Chechile, & Taylor, in press). Explanations of the memory enhancements due to generative encoding have involved distinctiveness (McDaniel, Waddill, & Einstein, 1988), cue–target relational and item-specific processing (Hirshman & Bjork, 1988), procedural processes (McNamara & Healy, 1995), and multiple cuing (Soraci et al., 1994, 1999). Because generative learning effects occur across a wide range of testing contexts and materials, they have theoretical significance and are potentially important for the design of educational interventions.

One type of generative learning concerns "aha" processes. Auble, Franks, and Soraci (1979) were the first to demonstrate the "aha" effect in memory. In their study, grammatically correct but incomprehensible sentences (e.g., "The haystack was important because the cloth ripped") were presented either with a post-sentence solution word that would disambiguate the context (i.e., "parachute") or with the solution meaningfully embedded within the sentence so that the sentence was naturally understandable. A subsequent recall test indicated that sentential recall was superior in the cue-after (i.e., "aha") condition. A recent study by Wills, Soraci, Chechile, and Taylor (2000) extended this "aha" phenomenon to visual stimuli. In this study, participants were shown pictures of connect-the-dot objects in three different manners. Subjects in the Dot condition had to complete a connect-the-dot object so that the full outline became visible. Subjects in the Trace condition had to trace the outline of an already completed connect-the-dot object. Finally, participants in the Scan condition had to visually scan the completed object while reading the sequence of connect-the-dot numbers. Pictorial recall was better in the Dot (i.e., generative) condition than in either of the comparison conditions, but only in the absence of foreknowledge of the to-be-completed picture. If participants were given previews of the pictures prior to completing the connect-the-dot task, the difference between the three conditions disappeared. Thus, the advantage observed occurred only in a generative learning condition in which an unknown figure was slowly revealed by connecting dots to form the outline of the object. This pure demonstration of an "aha" effect for visual stimuli attests further to the generalizability of generative learning effects.

The physiological processes that are engaged under experimental conditions that involve active, generative encoding have been investigated in several studies. Posner and Raichle (1994) have demonstrated, via positron emission tomography (PET) neuroimaging techniques, enhanced activation in multiple cortical areas when individuals are required to generate semantic associates to provided cues. Research involving phasic alertness and its relationship to indices such as cardiac rate has shown the importance of attentional enhancement as it relates to physiological functioning (Soraci et al., 1986). Yet there have been virtually no attempts up to this point to investigate the generative aspect of "aha" effects as it relates to underlying neural functioning.

In a recently completed study, Warren, Soraci, Chechile, and Holcomb (1999) extended Auble et al.'s (1979) original "aha" research utilizing an event-related potential (ERP) methodology. In this study, they presented participants with sentences of four types: normal sentences, anomalous-ending sentences (e.g., Kutas & Hillyard, 1980), easy "aha" sentences, and difficult "aha" sentences. Following each sentence, a cue consistent with the context was presented. Examples of the four conditions are shown in Table 3.3. Results indicated the

**Table 3.3**
Examples of the Four Types of Sentences Used by Warren et al. (2000)

| | Sentence | Cue |
|---|---|---|
| Normal | She cannot tell a lie because she is honest. | TRUTHFUL |
| Anomalous-Ending | He called the dentist because his tooth rang. | CAVITY |
| Easy "Aha" | Her toes were pointed because the dance started. | BALLET |
| Difficult "Aha" | The house grew small because the sun came out. | IGLOO |

presence of the typical N400 following presentation of the final word in the anomalous-ending sentences (e.g., Kutas & Hillyard, 1980). There were no such effects across the other three conditions. In a novel finding, there was an N400 following presentation of the disambiguating cue for each difficult "aha" sentence. The presence of this N400 to difficult "aha" cue words, and absence of this effect for all other conditions, indicates that full resolution of the difficult sentences still was occurring during the presentation of the cue words. In addition, the difficult "aha" condition was associated with the presence of a late positivity in the ERP data. This effect was not present in the other three conditions. Late positive components (LPC) have been linked to working memory. Thus, it appears that the LPC in the difficult "aha" condition suggests that these sentences are being re-integrated to achieve resolution. We intend to investigate this phenomenon in future studies involving an examination of intelligence-related differences in "aha" effects as they relate to memory enhancement.

Based on this program of research on generation (Chechile & Soraci, 1999; Soraci et al., 1994, 1999) and "aha" (Auble, Franks, & Soraci, 1979; Wills, Soraci, Chechile, & Taylor, 2000) effects in memory, we designed a generative encoding context to increase the free-recall rates of individuals with and without mental retardation (Carlin, Soraci, Dennis, Chechile, & Loiselle, 2001). Our goal in designing the task was to develop a method for presenting visual materials to individuals with mental retardation that would enhance free-recall rates without the need for extensive strategy training or verbal instructions, and that could be adapted easily for computerized learning environments. Our work in the areas of generative and "aha" learning indicated that an encoding context that was generative, induced activation of multiple potential retrieval routes, and involved resolution of a problem (i.e., an "aha" effect) would be most effective for enhancing free-recall rates. The manipulation we utilized to produce these effects was the fading of items using multiple stages of Gaussian blur. Clear images were blurred in equal increments to create eleven images for each picture. For most images, the first six steps of blur in the fading sequence were not identifiable. As the sequence progressed, however, the participant could properly

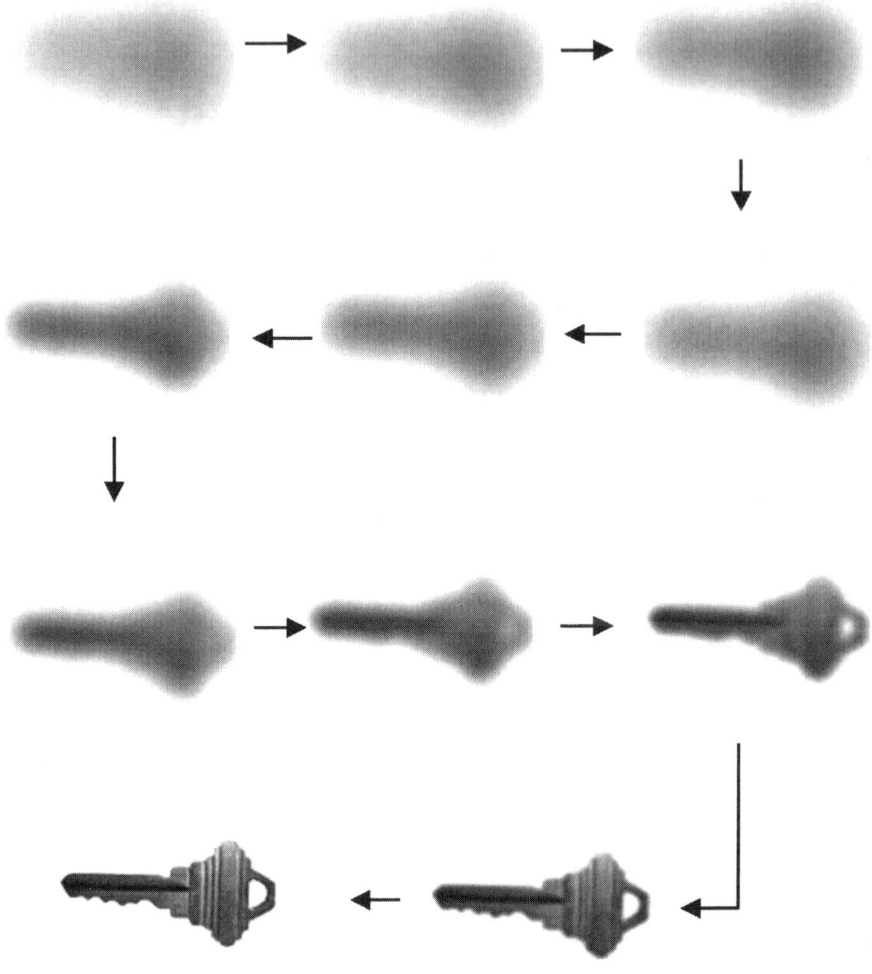

**Figure 3.6**
Demonstration of the Ten Stages of Blur Used in the Free-Recall Study. The sequence is designed to promote generation of possible solutions prior to identification of the object.

identify the object and provide a verbal label for it (see Figure 3.6 for an example sequence). This form of manipulation promoted the generation of possible solutions that could serve as retrieval cues at test, and induced the recognition ("aha") of solving an identification problem on each trial.

In our critical test, we compared a condition in which pictures were initially presented out of focus and slowly came into focus ("fade in" or "aha") to a control condition in which pictures were initially presented clearly and then were slowly "faded out." There were no explicit instructions to generate potential labels for the fading sequences. Rather, the nature of the visual presentation was designed to induce particular cognitive operations that enhance performance on a generative recall task. The participants were a group of individuals with mild mental retardation and groups of individuals without mental retardation that were matched according to chronological age (CA) and mental age (MA) as determined by the Peabody Picture Vocabulary Test (PPVT-III). Results indicated that the individuals with mental retardation and the CA-matched controls recalled the fade-in pictures at a higher rate than fade-out pictures. This supported our contention and demonstrated that focused manipulations of a learning context can significantly impact the processing and performance of individuals with mental retardation. The lack of an effect for the MA-matched individuals is consistent with the position of Engle and Nagle (1979) that younger individuals (< nine years of age) would not benefit as much from this type of manipulation due to a less-developed semantic network. For example, these individuals may not generate as many potential solutions during the fading sequence, thus limiting its effectiveness. However, perhaps other similar manipulations such as puzzle construction or connect-the-dot pictures (see Wills et al., 2000) that induce similar cognitive operations may be effective for these younger individuals.

## PHYSIOLOGICAL CORRELATES

The pattern of performance similarities and differences between the individuals with and without mental retardation discussed previously raises the question of the potential neural bases for the intelligence-related differences observed. Our view on this issue is in contradistinction to those that imply that such group differences are necessarily dependent upon structural, and relatively nonmodifiable, deficits associated with mental retardation. Rather, while remaining largely agnostic about the neurological basis of the observed group differences, we have focused on enhancing the performances of individuals with mental retardation by developing methods that perceptually increase the salience of critical stimulus relations (e.g., oddity, matching to sample, visual search) and promote adaptive attending behaviors (e.g., free recall, guided search). However, identifying the neurological loci associated with performance differences across individuals with and without mental retardation or with varying etiologies of mental retardation will be an important advance toward understanding mental retardation and developing the most effective interventions. Later, we discuss several potential bases for the group differences discussed previously. Some have been the focus of active research involving individuals with mental retardation (i.e., the magnocellular visual pathway), whereas others that are in need of experimental testing are merely interesting speculations at this time.

Given our consistent findings of lowered sensitivities to stimulus relations and reports of performance decrements on depth (Fox & Oross, 1988) and motion processing tasks (Fox & Oross, 1990, 1992), there has been speculation that one potential critical locus is the magnocellular visual pathway (Soraci & Carlin, 1992; Fox & Oross, 1992). This visual pathway is primarily responsible for the processing of motion, depth cues, and figure-to-ground relations in the human geniculocortical visual system (e.g., Livingstone & Hubel, 1988; Shapley, 1990). Further, there have been several reports associating inefficient magnocellular processing with developmental dyslexia (e.g., Cornelissen, Richardson, Mason, Fowler, & Stein, 1995; Demb, Boynton, & Heeger, 1998; Eden, Van Meter, Rumsey, Maisog, Woods, & Zeffiro, 1996; Livingstone, Rosen, Drislane, & Galaburda, 1991). However, some of our recent work (Carlin, Hobbs, Bud, & Soraci, 1999) has indicated that individuals with mental retardation perform well on motion-detection tasks if exposed to a brief training procedure. A recently completed study comparing motion-detection thresholds of individuals with and without mental retardation indicated that individuals with mental retardation performed much better than estimated by Fox and Oross (1990), and modest intelligence-related differences emerged only when the figure-to-ground relation was dramatically reduced (Carlin, Soraci, Strawbridge, & Loiselle, 2000). The large reduction in the magnitude of difference in detection thresholds across individuals with and without mental retardation after institution of a brief training procedure would seem to indicate that earlier differences were procedurally based and not neurologically based. However, that smaller group differences still did emerge at particular levels of the task leaves open the possibility that there may be a neurologically based difference across these populations. Perhaps small differences in detection thresholds do have significant behavioral consequences. Resolving the debate concerning magnocellular involvement in mental retardation will require use of a broader array of visual processing tasks and advanced imaging techniques that will provide more direct information about physiological functioning.

Particularly relevant to our work on visual search abilities and the regulation of attention are recent studies (Botvinick, Nystrom, Fissell, Carter, & Cohen, 1999; Carter, Braver, Barch, Botvinick, Noll, & Cohen, 1998; Carter et al., 2000) on the role of the anterior cingulate cortex (ACC). These studies have employed event-related fMRI to record regional brain activations during tasks such as a Stroop task, a flanker task, and a sustained-attention task. Based on their studies of normally functioning adult volunteers, these authors have contended that the ACC performs a response–conflict monitoring role. The ACC's role is to detect the level of response competition presented by a particular task, and perhaps to indicate the need for greater top-down control of performance to maximize accuracy and speed. Thus, the ACC would serve as part of a broader executive functioning network that likely includes coordinated activity in numerous areas of the brain including the frontal lobes (e.g., Dempster, 1991) and perhaps the locus coeruleus (Usher, Cohen, Servan-Schreiber, Rajkowski, & Aston-Jones, 1999). Focus on these structures and areas in the brain involves the pursuit of understanding the physiology underlying top-down control of behavior, whereas the earlier focus on magnocellular visual pathway functioning represented a concern with sensory functioning (i.e., bottom-up processing).

Toates (1998) presented evidence that the relative weightings of top-down and bottom-up controls of behavior vary as a function of task, developmental level, domain experience, and neurological functioning level. According to this model, understanding performance differences between individuals with and without mental retardation would require knowledge of individuals' sensory functioning and cognitive capabilities, each a function of age and neurological functioning, and how individuals' abilities interact with the nature of the experimental or educational task encountered. Many of the intelligence-related performance differences discussed in this chapter are related to this critical interaction between top-down and bottom-up processes. Further, the training programs described represent methods for addressing both types of processing deficiencies. In terms of bottom-up processing, our research has shown that individuals with mental retardation have decreased sensitivities to particular stimulus relations relative to individuals without mental retardation, and that these limitations often can be overcome when a learning task is structured to make the critical relations more salient. Matching behavior can be induced by utilizing pop-out arrays (i.e., arrays with many nonmatching, or *S*–, stimuli) initially, and slowly decreasing the perceptual guidance across training steps. With regard to top-down processing, our memory-enhancement program is an example of a manipulation type that may be used to overcome strategic deficiencies for particular individuals. If individuals do not spontaneously use memory-enhancement strategies (e.g., rehearsal, mnemonics), then structuring the learning task in a manner that induces memory-enhancing cognitive operations, which we assume is what occurs, can lead to increased memory rates. An understanding of the factors that control or limit behavior on a particular task, therefore, is critical for designing the most effective training procedures. We believe the top-down/bottom-up framework is helpful in this regard, and that increased knowledge of the underlying neurological bases of these forms of processing will lead to more advanced research into the causes of mental retardation, the heterogeneity of performances observed in this population, and the development of effective and efficient educational training programs.

## SUMMARY

The foregoing work demonstrates the utility of a program of research designed to augment the performances of individuals with and without mental retardation via manipulations of task and perceptual structure. The program of research includes initial use of broad-based assessments of perceptual and attentional skills to identify group patterns of behavior to inform the design of individualized instruction programs. The individualized instruction programs are designed to maximize the performances of individuals based on the knowledge of their current capabilities and perceptual or attentional strengths and weaknesses. The combination of basic and applied research, and melding of behavioral and cognitive methodologies, provides great promise for advancing understanding of the nature of mental retardation and developing methods for remediating various forms of performance deficiencies reported in the literature and encountered in the classroom. In particular, we believe these types of manipulations could easily be incorporated into many currently available computerized learning programs (e.g., Bransford et al., 1999; Sandford & Turner, 1999) to augment the performances of individuals with various forms of attentional problems.

## NOTE

This research was supported by the National Institute of Child Health and Human Development Grants HD23682, HD36832, and HD25995 awarded to the Shriver Center of the University of Massachusetts Medical School. Its contents are solely the responsibility of the authors and do not necessarily represent the official views of the NICHD.
Correspondence and requests for reprints should be sent to Michael Carlin, Psychological Sciences Division, Shriver Center, 200 Trapelo Road, Waltham, MA 02452; michael.carlin@umassmed.edu.

## REFERENCES

Aks, D.J., & Coren, S. (1990). Is susceptibility to distraction related to mental ability? *Journal of Educational Psychology, 82*(2), 388–390.

Auble, P.M., Franks, J.J. & Soraci, S.A. (1979). Effort toward comprehension: Elaboration or "aha!"? *Memory & Cognition, 7,* 426–434.

Belmont, J.M., & Butterfield, E.C. (1971). Learning strategies as determinants of memory deficiencies. *Cognitive Psychology, 2,* 411–420.

Borkowski, J.G. (1985). Signs of intelligence: Strategy generalization and metacognition. In S.R. Yussen (Ed.), *The growth of reflection in children* (pp. 105–144). New York: Academic Press.

Botvinick, M., Nystrom, L.E., Fissell, K., Carter, C.S., & Cohen, J.D. (1999). Conflict monitoring versus selection-for-action in anterior cingulate cortex. *Nature, 402,* 179–181.

Bransford, J.D., & The Cognition and Technology Group at Vanderbilt University. (1999). Designing environments to reveal, support, and expand our children's potentials. In S.A. Soraci, & W.J. McIlvane (Eds.), *Perspectives on fundamental processes in intellectual functioning: Vol. 1 A survey of research approaches.* Stamford, CT: Ablex.

Bray, N.W., Fletcher, K.L., & Turner, L.A. (1997). Cognitive competencies and strategy use in individuals with mental retardation. In W.E. MacLean, Jr. (Ed.), *Ellis' handbook of mental deficiency, psychological theory and research* (3rd ed., pp. 197–217). Mahwah, NJ: Erlbaum.

Bray, N.W., Reilly, K.D., Fletcher, K.L., Huffman, L.F., Grupe, L.A., Villa, M.F., & Anumolu, V. (1998). Memory competencies and deficiencies: A conceptual framework and the potential of connectionist models. In S.A. Soraci & W.J. McIlvane (Eds.), *Perspectives on fundamental processes in intellectual functioning:* Vol. 1. (3–44). Greenwich, CT: Ablex/JAI Press.

Brown, A.L. (1970). Subject and experimental variables in the oddity learning of normal and retarded children. *American Journal of Mental Deficiency, 75,* 142–151.

Brown, J.S., Collins, A., & Duguid, P. (1989). Situated cognition and the culture of learning. *Educational Researcher, 18,* 32–42.

Carlin, M.T., Hobbs, K.L., Bud, M.J., & Soraci, S.A. (1999). Detection of motion-defined forms by individuals with mental retardation and autism: Evidence of modifiability. *Intelligence, 27,* 141–156.

Carlin, M.T., Soraci, S.A., Dennis, N.A., Chechile, N.A., & Loiselle, R.C. (2001). Enhancing free recall rates of individuals with mental retardation. *American Journal on Mental Retardation, 106,* 314–326.

Carlin, M.T., Soraci, S.A., Dennis, N.A., Strawbridge, C.P., & Checile, N.A. (2002). Guided visual search in individuals with mental retardation. *American Journal on Mental Retardation, 107* (4).

Carlin, M.T., Soraci, S.A., Goldman, A.L., & McIlvane, W.J. (1995). Visual search in unidimensional arrays: A comparison between subjects with and without mental retardation. *Intelligence, 21,* 175–195.

Carlin, M.T., Soraci, S.A., Strawbridge, C.P., Dennis, N.A., & Loiselle, R. (in press). Detection of changes in naturalistic scenes: Comparisons of individuals with and without mental retardation. *American Journal on Mental Retardation.*

Carlin, M.T., Soraci, S.A., Strawbridge, C., & Loiselle, R. (2000, March). *Motion perception and mental retardation.* Poster presented at the 33rd Annual Gatlinburg Conference on Research and Theory in Mental Retardation and Developmental Disabilities, San Diego, CA.

Carter, C.S., Braver, T.S., Barch, D.M., Botvinick, M.M., Noll, D., & Cohen, J.D. (1998). Anterior cingulate cortex, error detection, and the online monitoring of performance. *Science, 280,* 747–749.

Carter, C.S., Macdonald, A.M., Botvinick, M., Ross, L.L., Stenger, V.A., Noll, D., & Cohen, J.D. (2000). Parsing executive processes: Strategic vs. evaluative functions of the anterior cingulate cortex. *Proceedings of the National Academy of Sciences, 97,* 1944–1948.

Cha, K.-H., & Merrill, E.C. (1994). Facilitation and inhibition in visual selective attention processes of individuals with and without mental retardation. *American Journal on Mental Retardation, 98,* 594–600.

Chechile, R.A., & Soraci, S.A. (1999). Evidence for a multiple-process account of the generation effect. *Memory, 7,* 483–508.

Cole, C.L., & Hughes, P.K. (1984). A field trial of attention and search conspicuity. *Human Factors, 26*(4), 299–313.

Cornelissen, P., Richardson, A., Mason, A., Fowler, S., & Stein, J. (1995). Contrast sensitivity and coherent motion detection measured at photopic luminance levels in dyslexics and controls. *Vision Research, 35,* 1483–1494.

Demb, J.B., Boynton, G.M., & Heeger, D.J. (1998). Functional magnetic resonance imaging of early visual pathways in dyslexia. *The Journal of Neuroscience, 18,* 6939–6951.

Dempster, F.N. (1991). Inhibition processes: A neglected dimension of intelligence. *Intelligence, 15,* 157–173.

Dinsmoor, J.A. (1985). The role of observing and attention in establishing stimulus control. *Journal of the Experimental Analysis of Behavior, 43,* 365–381.

Duncan, J., & Humphreys, G.W. (1992). Beyond the search surface: Visual search and attentional engagement. *Journal of Experimental Psychology: Human Perception and Performance, 18*(2), 578–588.

Eden, G.F., Van Meter, J.W., Rumsey, J.M., Maisog, J.M., Woods, R.P., & Zeffiro, T.A. (1996). Abnormal processing of visual motion in dyslexia revealed by functional brain imaging. *Nature, 382,* 6–69.

Ellis, N.R. (1970). Memory processes in retardates and normals. In N.R. Ellis (Ed.), *International review of research in mental retardation:* Vol. 4. (pp. 1–32). New York: Academic Press.

Ellis, N.R., & Sloan, W. (1959). Oddity learning as a function of mental age. *Journal of Comparative and Physiological Psychology, 52,* 228–230.

Engle, R.W., & Nagle, R.J. (1979). Strategy training and semantic encoding in mildly retarded children. *Intelligence, 3,* 17–30.

Folk, C.L., Remington, R.W., & Johnston, J.C. (1992). Involuntary covert orienting is contingent on attentional control setting. *Journal of Experimental Psychology: Human Perception and Performance, 18(4),* 1030–1044.

Fox, R., & Oross, S. III. (1988). Deficits in stereoscopic depth discrimination by mildly mentally retarded adults. *American Journal of Mental Retardation, 93,* 232–244.

Fox, R., & Oross, S., III. (1990). Mental retardation and perception of global motion. *Perception & Psychophysics, 48,* 252–258.

Fox, R., & Oross, S., III. (1992). Perceptual deficits in mildly mentally retarded adults. In N.W. Bray (Ed.), *International review of research in mental retardation:* Vol. 18. (pp. 1–27). New York: Academic Press.

Henderson, J.M., & Hollingworth, A. (1999). The role of fixation position in detecting scene changes across saccades. *Psychological Science, 10*(5), 438–443.

Hirshman, E., & Bjork, R.A. (1988). The generation effect: Support for a two-factor theory. *Journal of Experimental Psychology: Learning, Memory, and Cognition, 13,* 484–494.

Iarocci, G., & Burack, J.A. (1998). Understanding the development of attention in persons with mental retardation: Challenging the myths. In J.A. Burack, R.M. Hodapp, & E. Zigler (Eds.), *Handbook of mental retardation and development* (pp. 349–381). New York: Cambridge University Press.

Johns, E.E., & Swanson, L.G. (1988). The generation effect with nonwords. *Journal of Experimental Psychology: Learning, Memory, and Cognition, 14,* 581–593.

Kinjo, H., & Snodgrass, J.G. (2000). Does the generation effect occur for pictures? *American Journal of Psychology, 113,* 95–121.

Kutas, M., & Hillyard, S.A. (1980). Reading senseless sentences: Brain potentials reflect semantic incongruity. *Science, 207*(11), 203–205.

Livingstone, M., & Hubel, D. (1988). Segregation of form, color, movement, and depth: Anatomy, physiology, and perception. *Science, 240,* 740–749.

Livingstone, M.S., Rosen, G.D., Drislane, F.W., & Galaburda, A.M. (1991). Physiological and anatomical evidence for a magnocellular defect in developmental dyslexia. *Proceedings of the National Academy of Sciences, 88,* 7943–7947.

Mackay, H.A., Soraci, S.A., Carlin, M.T., Dennis, N.A., & Strawbridge, C.P. (2002). Guiding visual attention during acquisition of matching-to-sample. *American Journal on Mental Retardation, 107* (6), 445–454.

McDaniel, M.A., Waddill, P.J., & Einstein, G.O. (1988). A contextual account of the generation effect: A three-factor theory. *Journal of Memory and Language, 27,* 521–536.

McNamara, D.S., & Healy, A.F. (1995). A procedural explanation of the generation effect: The use of an operand retrieval strategy for multiplication and addition problems. *Journal of Memory and Language, 34,* 399–416.

Merrill, E.C., Cha, K.-H., & Moore, A.L. (1994). Suppression of irrelevant location information by individuals with and without mental retardation. *American Journal on Mental Retardation, 99*(2), 207–214.

Penn, N.E., Sindberg, R.M., & Wohlheuter, M.J. (1969). The oddity concept in severely retarded children. *Child Development, 40,* 154–161.

Peynircioglu, Z.F. (1989). The generation effect with pictures and nonsense figures. *Acta Psychologica, 70,* 153–160.

Posner, M.I., & Raichle, M.E. (1994). *Images of mind.* New York: Scientific American Library.

Ray, B.A. & Sidman, M. (1967). Reinforcement schedules and stimulus control. In W.N. Schoenfeld (Ed.), *The theory of reinforcement schedules.* (pp. 187–214). New York: Appleton-Century-Crofts.

Rensink, R.A., O'Regan, K., & Clark, J.J. (1997). To see or not to see: The need for attention to perceive changes in scenes. *Psychological Science, 8*(5), 368–373.

Sandford, J.A., & Turner, A. (1999). Captain's Log Personal Trainer [Computer software]. Richmond, VA: BrainTrain.

Schadler, M. (1973). Development of relational learning: Effects of instruction and delay of transfer. *Journal of Experimental Child Psychology, 16,* 459–471.

Scott, M.S., & House, B.J. (1978). Repetition of cues in children's oddity learning and transfer. *Journal of Experimental Child Psychology, 25,* 58–70.

Serna, R., & Carlin, M.T. ( in press). Guiding visual attention in individuals with mental retardation. To appear in *International Review of Research in Mental Retardation.* New York: Academic Press.

Shapley, R. (1990). Visual sensitivity and parallel retinocortical channels. *Annual Review of Psychology, 41,* 635–658.

Simons, D.J., & Levin, D.T. (1998). Failure to detect changes to people during a real-world interaction. *Psychonomic Bulletin & Review, 5*(4), 644–649.

Skinner, B.F. (1935). The generic nature of the concepts of stimulus and response. *Journal of General Psychology, 12,* 40–65.

Slamecka, N.J., & Graf, P. (1978). The generation effect: Delineation of a phenomenon. *Journal of Experimental Psychology: Human Learning and Memory, 4,* 592–604.

Soraci, S.A., Alpher, V.S., Deckner, C.W., & Blanton, R.L. (1983). Oddity performance and the perception of relational information. *Psychologia, 26,* 175–184.

Soraci, S.A., Barlean, J.L., Haenlein, M., & Baumeister, A.A. (1986). Lower sensitivity to alterations of auditory relational information in mentally retarded than in nonretarded adults. *Physiological Psychology, 14,* 146–149.

Soraci, S.A., Baumeister, A.A., & Carlin, M.T. (1993). Stimulus organization and stimulus detection: Intelligence-related differences. In D.K. Detterman (Ed.), *Current topics in human intelligence: Vol. 3. Individual differences and cognition.* (pp. 283–306). Norwood, NJ: Ablex.

Soraci, S.A., & Carlin, M.T. (1992). Stimulus organization and relational learning. In N.W. Bray (Ed.), *International review of research in mental retardation:* Vol. 18. (pp. 29–53). New York: Academic Press.

Soraci, S.A., Carlin, M.T., & Chechile, R.A. (1998). The detection of interstimulus relations: A locus of intelligence-related differences. In J. Kingma, & W. Tomic (Eds.), *Advances in cognition and education: Contextual issues in research on intelligence.* Greenwich, CT: Ablex.

Soraci, S.A., Carlin, M.T., Chechile, R.A., Franks, J.J., Wills, T.W., Watanabe, T. (1999). Encoding variability and cueing in generative processing. *Journal of Memory and Language, 41,* 541–559.

Soraci, S.A., Carlin, M.T., Deckner, C.W., & Baumeister, A.A. (1990). Detection of stimulus organization: Evidence of intelligence-related differences. *Intelligence, 14,* 435–447.

Soraci, S.A., Carlin, M.T., Sharp, D.L., Franks, J.J., Vye, N., & Bransford, J.D. (1993). It's what's up front that counts: Some thoughts on the information given. *Experimental Analysis of Behavior Bulletin, 10*(2), 26–30.

Soraci, S.A., Carlin, M.T., & Wiltse, L.J. (1998). Facilitating visual attention. In S.A. Soraci, & W.J. McIlvane (Eds.), *Perspectives on fundamental processes in intellectual functioning: Vol. 1. A survey of research approaches.* (pp. 45–65). Greenwich, CT: Ablex.

Soraci, S.A., Deckner, C.W., Baumeister, A.A., Bryant, J.T., Mackay, H.A., Stoddard, L.T., & McIlvane, W.J. (1991). Generalized oddity performance in preschool children: A bimodal training procedure. *Journal of Experimental Child Psychology, 51,* 280–295.

Soraci, S.A., Deckner, C.W., Haenlein, M., Baumeister, A.A., Murata-Soraci, K., & Blanton, R.L. (1987). Oddity performance in preschool children at risk for mental retardation: Transfer and maintenance. *Research in Developmental Disabilities, 8,* 137–151.

Soraci, S.A., Franks, J.J., Bransford, J.D., Chechile, R.A., Belli, R.F., Carr, M., & Carlin, M.T. (1994). Incongruous item generation effects: A multiple-cue perspective. *Journal of Experimental Psychology: Learning, Memory, and Cognition, 20,* 1–12.

Theeuwes, J. (1992). Perceptual selectivity for color and form. *Perception & Psychophysics, 51,* 599–606.

Toates, F. (1998). The interaction of cognitive and stimulus-response processes in the control of behaviour. *Neuroscience and Biobehavioral Reviews, 22,* 59–83.

Treisman, A.M. (1988). Features and objects: The fourteenth Bartlett memorial lecture. *The Quarterly Journal of Experimental Psychology, 40A,* 201–237.

Treisman, A.M., & Gelade, G. (1980). A feature-integration theory of attention. *Cognitive Psychology, 12,* 97–136.

Treisman, A.M., & Gormican, S. (1988). Feature analysis in early vision: Evidence from search asymmetries. *Psychological Review, 95,* 15–48.

Turnure, J.E. (1985). Communication and cues in the functional cognition of the mentally retarded. In N.R. Ellis & N.W. Bray (Eds.), *International review of research in mental retardation:* Vol. 13. (pp. 43–77). New York: Academic Press.

Usher, M., Cohen, J.D., Servan-Schreiber, D., Rajkowski, J., & Aston-Jones, G. (1999). The role of locus coeruleus in the regulation of cognitive performance. *Science, 283,* 549–554.

Warren, A.M., Soraci, S.A., Chechile, R.A., & Holcomb, P.J. (1999, November). *"Aha" sentences and ERPs.* Poster presented at the 40th Annual Meeting of the Psychonomic Society, Los Angeles, CA.

Wills, T.W., Soraci, S.A., Chechile, R.A., & Taylor, H.A. (2000). "Aha" effects in the generation of pictures. *Memory and Cognition, 28,* 939–948.

Wolfe, J.M., Cave, K.R., & Franzel, S.L. (1989). Guided search: An alternative to the feature integration model for visual search. *Journal of Experimental Psychology: Human Perception and Performance, 15,* 419–433.

Yantis, S. (1993). Stimulus-driven attentional capture and attentional control settings. *Journal of Experimental Psychology: Human Perception and Performance, 19*(3), 676–681.

Yantis, S., & Egeth, H.E. (1999). On the distinction between visual salience and stimulus-driven attentional capture. *Journal of Experimental Psychology: Human Perception and Performance, 25*(3), 661–676.

Chapter 4

# Stimulus Overselectivity and Observing Behavior in Individuals with Mental Retardation

*William V. Dube, Kristin M. Lombard, Kevin M. Farren, Debra S. Flusser, Lyn M. Balsamo, Thomas R. Fowler, and Gerson Y. Tomanari*

Stimulus overselectivity (or restricted stimulus control) is a widely acknowledged problem in the education of individuals with developmental disabilities such as mental retardation and autism. Overselectivity refers to stimulus control that is atypically limited with respect to range, breadth, or number of stimuli or stimulus features (Lovaas, Koegel, & Schreibman, 1979; Schreibman, 1997). Overselectivity may not always be obvious. For example, a student may learn to identify his printed name, John, on the basis of the initial letter only. Proficiency tests might include both expressive and receptive components: For the former, naming printed words aloud; for the latter, matching-to-sample tests in which the student selects the printed name from among several alternatives (the comparisons) conditionally upon hearing the spoken name (the sample). If the set of printed-name stimuli for these tests included only one name with the initial letter "J" (e.g., as might result if the set were composed of the names of the children in the class), then stimulus control by the "J" would be sufficient for high accuracy scores, and the overselectivity problem would go undetected. Overselectivity would become evident only if John were presented in a context that also included other names such as Jack and Jill (Kledaras, Dube, Flusser, McIlvane, & Potter, 1999).

Stimulus overselectivity was first described as such in the literature of operant clinical psychology (Lovaas, Schreibman, Koegel, & Rehm, 1971; reviewed in Lovaas et al., 1979). The evaluation methods generally involved initial discrimination training with multi-element stimuli followed by testing with the individual elements to determine how many of those elements had gained control of behavior. For example, if initial training established stimulus *ABC* as positive and *XYZ* as negative, tests would present the various combinations of individual elements, *A* versus *Y*, *B* versus *X*, and so forth. When given such tests, individuals with developmental disabilities may respond appropriately to some elements but not to others, and to fewer elements than nondisabled individuals. Although overselectivity is often associated with autism, it is also a common feature of the learning difficulties found in individuals with mental retardation (Allen & Fuqua, 1985; Dube & McIlvane, 1997; Huguenin, 1997; Litrownik, McInnis, Wetzel-Pritchard, & Filipelli, 1978; Meisel, 1981; Schneider & Salzberg, 1982; Stromer, McIlvane, Dube, & Mackay, 1993; Whiteley, Zaparniuk, & Asmundson, 1987; Wilhelm & Lovaas, 1976).

Although overselectivity has been demonstrated within the auditory modality (e.g., Schreibman, Kohlenberg, & Britten, 1986), the primary focus has been on visual stimuli. Overselectivity has been documented with several types of multi-element stimuli, including those with elements of the same stimulus dimension (e.g., arrays of discrete forms; Koegel & Wilhelm, 1973) and multidimensional elements (differences in color, form, etc.; Kovattana & Kraemer, 1974). The finding has been replicated with controls for discrimination of all elements presented individually (Dube, Kledaras, Iennaco, Stoddard, & McIlvane, 1990) and discrimination of individual stimuli when presented in multiple-stimulus arrays (Stromer et al., 1993).

The degree of overselectivity is often correlated with mental age-equivalent scores on standardized tests in both typically developing children (e.g., Eimas, 1969; Schover & Newsom, 1976) and individuals with developmental disabilities (e.g., Schover & Newsom, 1976; Wilhelm & Lovaas, 1976). For example, with three-element training and testing like that in the *ABC/XYZ* example above, Wilhelm and Lovaas (1976) found stimulus control by all three elements in typically developing eleven-year-old children, two of three elements in children with moderate mental retardation, and fewer than two elements (mean = 1.6) in children with severe mental retardation.

Overselectivity may be related to a wide range of deleterious effects on behavioral functioning. It can have a negative impact on observational learning, generalization, and transfer from prompts to teaching materials (reviewed in Lovaas et al., 1979). A potentially far-reaching effect may result if overselectivity with auditory (e.g., vocal speech) or visual cues (e.g., printed words) leads to deficits in language and communication skills (Koegel, Schreibman, Britten, & Laitinen, 1979; Schreibman, 1997). In addition to the obvious educational value of functional language and communication, such skills may also be essential for building social relationships, attaining optimal school placement, making choices, and achieving maximal independence.

## A LABORATORY MODEL OF STIMULUS OVERSELECTIVITY

Historically, the majority of studies that have examined stimulus overselectivity in individuals with moderate to severe autism or mental retardation used simple-discrimination procedures similar to those developed by Lovaas and colleagues and described previously (reviewed in Lovaas et al., 1979). More recently, our laboratory has adopted conditional-discrimination methodology—specifically, delayed matching to sample (DMTS) with multiple sample stimuli (Dube & McIlvane, 1997, 1999; Stromer et al., 1993). DMTS has been used previously to study divided attention in the animal cognition literature (e.g., Maki & Leuin, 1972), as well as breadth of attention across multiple stimulus dimensions such as form and color in individuals with developmental disabilities (Litrownik et al., 1978; Mackie & Mackay, 1982; Schneider & Salzberg, 1982; Whiteley et al., 1987). The DMTS procedure provides great flexibility for research in overselectivity. For example, it allows evaluation and training with both small stimulus sets (as in simple-discrimination procedures) or large stimulus sets. Many of our studies use a version of the DMTS task in which different stimuli are presented on every trial. This trial-unique version of the task provides the means to assess overselectivity in a context with a continuing requirement for careful observation of all stimuli. The trial-unique task also allows the effects of procedural modifications to be assessed independently of their effect on specific stimuli.

The evaluation procedure will be described briefly here; procedural details can be found in Dube and McIlvane (1997). Test sessions are presented by a computer with a touchscreen-equipped monitor. Preliminary tests familiarize participants with the matching-to-sample procedure and verify several prerequisite skills such as accurate performance when stimulus control by a single stimulus is sufficient. For individuals with mental retardation, most of the overselectivity evaluations are accomplished with a two-sample DMTS task, illustrated in Figure 4.1. The task consists of a series of matching-to-sample trials in which two sample stimuli are displayed on each trial. The samples remain available for observation until the participant touches them; the duration of the observing period is therefore determined by the participant. After a touch to the sample-stimulus display area, the samples disappear and an array of single-comparison stimuli is presented immediately (i.e., 0-s DMTS). On each trial, the correct comparison stimulus is identical to one of the sample stimuli. For example, given samples *A* and *B*, the comparisons may be *A*, *C*, and *D* (for experimental control, the actual stimuli are most often arbitrary visual forms like those shown in Figure 4.1). On each trial, only one of the two samples appears in the comparison array and, importantly, the participant cannot predict which one it will be because the correct comparison is identical to the left or right sample equally often but in an unpredictable order.

Accurate performance on two-sample DMTS (>90% correct) indicates no overselectivity with two sample stimuli; regardless of which sample appears as the correct comparison, the participant is able to match it. Overselectivity is operationally defined by intermediate accuracy scores (e.g., approximately 67% for the three-choice task): The participant is able to match only one of the two sample stimuli. On those trials where that stimulus appears as a comparison

**Figure 4.1**
Two-sample Delayed Matching-to-Sample. Trials begin with the presentation of two sample stimuli (*A*) on a touchscreen-equipped computer monitor. The samples remain available for observation until the participant touches the screen (*B*), whereupon the samples disappear and three comparison stimuli appear (*C*). One of the comparisons is identical to one of the samples, and the correct response is to touch it (*D*).

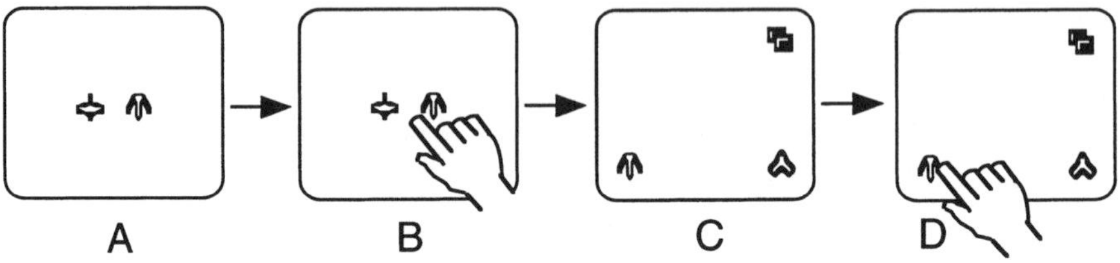


(half of the trials in the session), the participant is always correct. On the remaining trials, performance is at chance levels. The intermediate accuracy score for the entire session results from averaging scores from both types of trials (for a detailed analysis of stimulus control in two-sample DMTS, see Dube & McIlvane, 1997).

## OBSERVING BEHAVIOR AND OVERSELECTIVITY

Overselectivity is exhibited both in the classroom and in the laboratory as discriminative behavior. Examples include the vocal verbal responses that constitute reading aloud, and the manual selection responses in tasks such as matching-to-sample. To return to the example of the printed name John, John's teacher may conclude that his behavior is overselective after an error analysis of matching-to-sample selection responses on trials where printed-word discriminative stimuli do or do not share the common feature of the initial letter "J." The selection response, however, is actually the last link in a chain of behavioral events. Earlier links in that chain include observing behavior. Depending on the situation, observing behavior with visual stimuli may include a number of preliminary responses such as pressing a light switch or turning the page of a book. Ordinarily, observing behavior includes orienting the head and moving the eyes so that light reflected or emitted by the stimuli falls on the fovea. Effective observing behavior is a prerequisite for accurate visual discrimination (Dinsmoor, 1985; Schroeder, 1997).

Basic research on observing behavior indicates that it is maintained when its immediate consequence is the production of discriminative stimuli (Dinsmoor, 1983; Wyckoff, 1952, 1969). The results of research comparing observing behavior in two or more conditions with different reinforcement contingencies show that observing responses occur more frequently, have longer durations, and/or have shorter latencies when they are followed by the production of stimuli signaling a relatively higher magnitude or frequency of reinforcement. When observing responses are followed by the production of stimuli signaling a relatively lower magnitude or frequency of reinforcement, they occur less frequently, have shorter durations, and/or have longer latencies (Allen & Lattal, 1989; Case & Fantino, 1981, 1989; Case, Fantino, & Wixted, 1985; Case, Ploog, & Fantino, 1990; Dinsmoor, Mulvaney, & Jwaideh, 1981; Fantino & Case, 1983; Fantino, Case, & Altus, 1983; Mueller & Dinsmoor, 1984; Mulvaney, Hughes, Jwaideh, & Dinsmoor, 1981). For a recent summary of research on observing-behavior, see Fantino (1998; cf. Lieberman, Cathro, Nichol, & Watson, 1997).

Two recent research findings suggested that deficient observing behavior may play a role in overselectivity. The first finding was that the specific stimuli discriminated in overselectivity can be predicted and controlled by manipulating reinforcement probabilities (Dube & McIlvane, 1997). Participants were teenagers with moderate to severe mental retardation who, on pretests with the two-sample DMTS task described previously, had intermediate accuracy scores indicative of overselectivity. The experimental stimuli were four arbitrary forms, here designated *A, B, X,* and *Y.* Initial training consisted of a standard DMTS task with only one sample per trial, a task on which all participants had high accuracy. During initial training, the reinforcement schedules were such that correct matching with two of the stimuli (e.g., *A* and *B*), produced a high rate of reinforcement, and matching the other two (*X* and *Y*) produced a substantially lower rate of reinforcement. The stimuli were then presented in the two-sample DMTS task. Data analyses from trials where the samples consisted of one high-rate stimulus and one low-rate stimulus (e.g., *A* and *X*) showed that the overselective stimulus control was restricted to the high-rate stimuli (*A* and *B*). Next, participants returned to the standard DMTS task, but now the reinforcement schedules were reversed (e.g., low rates for *A* and *B,* and high rates for *X* and *Y*). After this training, the two-sample tests were repeated. The results showed that stimulus control had shifted so that the new high-rate stimuli (e.g., *X* and *Y*) were now the ones that were discriminated. The participants remained overselective throughout the experiment in that they could match only one of the two sample stimuli; the specific stimuli that were discriminated, however, were predicted and controlled by manipulating the reinforcement schedules. This sensitivity of overselectivity to relative rates of reinforcement parallels the research findings for observing behavior.

The second finding suggesting deficient observing behavior was that overselectivity was substantially reduced by differential observing responses (DORs). DORs control observing behavior and verify discrimination. For example, given the printed-word stimulus John, a DOR procedure might require the student to spell the name aloud. To produce the responses, "J, O, H, N," the student must observe each letter, and the forms of the responses verify that each letter was appropriately discriminated. With matching-to-sample procedures, the DOR may be contrasted with the more common nondifferential observing response of simply touching the sample stimulus. Gutowski, Geren, Stromer, and

**Figure 4.2**
Individual Session Accuracy Scores for the Standard Two-Sample DMTS Task (No DOR) Shown in Figure 4.1, and a Variation Incorporating a Differential Observing-Response Procedure (DOR) Into the Sample Observation Period. Gaps in the data for Cathy indicate sessions where the DOR matching procedure was reviewed in isolation.

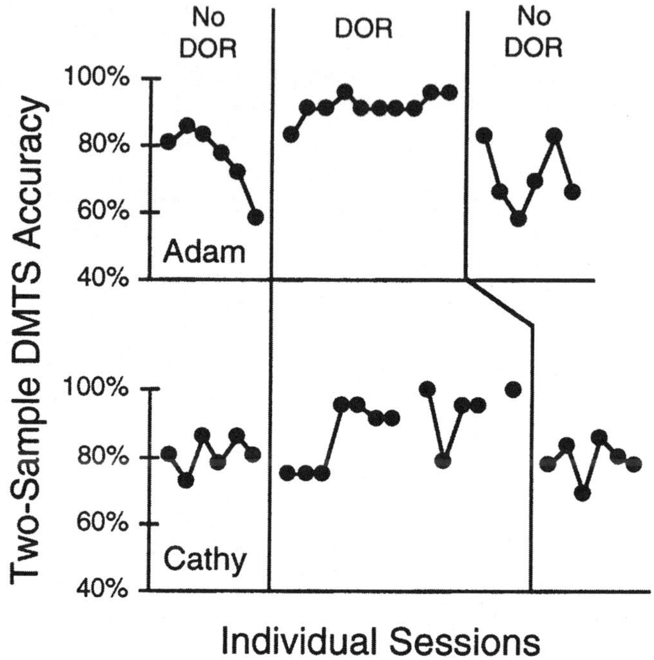


Mackay (1995) showed that stimulus overselectivity was reduced in two individuals with moderate mental retardation when they were required to name the stimuli (pictures of common objects) in a two-sample DMTS task.

Because naming is not always available as a differential observing response (especially in special-education situations), we have examined generalized, nonverbal differential observing-response procedures. Participants in our studies were teenagers with moderate to severe mental retardation who had intermediate accuracy scores on the trial-unique version of the two-sample DMTS task described before (i.e., different stimuli appeared on every trial). In the standard procedure (Figure 4.1), the sample stimuli remained displayed until the participant touched the sample-display area (a nondifferential observing response), and then the samples disappeared as the comparison stimuli were presented. The experimental manipulation was to insert one or more simultaneous matching tasks into the sample-observation period. In one type of DOR procedure, sets of individual comparison stimuli appeared on the computer screen, and each set provided the opportunity to match one of the sample stimuli. These DORs were never followed by differential consequences; after the participant completed them, the DMTS trial simply continued (the sample disappeared and a comparison display was presented). Figure 4.2 shows the results with two individuals. In both cases, DMTS accuracy improved with the DOR procedure and returned to intermediate levels when DORs were no longer prompted.

In another variation of the procedure, the stimuli for the DOR components were arranged so that discrimination of both of the sample stimuli could be verified with a single response (Dube & McIlvane, 1999). For example, given sample *AB*, the DOR comparison stimuli might be *AB* (correct), *AC*, and *DB* (incorrect). Note that both of the incorrect alternatives had one stimulus in common with the sample, and thus consistently correct DORs could not be based on one sample only (Allen & Fuqua, 1985; Burke & Cerniglia, 1990; Koegel & Schreibman, 1977; Schreibman, Charlop, & Koegel, 1982). The imposition of this DOR procedure also produced increases in DMTS matching accuracy comparable to those shown in Figure 4.2 (see Dube & McIlvane, 1999).

In summary, this research indicated that (a) overselectivity and observing behavior were similarly sensitive to reinforcement contingencies, and (b) experimental manipulations that controlled observing behavior also reduced overselectivity. Findings such as these suggested the value of a more direct approach for studying observing behavior in individuals with mental retardation. In the remainder of this chapter, we will describe research methods that use eye tracking to examine the relation between overselectivity and the topography of observing behavior; we will illustrate this approach with some of our initial results.

## DIRECT ASSESSMENT OF OBSERVING BEHAVIOR VIA EYE TRACKING

Recent advances in eye-tracking technology allow direct assessment of observing behavior in at least some cases of overselectivity in individuals with mental retardation. Our laboratory is equipped with an ISCAN head-mounted eye-tracking system (ISCAN Corp., Burlington, MA). The imaging components of the corneal reflection system include miniature video cameras attached to a headband by a supporting arm, as shown in Figure 4.3. Because the cameras are head-mounted, it is not necessary to immobilize the participant's head during recording—whenever the head moves, the cameras move with it. This feature of the apparatus makes it possible to accomplish eye tracking with individuals who would be unable or unwilling to participate in sessions involving mechanical head restraints such as a chin rest or a bite bar. Further, elimination of the requirement for head restriction also eliminates concerns that partici-

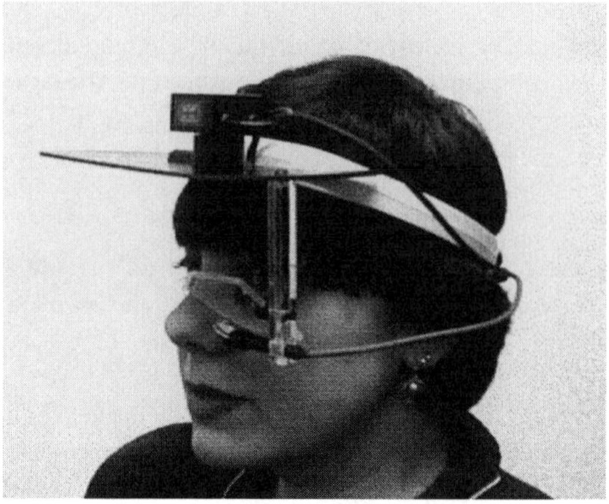

**Figure 4.3**
Headgear for the ISCAN Head-Mounted Eye-Tracking System.

pants may not perform tasks in the same ways that they would when given free head movement; for example, Creedon (1999) recently reported that children with autism tended to move the head and eyes together when target locations changed.

A plastic replica of the headgear is used to train participants with mental retardation. During training sessions, participants learn to (a) wear the headgear and request its removal appropriately, (b) perform a brief calibration routine, and (c) tolerate wearing the headgear for at least ten minutes while performing the calibration routine and the two-sample DMTS task. Training has required as little as one session for some individuals with mild mental retardation and good receptive language skills, and more than a dozen sessions for most individuals with more severe disabilities (for details, see Dube, Lombard, Farren, Flusser, Balsamo, & Fowler, 1999).

The relation between observing behavior and overselectivity will be illustrated with eye-tracking data for Participant DTM, a twelve-year-old young man with moderate mental retardation (Peabody Picture Vocabulary Test [PPVT] mental age equivalent score, 4.2 yr.). DTM's preliminary assessment was consistent with overselectivity: Accuracy scores on tests that required stimulus control by individual stimuli were always above 90 percent, and the mean accuracy score for twelve sessions of the two-sample DMTS task was 62 percent (range 47%–72% with no trend). Training to use the eye-tracking equipment required sixteen sessions (described in Dube et al., 1999).

For contrast, we will also present eye-tracking data for a nonclinical adult, LCN, whose accuracy scores were above 95 percent for all tasks, including two-sample DMTS. For both participants, the data are from their first eye-tracking session with the two-sample DMTS task; in these sessions, overall accuracy scores for LCN and DTM were 96 percent and 61 percent, respectively.

Figures 4.4 and 4.5 show trial-by-trial records of observing behavior for LCN and DTM, respectively. Trial numbers appear to the left of each plot. To allow some time for performance to stabilize, the analyses begin with DMTS trial number 12. In each plot, the upper and lower horizontal bars show observing durations for the left and right sample stimuli, respectively. Each plot begins with the first observation of a sample stimulus and continues to the end of the last observation. Gaps between bars indicate times during which neither of the sample stimuli were observed. Bars for the sample stimulus that subsequently appeared as the correct comparison are shown in black, and bars for the sample that did not appear in the comparison array are striped. During the sample observation period, of course, the participant could not predict which sample would appear as the correct comparison; the distinction is made in this presentation to assist in the interpretation of results. Letters to the right of each plot indicate whether the participant's response to the comparison array was correct (C) or an error (X). For example, in Figure 4.4, Trial 12 began with an observation of the left sample stimulus for 0.50 seconds (s), then the right sample for 0.27 s, and then back to the left sample for 0.47 s. Total observing duration was 1.24 s (the reader's estimation of approximate durations will be sufficient for the present discussion). When the comparisons were displayed, the stimulus that had been the left sample was the correct comparison (black bars), and the participant selected it (C).

Figure 4.4 shows the data for LCN; her observing behavior is typical of what we have found in eye-tracking sessions with eight nonclinical adults performing the two-sample DMTS task. Observing patterns were highly regular from trial to trial; every trial except Trial 24 began with a left-right-left sequence. Every sample stimulus was observed at least one time on every trial, and there were only two trials (16 and 23) with total observing durations of less than 1 s. LCN's only error occurred on Trial 18, but the data in Figure 4.4 do not suggest any relation between her observing behavior and this error.

Figure 4.5 shows DTM's data. Observing patterns were more irregular than LCN's, with substantial variation from trial to trial. On twelve trials, he observed only one of the sample stimuli. On four of these trials (17, 19, 30, 31), the observed stimulus was the correct comparison, and DTM was correct on all of them. On the remaining 8 trials (12, 13, 14, 16, 22, 24, 33, and 36), the observed stimulus did not appear in the comparison array, and he made errors on six of these eight trials. Observing durations were generally shorter than LCN's. There were fifteen trials on which total observing duration was 1 s or less. There were four trials on which DTM observed both sample stimuli and still made errors, and observing durations for the sample that was the correct comparison on three of these trials (black bars in Trials 15, 20, and 32) were relatively brief, 0.19 to 0.26 s. Taken together, observing failures and brief observing durations occurred on nine of the ten trials with errors.

**Figure 4.4**

Trial-by-Trial Record of Observing-Behavior for Participant LCN, a Nonclinical Adult. Trial numbers appear to the left of each plot. Upper (L) and lower (R) horizontal bars show durations of observing the left and right sample stimuli, respectively, in seconds according to the scale at the bottom of the figure. Black bars indicate the sample stimulus that was the correct comparison, and striped bars indicate the sample that did not appear as a comparison. C and X to the right of each plot indicate correct and incorrect responses to the comparison array, respectively. (From Dube et al., 1999.)

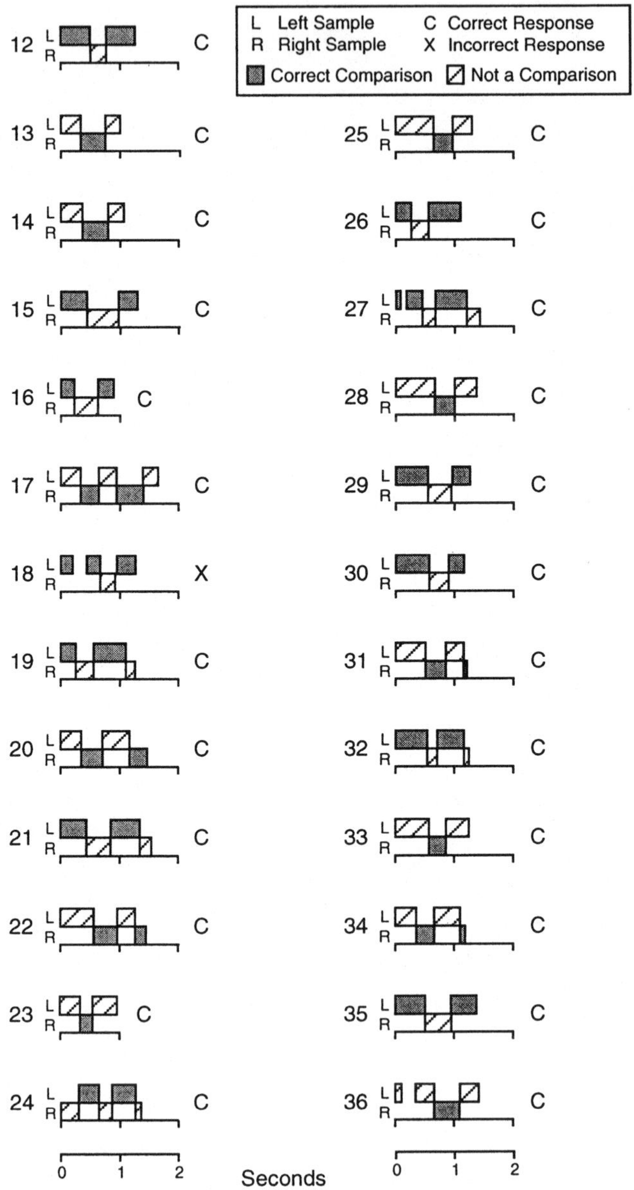

**Figure 4.5**
Trial-by-Trial Record of Observing-Behavior for Participant DTM, a Young Man with Moderate Mental Retardation. See the caption for Figure 4.4. (From Dube et al., 1999.)

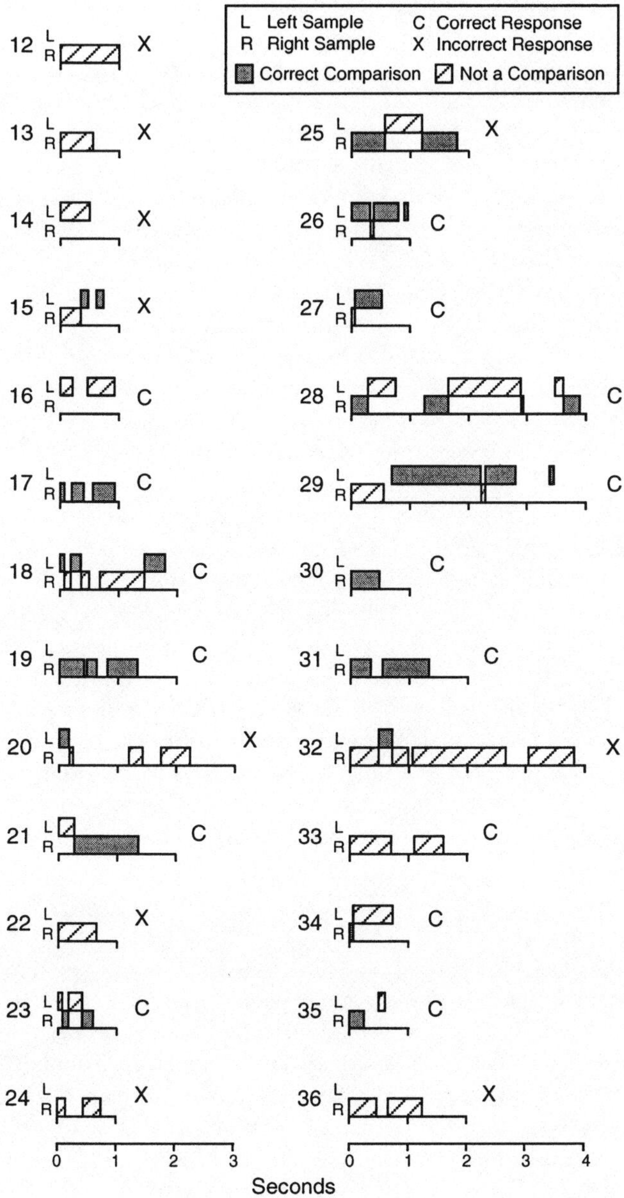

## IMPLICATIONS FOR FURTHER RESEARCH

The eye-tracking data show that overselectivity may be accompanied by failures to observe all of the relevant stimuli. On trials in which DTM observed only one sample stimulus, he was always correct if that stimulus was the correct comparison, but he was correct only at chance levels if it was not. As of this writing, we have obtained similar data from four individuals with mental retardation who responded overselectively on the two-sample DMTS task. These results suggest that stimulus overselectivity may be related to deficient observing behavior.

Evidence linking overselective performances to specific irregularities in observing behavior raises the question of whether overselectivity could be remediated by modifying observing. Previous research indicates that observing behavior is open to modification by manipulating its consequences (e.g., Rosenberger, 1973; Schroeder & Holland, 1968a,b; 1969). One form of the question is: If behavior-modification techniques were used to change the observing patterns in individuals such as DTM, so that each stimulus were observed on every trial, would two-sample DMTS accuracy increase to a high level? That is, if observing were normalized, would performance also be normalized? On the one hand, if the underlying problem were due primarily to a deficit in cognitive attentional resources, then arranging for visual fixations on stimuli would be insufficient. On the other hand, if the problem were not a lack of capacity for discrimination of multiple stimuli, but rather an impoverished observing repertoire for complex stimulus displays, then arranging for adequate scanning patterns could produce improved discrimination accuracy.

Because the eye-tracking apparatus produces a real-time video image, it provides the information necessary for the experimenter to arrange contingencies for observing responses. For example, the stimulus–observation period for the two-sample DMTS task may be extended until both sample stimuli have been observed. An important goal for continued research is to examine the effects of such interventions.

## NOTE

Data collection and manuscript preparation were supported by NICHD grants P01 HD 25995 and R01 HD37055. Lyn Balsamo is currently at the American University, Washington, DC, and Tom Fowler is at Hartford University, Hartford, CT. We thank Aranya Albert, Jason Krienke, and Jason Langlois for their help with data analysis; Steve Oross and Bill McIlvane for advice on the project; Bill McIlvane and Richard Serna for comments on the manuscript; Lynn Doucette and Joanne Kledaras for photography; and the staff and participating students of The Protestant Guild Learning Center for their cooperation. Special thanks to Rikki Razdan and Alan Kielar of the ISCAN Corp. for technical support and assistance. Address correspondence to William V. Dube, Psychological Sciences Division, E.K. Shriver Center, 200 Trapelo Road, Waltham, MA 02452, or william.dube@umassmed.edu.

## REFERENCES

Allen, K.D., & Fuqua, R.W. (1985). Eliminating selective stimulus control: A comparison of two procedures for teaching mentally retarded children to respond to compound stimuli. *Journal of Experimental Child Psychology, 39,* 55–71.

Allen, K.D. & Lattal, K.A. (1989). On conditional reinforcing effects of negative discriminative stimuli. *Journal of the Experimental Analysis of Behavior, 52,* 335–339.

Burke, J.C., & Cerniglia, L. (1990). Stimulus complexity and autistic children's responsivity: Assessing and training a pivotal behavior. *Journal of Autism and Developmental Disorders, 20,* 233–253.

Case, D.A. & Fantino, E. (1981). The delay-reduction hypothesis of conditioned reinforcement and punishment: Observing behavior. *Journal of the Experimental Analysis of Behavior, 35,* 93–108.

Case, D.A. & Fantino, E. (1989). Instructions and reinforcement in the observing behavior of adults and children. *Learning and Motivation, 20,* 373–412.

Case, D.A., Fantino, E., & Wixted, J. (1985). Human observing maintained by negative informative stimuli only if correlated with improvement in response efficiency. *Journal of the Experimental Analysis of Behavior, 43,* 289–300.

Case, D.A., Ploog, B.O., & Fantino, E. (1990). Observing behavior in a computer game. *Journal of the Experimental Analysis of Behavior, 54,* 185–199.

Creedon, M.P. (1999). AcuVision 1000™ practice and functional visual skills performance by students with autism [Abstract]. *Proceedings of the 32nd Annual Gatlinburg Conference on Research and Theory in Mental Retardation and Developmental Disabilities,* 150.

Dinsmoor, J.A. (1983). Observing and conditioned reinforcement. *Behavioral and Brain Sciences, 6,* 693–728, including commentaries.

Dinsmoor, J.A. (1985). The role of observing and attention in establishing stimulus control. *Journal of the Experimental Analysis of Behavior, 43,* 365–382.

Dinsmoor, J.A., Mulvaney, D.E., & Jwaideh, A.R. (1981). Conditioned reinforcement as a function of duration of stimulus. *Journal of the Experimental Analysis of Behavior, 36,* 41–49.

Dube, W.V., Kledaras, J.B., Iennaco, F.M., Stoddard, L.T., & McIlvane, W.J. (1990). Observing complex visual stimuli: Effects of component pretraining. *Experimental Analysis of Human Behavior Bulletin, 8,* 7–11.

Dube, W.V., Lombard, K.M., Farren, K.M., Flusser, D.S., Balsamo, L.M., & Fowler, T.R. (1999). Eye tracking assessment of stimulus overselectivity in individuals with mental retardation. *Experimental Analysis of Human Behavior Bulletin, 17,* 8–14.

Dube, W.V., & McIlvane, W.J. (1997). Reinforcer frequency and restricted stimulus control. *Journal of the Experimental Analysis of Behavior, 68,* 303–316.

Dube, W.V., & McIlvane, W.J. (1999). Reduction of stimulus overselectivity with nonverbal differential observing responses. *Journal of Applied Behavior Analysis, 32,* 25–33.

Eimas, P.D. (1969). Multiple-cue discrimination learning in children. *The Psychological Record, 19,* 417–424.

Fantino, E. (1998). Judgment and decision making: Behavioral approaches. *The Behavior Analyst, 21,* 203–218.

Fantino, E., & Case, D.A. (1983). Human observing: Maintained by stimuli correlated with reinforcement but not extinction. *Journal of the Experimental Analysis of Behavior, 40,* 193–210.

Fantino, E., Case, D.A., & Altus, D. (1983). Observing reward-informative and uninformative stimuli by normal children of different ages. *Journal of the Experimental Analysis of Behavior, 36,* 437–452.

Gutowski, S.J., Geren, M., Stromer, R., & Mackay, H.A. (1995). Restricted stimulus control in delayed matching to complex samples: A preliminary analysis of the role of naming. *Experimental Analysis of Human Behavior Bulletin, 13,* 18–24.

Huguenin, N.H. (1997). Employing computer technology to assess visual attention in young children and adolescents with severe mental retardation. *Journal of Experimental Child Psychology, 65,* 141–170.

Kledaras, J.B., Dube, W.V., Flusser, D.S., McIlvane, W.J., & Potter, M.J. (1999, May). *From the research setting to the classroom: Applying a behavioral technology for teaching sight words in a school setting.* Presented at the Annual Conference of the Association for Behavior Analysis, Chicago, IL.

Koegel, R.L., & Schreibman, L. (1977). Teaching autistic children to respond to simultaneous multiple cues. *Journal of Experimental Child Psychology, 24,* 299–311.

Koegel, R.L., Schreibman, L., Britten, K., & Laitinen, R. (1979). The effects of schedule of reinforcement on stimulus overselectivity in autistic children. *Journal of Autism and Developmental Disabilities, 9,* 383–396.

Koegel, R.L., & Wilhelm, H. (1973). Selective responding to the components of multiple visual cues by autistic children. *Journal of Experimental Child Psychology, 15,* 442–453.

Kovattana, P.M., & Kraemer, H.C. (1974). Response to multiple visual cues of color, size, and form by autistic children. *Journal of Autism and Childhood Schizophrenia, 4,* 251–261.

Lieberman, D.A., Cathro, J.S., Nichol, K., & Watson, E. (1997). The role of S– in human observing behavior: Bad news is sometimes better than no news. *Learning and Motivation, 28,* 20–42.

Litrownik, A.J., McInnis, E.T., Wetzel-Pritchard, A.M., & Filipelli, D.L. (1978). Restricted stimulus control and inferred attentional deficits in autistic and retarded children. *Journal of Abnormal Psychology, 87,* 554–562.

Lovaas, O.I., Koegel, R.L., & Schreibman, L. (1979). Stimulus overselectivity in autism: A review of research. *Psychological Bulletin, 86,* 1236–1254.

Lovaas, O.I., Schreibman, L., Koegel, R.L., & Rehm, R. (1971). Selective responding by autistic children to multiple sensory input. *Journal of Abnormal Psychology, 77,* 211–222.

Mackie, R., & Mackay, C.K. (1982). Attention vs. retention in discrimination learning of low-MA retarded adults and MA-matched retarded children. *American Journal of Mental Deficiency, 86,* 543–547.

Maki, W.S., & Leuin, T.C. (1972). Information processing by pigeons. *Science, 176,* 535–536.

Meisel, C.J. (1981). Stimulus overselectivity in mentally retarded adolescents: Effects of pretraining on cue identification. *American Journal of Mental Deficiency, 86,* 317–322.

Mueller, K.L., & Dinsmoor, J.A. (1984). Testing the reinforcement properties of S-: A replication of Lieberman's procedure. *Journal of the Experimental Analysis of Behavior, 41,* 17–25.

Mulvaney, D.E., Hughes, L.H., Jwaideh, A.R., & Dinsmoor, J.A. (1981). Differential production of positive and negative discriminative stimuli by normal and retarded children. *Journal of Experimental Child Psychology, 32*, 389–400.

Rosenberger, P.B. (1973). Concurrent schedule control of human visual target fixations. *Journal of the Experimental Analysis of Behavior, 20*, 411–416.

Schneider H.C., & Salzberg, C.L. (1982). Stimulus overselectivity in a match-to-sample paradigm by severely retarded youth. *Analysis and Intervention in Developmental Disabilities, 2*, 273–304.

Schover, L.R., & Newsom, C.D. (1976). Overselectivity, developmental level, and overtraining in autistic and normal children. *Journal of Abnormal of Psychology, 4*, 289–298.

Schreibman, L. (1997). The study of stimulus control in autism. In D.M. Baer & E.M. Pinkston (Eds.), *Environment and behavior* (pp. 203–209). Boulder, CO: Westview.

Schreibman, L., Charlop, M.H., & Koegel, R.L. (1982). Teaching autistic children to use extra-stimulus prompts. *Journal of Experimental Child Psychology, 33*, 475–491.

Schreibman, L., Kohlenberg, B.S., & Britten, K.R. (1986). Differential responding to content and intonation components of a complex auditory stimulus by nonverbal and verbal autistic children. *Analysis and Intervention in Developmental Disabilities, 6*, 109–125.

Schroeder, S.R. (1997). Selective eye fixations during transfer of discriminative stimulus control. In D.M. Baer & E.M. Pinkston (Eds.), *Environment and behavior* (pp. 97–110). Boulder, CO: Westview.

Schroeder, S.R., & Holland, J.G. (1968a). Operant control of eye movements. *Journal of Applied Behavior Analysis, 1*, 161–166.

Schroeder, S.R., & Holland, J.G. (1968b). Operant control of eye movements during human vigilance. *Science, 161*, 292–293.

Schroeder, S.R., & Holland, J.G. (1969). Reinforcement of eye movement with concurrent schedules. *Journal of the Experimental Analysis of Behavior, 12*, 897–903.

Stromer, R., McIlvane, W.J., Dube, W.V., & Mackay, H.A. (1993). Assessing control by elements of complex stimuli in delayed matching to sample. *Journal of the Experimental Analysis of Behavior, 59*, 83–102.

Whiteley, J.H., Zaparniuk, J., & Asmundson, G.J.G. (1987). Mentally retarded adolescents' breadth of attention and short-term memory processes during matching to sample discriminations. *American Journal of Mental Deficiency, 92*, 207–212.

Wilhelm H., & Lovaas, O.I. (1976). Stimulus overselectivity: A common feature in autism and mental retardation. *American Journal of Mental Deficiency, 81*, 26–31.

Wyckoff, L.B., Jr. (1952). The role of observing responses in discrimination learning: Part I. *Psychological Review, 59*, 431–442.

Wyckoff, L.B., Jr. (1969). The role of observing responses in discrimination learning. In D.P. Hendry (Ed.), *Conditioned reinforcement* (pp. 237–260). Homewood, IL: Dorsey.

Chapter 5

# Arousal Modulation of Neonatal Visual Attention: Implications for Development

*Judith M. Gardner, Bernard Z. Karmel, and Michael J. Flory*

## INTRODUCTION

Infants cannot gain information, learn about the world, or interact socially unless they first actively attend to relevant features of their environment. Typically, term infants are born with most of their physiological systems coordinated and well-adapted to the immediate environment. This structural and functional organization is evident in the fundamental processes of state modulation and attention to stimulation. For neonates, arousal and attention work interdependently as a single, quantitatively organized, homeostatic system that combines the effects of internal and external factors to specify attention to particular stimuli (Gardner & Karmel, 1984, 1995; Gardner, Karmel, & Magnano, 1992; Gardner, Lewkowicz, Rose, & Karmel, 1986; Gardner & Turkewitz, 1982). Clear distinction between arousal and attention appears to occur only later in development after the newborn period.

As infants mature, they gain more control over their internal and external worlds so that their arousal and attention become more independent from one another. That is, with development, which stimuli are attended to become more dependent on experience and meaning than on any specific set of physical features, and the infant becomes less dependent on variations in arousal whether or not produced by internal and external sources of stimulation (Gardner & Karmel, 1995; Geva, Gardner, & Karmel, 1999). Thus, arousal and attention act dynamically to provide the basic psychological and neurofunctional interface between an organism and its environment throughout development. How the transformation in this transaction comes about is a fundamental question for developmental scientists; it appears to be a necessary condition for development of normal perceptual, cognitive, and autoregulatory organization. In this chapter, we will describe our research findings on which these ideas are based and some new studies that helped us extend and refine our predictions to older ages.

## AROUSAL AND ATTENTION IN CNS-STRESSED NEONATES

Much evidence, including our own work, demonstrates that arousal and attention in early infancy depend upon the integrity and maturation of the central nervous system (CNS). Since early arousal and attention influence emerging perceptual, cognitive, and autoregulatory functioning in later infancy, injury to CNS structure and function should disrupt or alter development through damage to mechanisms controlling arousal and attention. We further hypothesize that arousal and attention will develop differently in different risk populations, depending on the type of CNS structural or neurofunctional perturbance that has occurred. Concurrently, we have chosen to study both normal and aberrant populations in which arousal and attention potentially are affected, such as high-risk infants assigned to the neonatal intensive care unit (NICU) or infants exposed chemically in utero to a neurotoxic substance such as cocaine. Using this strategy, we not only expand the range of potential responses that can be measured, but at the same time try to provide some practical insight into why certain risk populations develop as they do.

It has long been recognized that CNS insult can compromise arousal, attention, and state-modulating capacities in the developing infant (see Brazelton, Nugent, & Lester, 1987; Field, 1981; Korner, 1972; Parmelee, 1975; Prechtl, 1974; Thoman, 1981). In addition, disturbed arousal and attention, especially as precursors to later motivation and temperament characteristics, have been suggested to underlie certain deficits in performance in prenatally cocaine-exposed infants (see Lester et al., 1991; Mayes, Granger, Frank, Schottenfeld, & Bornstein, 1993; Mayes, Grillon, Granger, & Schottenfeld, 1998). We have attempted to determine the specific characteristics of brain function associated with these adverse outcomes through experimental manipulations of arousal (e.g., Gardner & Karmel, 1983, 1995; Gardner et al., 1992; Geva et al., 1999; Karmel, Gardner, & Magnano, 1991). Rather than viewing state as a confound, we find that manipulation of arousal can provide a wealth of information about the infant.

Mechanisms of arousal and attention are likely to be mediated by subcortical systems, especially prior to two or three months of age (see Bronson, 1982; Gardner & Karmel, 1983; Hoffmann, 1978; Johnson, 1990; Karmel et al., 1991; Karmel & Maisel, 1975; Morton & Johnson, 1991; Turkewitz, Gardner, & Lewkowicz, 1984; Turkewitz, Lewkowicz, & Gardner, 1983). Brainstem mechanisms involving early arousal modulation are assumed to underlie later development through the formation of brainstem–striatal–cortical pathways. In very low birthweight or severely asphyxiated neonates, CNS damage tends to occur in subcortical regions because those areas have the highest metabolic rate and blood supply, with the frailest capillary systems. In both preterm and term infants, CNS damage produced by hypoxia (e.g., hemorrhage, infarct, encephalopathy) often occurs to subcortical structures (Fawer, Dubowitz, Levene, & Dubowitz, 1983; Volpe, 2001) that are near areas or pathways classically thought to be involved in arousal and attention, such as the ascending and descending pathways of the reticular formation. Advances in medical care have produced changes in cohorts. Younger and sicker infants are surviving and there are increased numbers of very low birth weight infants in multiple gestations due to in vitro fertilization. Although there is decreased severe intracranial hemorrhage, there is no indication of any decrease in the suspected involvement of brainstem and other subcortical regions (Volpe, 2001). These regions are difficult to visualize structurally by cranial ultrasound (CUS; Levene, Wigglesworth, & Dubowitz, 1983), but easily studied functionally by auditory brainstem evoked responses (ABRs; Volpe, 2001). ABRs evaluate brainstem integrity by assessment of brainstem neuronal transmission speeds. We, along with others (reviewed in Hall, 1992, p. 488) have demonstrated that infants with documented structural CNS injury display abnormal ABR waveforms (90% or more of the time) in a manner similar to adults with traumatic head injury or compression from tumor in the posterior fossa (Karmel, Gardner, Zappulla, Magnano, & Brown, 1988; Zappulla, Malis, Greenblatt, & Karmel, 1984). As the infant recovers from CNS insult, both ABRs (Hall, 1992; Karmel et al., 1988) and CUSs tend to normalize. However, poor developmental outcome may be mediated by deviant arousal mechanisms arising from inadequate brainstem function not necessarily detected in structural abnormalities. For this reason, infants with brainstem neurofunctional abnormalities should be considered at risk for developmental problems (see also Cox, Hack, Aram, & Borawski, 1992; Murray, 1988).

## AROUSAL AND ATTENTION IN INTRA-UTERINE COCAINE-EXPOSED INFANTS

Both direct and indirect effects of cocaine on human fetal development and CNS organization have been reported, with indirect effects more widely noted. For instance, cocaine exposure in utero produces a variety of adverse outcomes in the fetus and newborn infant, including greater incidence of intrauterine growth retardation and overall decreases in birthweight, head circumference and body length (Frank, Bresnahan, & Zuckerman, 1993; Volpe, 1992; Zuckerman et al., 1989), as well as greater incidence of perinatal hypoxemia. These effects are likely associated with the known vasoconstrictive properties of cocaine that increase the likelihood of perinatal or birth asphyxia, leading to a higher incidence of intraventricular hemorrhage, hypoxic encephalopathy, and CNS ischemia. Such CNS insults all have been associated with deficits in neurobehavioral development and subsequent behavioral performance (Singer, Garber, & Kliegman, 1991; Singer, 1999).

In addition, studies have been reported with respect to cocaine's more direct influence as a neurotoxin through its action on developing CNS tissue. Many cocaine-exposed infants have problems with attention, arousal and state-regulation capacities shortly after birth (Field, 1995; Freedland, Karmel, Gardner, & Lewkowicz, 1998; Karmel & Gardner, 1996; Karmel, Gardner, & Freedland, 1996, 1998). Although these effects can be further confounded by the higher incidence of poorer nutrition, increased disease, inadequate prenatal care, and polysubstance abuse that characterize the lifestyles of habitual cocaine-using mothers (Olson, Grant, Martin, & Streissguth, 1995), nonetheless, direct effects on neonatal arousal attention modulation (Karmel & Gardner, 1996) and neurobehavioral performance (Gardner, Karmel, & Freedland, 2001; Karmel et al., 1998; Singer, Arendt, Minnes, Farkas, & Salvator, 2000) as well as indications of long-term sequelae (see Chasnoff, Anson, Hatcher, Stenson, Iuakea, & Randolph, 1998; Karmel et al., 1998; Koren, Nulman, Greenbaum, Loebstein, & Einarson, 1998; Lester, 1998; Mayes et al., 1998; Singer et al., 2002) have been noted. Azuma and Chasnoff's finding (1993) that cocaine exposure has both direct and indirect effects on the cognitive development of children at three years has been replicated in four- and six-year-olds (Chasnoff et al., 1998). Also, Singer and colleagues (Singer, Arendt, Song, Warshawsky, & Kliegman, 1994; Singer, Yamashita, Hawkins, Cairns, Baley, & Kliegman, 1994) confirmed this result on a sample of low birth-weight, cocaine-exposed infants. Language delay and higher frequency of autism also have been reported (Davis, Fennoy, Laraque, Kanem, Brown, & Mitchell, 1992). More recently, Mayes has reported a number of studies showing differences in cocaine-exposed infants and children, specifically with respect to arousal and attention regulation, that helps to put the varied findings into perspective, especially as they relate to later functioning (e.g., Mayes et al., 1998).

Thus, it appears that both early CNS injury and exposure to neurotoxins can alter, modify or deviate the course of normal neurofunctional and neurostructural development, affecting arousal and attention systems, at a minimum through a disruption of monoaminergic pathways. Dopaminergic neurons with cell bodies located in the lower and mid-brainstem innervate a number of higher structures, including the frontal cortex, the nucleus accumbence and the corpus striatum (Diamond, 1990, 2000; Kolb & Whishaw, 1990). In addition, serotonergic projections, which have been thought to act in the CNS to modulate and maintain behavior, arise from the dorsal nucleus of the raphe in the midbrain and project collaterals to both the striatum and substantia nigra, forming an ascending functional relationship with the reticular activating system and the limbic system (Carpenter, 1993). Although studies have looked at behavioral deficits in children after prematurity, brain injury, or cocaine exposure (Hack, Klein, & Taylor, 1995), most of these have been descriptive population studies and have not dealt specifically with mechanisms such as arousal and attention systems that may underlie these later deficits.

In order to study the influence of neonatal CNS pathology or intrauterine cocaine exposure on autoregulation and attention behavior in more detail, we classify infants into different groups based on their CNS risk due to cocaine exposure and structural and functional brain injury identified by CUS and ABR findings.

## INTERDEPENDENCE OF AROUSAL ON ATTENTION IN THE NEONATAL PERIOD

Our studies indicate that in neonates, arousal and attention to stimulation are inseparable, interdependent processes that we refer to as arousal-modulated attention (AMA). Exogenous effects of stimuli mutually interact with other exogenous effects and, as a set, interact with endogenous effects such as feeding. These endogenous and exogenous effects combine to reflect an optimal level of arousal maintained in a self-regulating, homeostatic manner. This can be conceptualized as an inverted U-shaped function with the optimal level at the top. The contributions of internal and external stimulation are balanced so that, for example, if internal levels are high, the infant will seek lower levels of external stimulation to achieve the same overall optimal level (i.e., homeostasis). Stimuli are attended relative to context and not relative to specific attributes. We interpret these effects to be sensory-nonspecific, most likely reflecting general central arousal and attention mechanisms rather than peripheral receptor or sensory-specific effects. Thus, any activation that produces orienting to specific stimuli also will include input from the nonspecific arousal system. Arousal influences not only tonic activity, but also gates specific sensory processing when internal or external factors are manipulated through feeding or stimulation (Gardner & Karmel, 1983; Karmel et al., 1991). This self-limiting homeostatic, behavioral system helps neonates maintain their arousal at the same overall level and maintains equilibrium (Zeskind & Marshall, 1991), with distinctions between internal and external changes in environmental energy probably arbitrary and artificial for the neonate.

We have operationally defined AMA in a series of studies looking at visual preference behavior using both spatial and temporal stimuli (Gardner & Karmel, 1983, 1984, 1995; Gardner et al., 1986, 1992; Karmel & Gardner, 1996; Karmel, Gardner, & Lewkowicz, 1983; Karmel et al., 1991, 1996). In this procedure, infants view all possible pairs of stimuli varying along some dimension when they are in three different arousal conditions: (1) less aroused—after feeding; (2) more aroused—endogenous (increased internal stimulation, i.e., before feeding); and (3) more aroused—exogenous (increased external stimulation, i.e., additional visual stimulation prior to each trial). For example, using different temporal frequencies, normal newborn and one-month-old infants prefer (look longer at) faster frequencies when less aroused after feeding and prefer slower temporal frequencies when more aroused due to either increased internal or external arousal (see Figure 5.1a). This ability to modulate attention with a clear interaction between arousal level and amount of stimulation is not

**Figure 5.1**
(a) Proportion of Time Normal Newborn Infants Looked at Stimulus Temporal Frequencies (in Hz) When They Were in the More and Less Aroused Conditions (b) Proportion of Time Newborn Infants with Severe Brain Injury Looked at Stimulus Temporal Frequencies (in Hz) When They Were in the More and Less Aroused Conditions (c) Proportion of Time Cocaine-Exposed Newborn Infants Looked at Stimulus Temporal Frequencies (in Hz) When They Were in the More and Less Aroused Conditions.

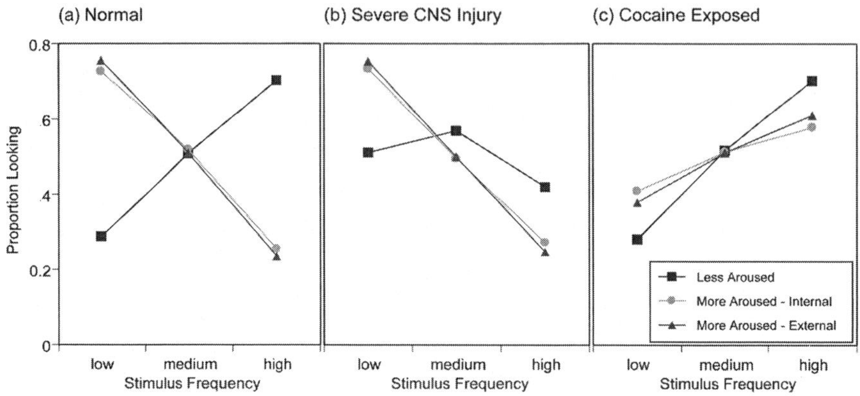


seen in infants with more severe brain injury. Infants with severe brain injury tend to prefer less stimulation even when less aroused (see Figure 5.1b). The result is that these infants are stimulus avoiders. By four months, normal infants prefer faster frequencies when both more and less aroused. Infants with severe CNS injury tend to shift toward faster frequencies when less aroused but still not prefer more intense stimuli when more aroused. Thus, the less the CNS involvement, the more looking at increased stimulation when less aroused, the larger the interaction at newborn and one month, and the less the interaction at four months between arousal condition and stimulus intensity preferences. Infants prenatally exposed to cocaine are also poor modulators with little interaction between more and less aroused conditions, but in the opposite direction. They tend to prefer more stimulation even when more aroused (see Figure 5.1c), resulting in infants who are stimulus seekers (Karmel & Gardner, 1996).

A converging measure of AMA uses a one-trial visual recognition memory (VRM) procedure (Geva et al., 1999). When infants are presented with two different stimuli, one novel and the other previously seen, they tend to look longer at the novel stimulus. This preference for novelty operationally defines VRM. Testing infants before and after feeding (more vs. less aroused) at newborn, one, and four months indicated that VRM behaved analogously to visual preferences, both in terms of shifting as a function of arousal, as well as reflecting differences due to brain injury and cocaine exposure. At newborn and one month, normal infants show familiarity preferences when more aroused and novelty preferences when less aroused (see Figure 5.2a). Infants with CNS in-

**Figure 5.2**
(a) Percentage of Time Normal Newborn Infants Looked at the Novel Stimulus When More and Less Aroused (b) Percentage of Time Newborn Infants with Severe Brain Injury Looked at the Novel Stimulus When More and Less Aroused (c) Percentage of Time Cocaine-Exposed Newborn Infants Looked at the Novel Stimulus When More and Less Aroused.

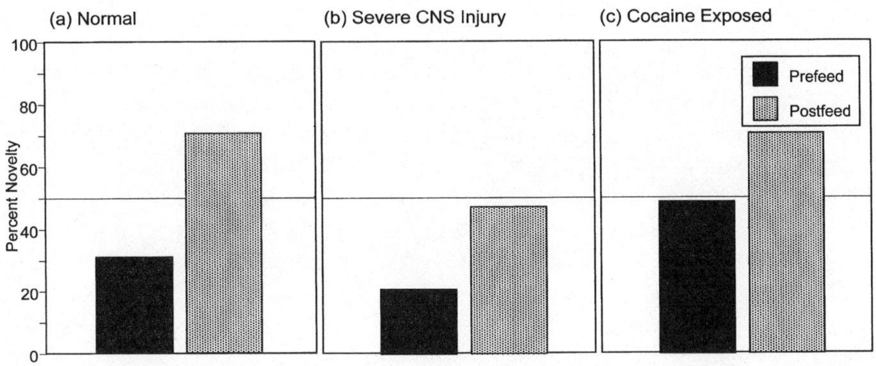


volvement do not shift as much to novelty preferences when less aroused (see Figure 5.2b) while cocaine-exposed infants tend to shift away from the familiar and more toward the novel even when more aroused (see Figure 5.2c). At four months, normal infants prefer the novel stimulus both before and after feeding. Infants with CNS damage now shift from marginal novelty responses before feeding to novelty preferences after feeding, and cocaine-exposed infants tend to show greater novelty preferences in general. Although the results have practical implications for high-risk studies, they also have broader theoretical significance because VRM is presumed difficult to obtain during the neonatal period. Understanding this paradigm in terms of AMA indicates that memory-related functions in young infants are strongly affected by arousal.

What appears to be simple visual processing in young infants actually is affected by a variety of factors. Both visual preferences to greater stimulation and preferring novelty are decreased by higher arousal, increasing CNS injury, and decreasing age. Therefore, using converging evidence from both visual preference and VRM techniques, systematic differences in neonatal attention can be obtained with controlled manipulations of arousal due to changes in internal (endogenous) and external (exogenous) sources of stimulation (Gardner & Karmel, 1983, 1984, 1995; Gardner et al., 1986, 1992; Gardner & Turkewitz, 1982; Geva et al., 1999; Karmel & Gardner, 1996; Karmel et al., 1983). Normal infants maintain equilibrium in their arousal and attention, while CNS pathology and neurotoxity due to cocaine exposure adversely affect the homeostatic regulatory mechanism that maintains this equilibrium and thus affect the ability to control arousal and attention. By four months, only infants with ABR but not CUS abnormalities or with the most severe CNS pathology still show arousal modulation effects by preferring lower levels of stimulation when more aroused (Karmel et al., 1996) or by preferring novelty after but not before feeding, thereby failing to mature from patterns of performance more typical of neonates.

## DEVELOPMENT OF AROUSAL AND ATTENTION AFTER THE NEONATAL PERIOD THROUGH THE FIRST YEAR

What are the developmental consequences of a unitary, early, homeostatic system for later development? Theories about brain-behavior relations in infants and children have been put forth with respect to visual attention (e.g., Braddick & Atkinson, 1988; Hoffmann, 1978; Johnson, 1996; Karmel, Kaye, & John, 1978; Karmel, Lester, McCarvill, Brown, & Hofmann, 1977; Karmel & Maisel, 1975; Richards, 2000; Richards & Hunter, 1998, 2002), emotional regulation (Calkins, Fox, & Marshall, 1996; Fox, 1994; Kagan, Reznick, & Snidman, 1987), and inhibitory control and cognition (Diamond, Prevor, Callender, & Druin, 1997). Likewise, there has been a long tradition of theories to explain both the development of arousal and attention systems in the first few years of life and their interaction in adults (Berlyne, 1958; Dember & Earl, 1957). But these theories tend to start after the neonatal period and do not deal with the earliest modes of behavior. For example, information about the neural mechanisms underlying attention stemming from Posner's (1988) studies of adult patients with lesions shows that orienting attention is influenced by adequate functioning of the parietal lobe, colliculus, and thalamic areas with each having its own unique contribution. Ruff and Rothbart (1996) and Rothbart, Posner, and Rosicky (1994) reviewed the developmental work in attention and argued for the use of "marker tasks" that specify underlying neural mechanisms, particularly in the first year or two. A similar argument was made by Karmel et al. (1978) for understanding neural mechanisms underlying developmental transitions in attention and cognition.

A number of changes occur in development after the neonatal period that can be linked to underlying neural systems. At about one month, the pathway from the basal ganglia to the superior colliculus provides inhibitory control over eye movements. A behavioral example of this new pathway is the difficulty infants have in breaking their gaze, even when continued orienting to the stimulus produces distress. This form of sustained attention has been referred to as "obligatory looking." Such sustained attention ends by about four months, probably due to the development of a pathway from the frontal eye fields to the colliculus and other parietal structures in conjunction with an increase in peripheral vision (Posner, 1988). Further evidence for developmental changes in brain organization after the neonatal period may be provided by the observed shifts in amplitude and latency distributions of visual evoked potentials (VEPs). For example, Hoffmann (1978) demonstrated that at six to eight weeks, visual preferences corresponded to diffuse amplitudes and latencies that were synchronous and widespread throughout the cortex, reflecting a common, probably subcortical, source projecting to multiple cortical regions. In contrast, visual preferences after ten weeks were associated with localized occipital pole VEP component amplitudes and latencies. Other forms of attention (possibly under inhibitory control, such as inhibition of return and object habituation) undergo substantial development between three and six months and are influenced by collicular function and the development of the posterior attention system (Posner, 1988). The posterior attention system is involved with orienting and investigative functions that we refer to as the sensory-specific period of attention because many studies show a developmental shift in behaviors (see Gardner & Karmel, 1983, 1995; Geva et al., 1999; Karmel et al., 1991; Lewis et al., 1998; Maurer & Lewis, 1979; Morton & Johnson, 1991) and VEPs become cortically localized as a function of stimulus modality mostly during this period (Ellingson, 1967; Hoffmann, 1978; Woodruff, 1978).

Rothbart and colleagues have investigated aspects of distress and vigilance with respect to attention in normal infants. Harman, Posner, Rothbart, and Thomas-Thrapp (1994) showed the relation between soothing of distress and attention to a distractor in normal infants. Rothbart, Ziaie, and O'Boyle (1992) showed that infants more prone to distress were less likely to look away from arousing stimuli, and Johnson, Posner, and Rothbart (1991) demonstrated a relation between soothability and the ability to disengage, and less distress to novel or intense stimuli. The coordination of vigilance and distress is important for normal development, with orienting attention necessary for modulating distress and vigilance necessary for attention to relevant stimuli. Deficits in such inhibitory autoregulatory control have been found in risk infants. Field (1982) showed differences in physiological and attention responses to animate and inanimate stimuli differing in arousal content. Eckerman and colleagues (Eckerman, Oehler, Hannan, & Molitor, 1995; Eckerman, Hsu, Molitor, Leung, & Goldstein, 1999) also reported differences in arousal control and distress while attending arousal-producing stimuli during a social interaction, and Karmel et al. (1996) found differences in AMA as a function of CNS pathology and cocaine exposure. Intact functioning of neurotransmitters, namely norepinephrine and dopamine, arising from the reticular activating system and locus ceruleus in the brainstem, is necessary for many functions related to arousal and attention. Orienting depends on the parietal lobe, pulvinar, and superior colliculus, each of which receives large amounts of norepinephrine. The relation between distress and visual attention probably involves the superior colliculus integrating these norepinephrine systems in the ability to disengage or be soothed by attending to a distractor.

Thus, CNS development, most likely cortical in origin, appears to modify the early basic subcortical influences of arousal on attention to generate the behavioral differentiation into two separate arousal and attention systems observed after the neonatal period. If cortical influences are inhibitory, developmentally-emerging cortical activity could act as negative feedback to a subcortical arousal system as this system projects to the cortex, thereby producing a higher-order homeostasis. Consequently, subcortical influences on attention are reduced and specific sensory system information is enhanced. The importance of the developing inhibitory system lies in the fact that perceptual and cognitive organization would be proportionally influenced to the extent to which the generalized AMA effect was inhibited in its interaction with sensory-specific information. Our recent findings indicate that the sensory-specific effects may be more intractable or canalized, less affected by damage, and less related to later functioning, than the arousal and inhibitory system effects. This type of functioning would be very adaptive for learning about the world just at the time when the infant is accumulating large amounts of detailed information and trying to integrate it all together.

As stated above, our four-month AMA data do not show differential effects for infants with only mild to moderate CNS involvement. In addition, at seven months, we evaluated VRM using both the six-trial (3 faces, 3 patterns) procedure developed by Rose and colleagues (Rose, 1983, 1988; Rose, Gottfried, & Bridger, 1979) and the ten-trial procedure normed by Fagan for the Infantest (Fagan, 1988; Fagan & Singer, 1983). In general, infants showed significant novelty preferences that were not differential across brain injury or prematurity in either procedure. We also analyzed the length of look on the Infantest as suggested by Jacobson (Jacobson, O'Neill, Padgett, Frankowski, & Bihun, 1992; Jacobson & Jacobson, 1995), who found increased length of looks in alcohol-exposed infants and decreased length of looks in cocaine-exposed infants. We found length of look decreased across age and increased with brain injury, but the associations to AMA and other attention variables were not as consistent. At these ages, after the hypothesized divergence into separate arousal and attention systems, length of look seems to be the more consistent variable. If novelty preference is more related to the attention process and length of look is more related to the arousal process, the effects seem stronger to arousal than to attention during this period.

## THE TRANSITION TO AUTOREGULATORY BEHAVIORS IN ONE- TO TWO-YEAR-OLDS

Development of higher levels of attentional control requires greater inhibitory control from frontal areas that first appear during the latter part of the first year, and the integration of these higher-level frontal control systems with sensory-specific systems (Diamond, 1990; Ruff & Rothbart, 1996). This higher-level attention control system appears to involve the midfrontal lobe including the anterior cingulate gyrus, regions also thought to be involved in the coordination of distress and attention. It has close connections to the dorsolateral frontal cortex, limbic system, adjacent motor areas, and posterior parietal lobe believed to be involved in inhibitory control behavior. Activity of the cingulate is modified by dopamine input from the underlying basal ganglia, which frequently are involved when early CNS damage occurs. Consequently, higher-level organization and control of behavior in the latter part of the first year and into the second year may be even more susceptible to aberrant development from damaged lower-level systems. High-risk infants frequently have deficits in focused attention, which we hypothesize is related to early deficiencies in arousal/attention modulation. Ruff (1988) provided support for this position by showing poor arousal control over external stimulation. Indeed, the inability to shift attention away from a stimulus with shifts in internal or external conditions may be the root cause linking some forms of CNS injury to later cognitive and emotional problems (Ruff & Rothbart, 1996). The ability to shift attention appropriately between stimuli, sustain attention, and resist distraction represents cognitive control over external events and impacts on subsequent learning and social relations (Ruff & Lawson, 1991). We propose that as infants with CNS pathology or neurotoxity due to cocaine exposure develop, their early difficulties regulating arousal may be manifested by attention-regulation problems in distracting environments. The period from ten to sixteen months is especially interesting, as focused attention and distractibility changes from a sensory-specific system of attention to one under greater frontal cortical control.

We used a modification of Ruff's procedure in which we computerized components to enhance stimulus presentation and facilitate analysis. Infants were videotaped for three two-minute trials beginning when the infant was presented with one of three different age-appropriate toys. Distractors (computer-generated animated pictures preceded by a beep) were presented for 4-second durations at random intervals of 3, 5, or 7 seconds contingent on an experimenter judging that the child was looking at the toy (Feldman, 1999; Feldman, Freedland, Gardner, & Karmel, 1996; Feldman, Freedland, Gardner, Karmel, & Gartenberg, 1999; Feldman, Gardner, Karmel, & Freedland, 1999).

We examined the effects of structural and functional brain injury from ten to sixteen months, enabling us to see how focused attention develops as the infant becomes less stimulus-bound and more able to voluntarily direct attention, and also how CNS injury affects this development (Feldman, 1999; Feldman, Gardner et al., 1999). At ten months, all infants were able to differentially inhibit attention to a distractor when in focused attention, as compared to casual attention. However, healthy term infants spent a larger proportion of their looking time in focused attention due to more episodes of focused attention rather than longer focused looks. This finding is consistent with early theories of attention, where attention to toys at ten months is highly stimulus-directed and characterized by short looks.

At sixteen months, the differences between healthy, term infants and infants with structural and functional damage emerged more clearly. Infants showed no differences as a function of brain injury in the number of times they engaged in focused attention. What now changed significantly was the duration of each focused epoch, with healthy, term infants sustaining focused attention for significantly longer periods of time than infants with more severe structural as well as functional-only (abnormal ABR but normal CUS) damage. Normal infants showed significantly less distractibility during focused attention than the CNS-involved infants. The increase in the length of each focused epoch and the decrease in distractibility during focused attention are indicative of a shift for the normal infants to the second system of attention, with increased voluntary control over attention and improved inhibitory control. This finding is important because it suggests that infants with structural and functional CNS damage may fail to transition to higher-level controls of attention, and dictates that these infants should continue to be followed to see if they do make this transition. It also suggests a possible way to remediate attentional difficulties that may be at the root of later deficits in school performance. If the focused epochs of these infants can be increased, their later difficulties may be averted.

Similarly, healthy term infants who were prenatally exposed to cocaine were compared to similar infants not exposed to cocaine (Feldman, Freedland et al., 1999). Although cocaine-exposed and non-exposed toddlers spent the same amount of time looking at the toy in general, and had the same number of focused attention episodes, cocaine-exposed toddlers' performance indicated specific difficulties in the ability to organize and sustain attention. Over trials, non-exposed toddlers showed decreased frequency of shifts in attentive behaviors, increased time spent in focused attention, and increased latency to turn to the distractor while in focused attention. On the first trial, cocaine-exposed toddlers tended to take longer to organize a focused attention response. Although by the third trial they were able to organize a focused attention response at the same rate as non-exposed toddlers, their ability to sustain attention deteriorated after the second trial. This was evidenced by reduced time spent in focused attention, failure to increase length of focused epochs, and decreased latency to turn to the distractor while in focused attention, all of which were in the opposite direction from non-exposed toddlers. Thus, the performance of cocaine-exposed toddlers on this early task of sustained attention suggests a shorter attention span with higher distractibility and shorter periods of sustained attention, which may underlie a number of behavioral and academic difficulties in later years.

The performance of healthy, term infants on tasks of focused attention supported previous normative research, with infants showing less distractibility when in focused attention than when casually looking at a toy. Healthy, term infants also showed attentional performance that indicated transition to higher-level cortical controls of attention. Infants with CNS injury and cocaine exposure both showed attentional processes that indicated either total or partial failure to attain attentional control indicative of higher cortical control by sixteen months but with different patterns of deficits.

## EXECUTIVE FUNCTION AND HIGHER LEVELS OF CONTROL AFTER TWO YEARS

High-risk infants have an increased likelihood of later school-related learning and attention problems, with neonatal CNS injury constituting a higher risk factor for poor neurodevelopmental outcome (Volpe, 2001). Although the global adverse effects of severe brain insult have been documented, less is known about "the lesser degrees of brain damage ... thought to be responsible for fine motor impairments, visuoperceptual and math difficulties, and hyperactivity" (Hack et al., 1995). Nevertheless, direct linkages between subcortical arousal systems and cortical frontal inhibitory control (IC) systems support our position on the importance of intact AMA in the neonatal period as a precursor to later executive function.

From the second year on, higher-level controls develop into regulatory mechanisms mediated by continued maturation of the frontal cortex (Diamond, 1990). The prefrontal cortex, involved in IC in adults (Lezak, 1983, 1989; Luria, 1973), becomes richly connected with the upper parts of the brainstem and thalamus and with other cortical zones, as well as with subcortical arousal systems (Carpenter, 1993). IC is necessary for planning and developing, as well as maintaining and shifting, behavioral actions or mental sets. Deficits in IC are reflected in difficulties making mental or behavioral shifts, which may result in perseverative, impulsive, or distractible behavior. Investigators have long suspected insufficient IC as the underlying cause of ADD/H symptoms (Aman, Roberts, & Pennington, 1998; Satterfield & Dawson, 1971). ADD/H children have been found to be slower and have more errors reflecting disinhibition (consistent with deficiencies in the frontal lobe inhibitory system) on reaction time tasks requiring visual sustained attention (Hynd & Willis, 1988). We found that infants with certain CNS problems showed deficits in IC in the early preschool years in regulating both verbal and motor output (Geva, 1995).

We (Geva, 1995; Geva, Gardner, & Karmel, 1998) used a modification of the original Rapid Sequential Automated Naming Tasks (RSANT; Eakin & Douglas, 1971; Denckla & Rudel, 1976; Wolf, 1984) to study IC at thirty-four months. The child was shown two stimulus boards sequentially, each containing a series of illustrations of common objects (in this case a fish and a cup). Arousal was manipulated by introducing additional extraneous stimuli (equal numbers of bubbles) around both objects on the second board. After determining vocabulary ability, children were asked to name the objects sequentially as fast as possible (with the assistance of a transparent pointer guide). Time to completion (reflecting mental speed) and number of perseverative errors (reflecting IC) were measured. Preliminary results revealed that neonatal CNS injury affected IC skills in regulating verbal output such that children with severe brain injury took longer and made more perseverative errors. An interaction was seen between the order of presentation of the arousing stimulus board and brain injury. When the card with bubbles was presented first, children with abnormal ABRs but normal structure on CUS performed like normals, but when it was presented second, they performed like the more severely damaged children. Effects were independent of language skills, and remained significant when general cognitive measures were treated as covariates.

We (Geva, 1995; Geva et al., 1998) also used a simplified version of the Bender Visual Motor Gestalt test at thirty-four months. The test requires copying of geometric figures and has been proposed to be sensitive to diffuse cortical disease or subcortical lesions (Bender, 1938; Lyle & Gottesman, 1977; Lyle & Quest, 1976; Riklan & Diller, 1961). Children were presented with two exemplars to copy and then trace, a straight line (3–7 in. long) and a dotted line (4–6 dots). The four tasks represented an ordered set of increasing difficulty. Performance was scored on a three-level scale for each condition from performance conforming to criteria, to perseverative performance (such as repeated marking of the same location or failing to cease activity and lifting the pencil within the page boundaries), to non-compliance with the task. We found that the more severely injured children were more likely not to perform adequately even on the simplest draw-a-line task. Tracing a solid line, and drawing and tracing a dotted line, were equally difficult across all levels of brain injury. Grouping children by performance on this task indicated normal neonatal AMA functioning was positively associated with fewer perseverative errors. Effects were independent of performance and eye–hand coordination skills measured by subscales on standardized cognitive measures.

## PREDICTING LATER BEHAVIOR USING AROUSAL AND ATTENTION MEASURES

To better understand the continuity of the arousal and attention behavior over age, we constructed a structural model to assess the predictive power and causal paths of our neonatal arousal and attention measures. Figure 5.3 demonstrates the predictive validity of these measures at stages of development from birth through sixteen months. (The numbers in Figure 5.3 are analogous to regression coefficients and show the degree to which a change in early attention predicts a change in attention at a later time.) The four- and seven- to thirteen-month measures, while associated with each other, did not predict attention at other ages.

The four-month measures show an effect only on the next time period (which is dominated by the seven-month measures). AMA at zero to one month, however, has a strong relationship to attention at sixteen months that

**Figure 5.3**
Simplified Structural Model of Influences of Neonatal Attention on Attention at Subsequent Ages. The numbers are analogous to regression coefficients and show the degree to which a change in early attention predicts a change in attention at a later time. The 7–13-month time period is dominated heavily by the 7-month measures.

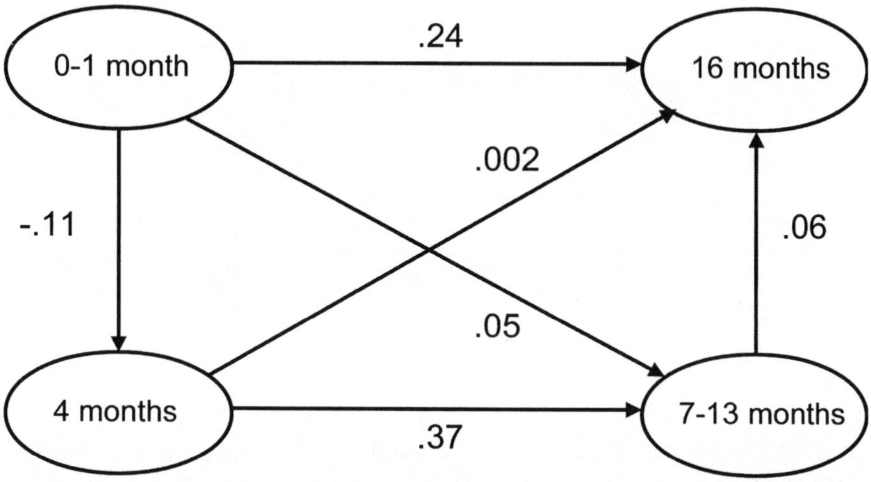


is unmediated by intervening time periods, suggesting that neonatal AMA has a lagged effect on later attentional capacity, or possibly that the characteristics of attention that we have captured at the intervening ages represent a different facet of the attentional system.

Figure 5.4 shows the relation of arousal and attention measures across age that contributed to Figure 5.3. Each of the tasks relates very well to the same task across age. AMA at newborn and one month is related within and across age, and to arousal-mediated VRM at one month and AMA at four months. AMA also relates better to measures after one year and has only limited relationships to measures during the sensory-specific period in the first year, that is, at four and seven months. This finding required us to revise our hypothesis that deficits in neonatal AMA underlie all types of later attention deficits. The findings suggest that after the newborn period, arousal and attention may emerge as two separate quasi-independent but related processes, each carrying different weight for predicting later behaviors. If each of our attention measures can be considered to contain two components, one reflecting arousal and one reflecting attention or cognition, we can evaluate the pattern of relations of those two components within and across measures. Thus, for example, we can examine VRM for both average length of look (arousal component) and percent novelty (attention or cognitive component), or focused attention for both latency to organize the first focused attention response (arousal) and percent time in focused versus nonfocused attention (attention). When we did this, the

**Figure 5.4**
Relation of Arousal and Attention Measures Across Age

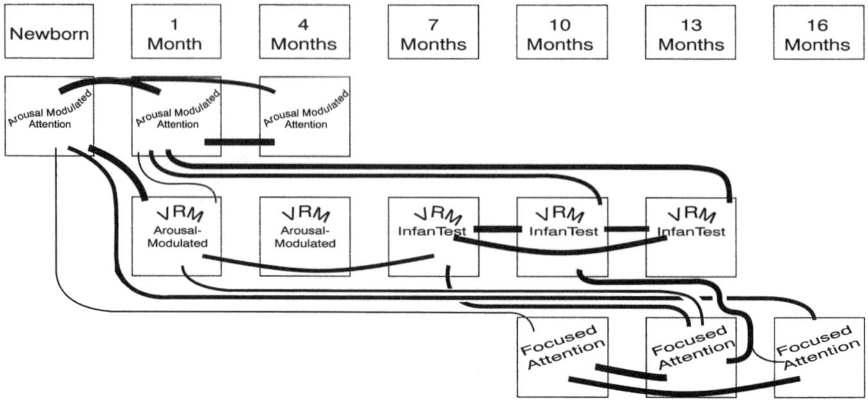


Note: Increasing strength of relationship is scaled by increasing line thickness.

relationships seemed more to reflect arousal than attentional mechanisms, although depending on the measure and age, different components were related.

## RELATIONSHIPS OF AROUSAL AND ATTENTION TO CNS PATHOLOGY AND STANDARDIZED MEASURES

To determine whether attentional abilities predicted later behavioral outcomes, we administered the Bayley Scales of Infant Development (BSID; Bayley, 1969; 1993) and the Griffiths Mental Development Scales (GMDS; Griffiths, 1984), two standardized cognitive measures that provide information on developmental level. Because the scaled scores were not uniform over age for either the BSID-I or II in our sample, we recalibrated the BSID-I and BSID-II to ensure that spurious age effects did not confound analyses (see Figure 5.6). Recalibration was accomplished by taking the raw scores across all tests for all infants considered normal in our population (1934 tests for 376 infants) and regressing on age (corrected) using LOCFIT (Loader, 1999), a refinement of the LOESS local regression technique (Cleveland & Devlin, 1988), to produce a smoothed raw score estimate over irregularly varying data. A second local regression was performed to estimate the standard deviation (SD) at each age. Residual values from the expected raw scores were then divided by SDs and adjusted to a mean of 100 and SD of 15 (like the scaled score of the BSID-II) across age. This allowed us to extend estimates below 50 (the artificial floor of the BSID) and improved our ability to detect real differences across different amounts of brain injury while discarding spurious ones. Figure 5.5 shows the

**Figure 5.5**
Effect of Recalibration of the BSID-I and II MDI as Shown by Original and Recalibrated Scores Across Age for Normal Infants

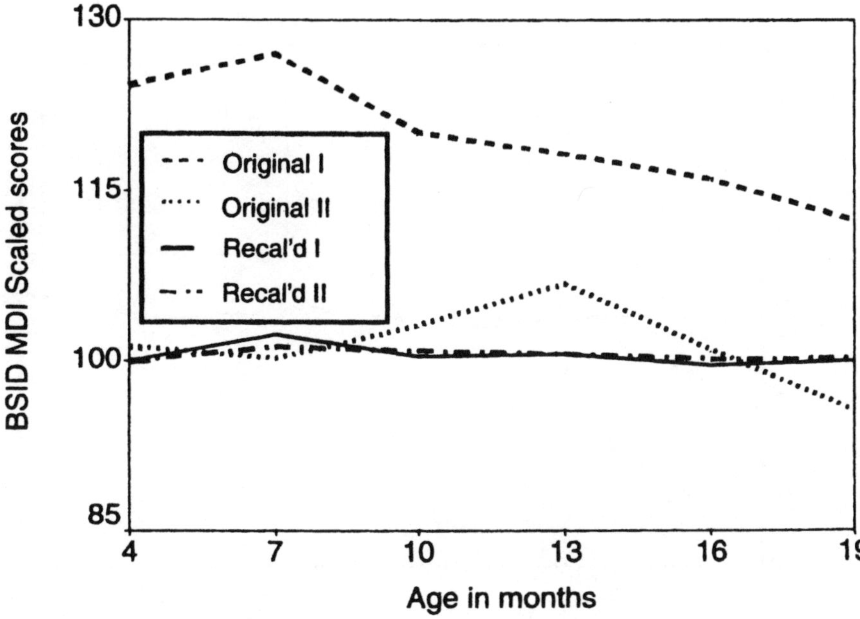


effect of recalibration on the Mental Development Index (MDI) for BSID-I and II from four to nineteen months for normal infants in our population.

Table 5.1 shows the relationship of arousal and attention tasks for each age to brain injury (excluding cocaine-exposed infants) and recalibrated BSID-II MDI and Psychomotor Development Index (PDI) at nineteen months. *R*s are presented and statistically significant effects are in bold. As can be seen, AMA was strongly associated with brain injury. AMA also was associated with BSID (Karmel et al., 2002). Figure 5.6 shows the predictive validity of AMA to BSID for all normal and CNS-involved infants but excluding cocaine-exposed infants. The upper and lower quadrants were plotted for the strength of the preference function in the optimal less aroused (after feeding) condition—the more positive, the stronger the preference for faster frequencies. When less aroused, infants with the uppermost 25 percent of positive slopes had higher BSID-II MDI across age than infants with the lowest 25 percent of positive slopes. These infants also were more likely to be normal. The utility of the newborn measure also was seen by its predictive validity to thirty-four months. The slope of the preference function in optimal arousal was associated significantly with thirty-four-month GMDS. We have recently also found a relationship between

**Figure 5.6**
Predictive Validity of AMA to BSID as Seen by the Relation Between AMA (the Slope of the Preference Function in the Less Aroused Condition) and the BSID-II MDI (Recalibrated) Across Age for Normal and Brain-Injured Infants (Excluding Cocaine-Exposed)

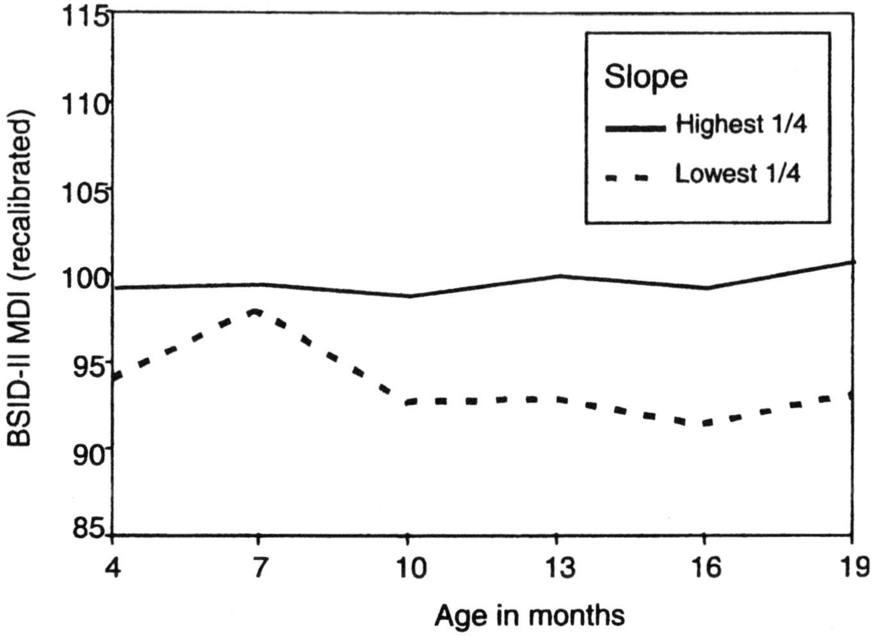


neonatal AMA and language inhibition at four to seven years of age (Belser, Sudhalter, & Gardner, 2000; Sudhalter, Belser, Gardner, & Karmel, 2001).

In general, findings at four and seven months were weak or inconsistent. These findings could be due to a number of reasons. Infants' behavior might be so sensory-specific that it is in some way canalized or resilient to many extraneous factors, and less affected by brain injury. This insulation from other effects probably is very adaptive because the job for infants at this time is to learn about their environment. Thus, this period is probably ideal for studying normal sensory and perceptual development but may not be for differentiating type and severity of risk, making it difficult to test processes associated with early AMA. The autoregulatory control component in modulating attention appears to be most important for this purpose both in the neonate and after a year, and may have been poorly probed in our tasks, which were heavily biased toward perceptual processing. Arousal-based rather than perceptual tasks might show better relations to early neonatal functioning, be more differentially affected by CNS involvement, and have better predictability to later behavior.

**Table 5.1**
Relationship of Arousal and Attention Tasks to CNS Injury and BSID-II

| | | CNS Injury | 19-month MDI | 19-month PDI |
|---|---|---|---|---|
| Age (mo) | Task | R | R | R |
| 0 | AMA | **.43** | **.16** | **.18** |
| 1 | AMA; VRM | **.46** | **.18** | **.22** |
| 4 | AMA; VRM | .05 | .07 | **.14** |
| 7 | AMA; VRM | .10 | **.14** | **.13** |
| 10 | Focused Attention; VRM | **.20** | **.26** | **.28** |
| 13 | Focused Attention; VRM | .13 | .08 | .12 |
| 16 | Focused Attention | **.17** | **.25** | **.15** |

*Note.* Significant *Rs* < .05 are bolded. BSID-II = Bayley Scales of Infant Development; Revised (Bayley, 1993); MDI = Mental Development Index; PDI = Psychomotor Development Index; AMA = arousal-modulated attention; VRM = visual recognition memory.

Measures at ten months (principally our focused-attention task) and at sixteen months related to brain injury and to BSID. Measures at thirteen months were not related to either and thus appeared inconsistent with ten- and sixteen-month findings, possibly reflecting transitional processes to higher-level attention. Sixteen-month measures were more stable and showed the strongest relations after the neonatal period to brain injury and BSID. Therefore, arousal and attention measures from ten and sixteen months demonstrated the most stable relationships to brain injury and BSID.

## SUMMARY

Arousal and attention form an inseparable, dynamically transacting process during the neonatal period. Normal neonates modulate their attention according to their arousal. They prefer more stimulation when less aroused and less stimulation when more aroused. Brain injury is differentiated by specific disruptions of this process. For this reason, infants with different types and severity of CNS problems modulate their attention differently. Infants with functional brainstem problems and those with more severe CNS injury or seizures are poor modulators. They tend to avoid higher levels of stimulation and prefer lower levels of stimulation even when less aroused. Infants prenatally exposed to cocaine also are poor modulators, but in the opposite direction. They tend to seek higher amounts of stimulation even when more aroused. Such different modes of early response to stimulation appear to have repercussions for information processing and social interactions not only during the neonatal period but also later in development. During the neonatal period, information is attended based on the amount of input from all sources, both internal and external. For normal neonates, who are excellent modulators, the range of information to which they are exposed would be very large. They would attend to and potentially process all types and amounts of information from all modalities, and they would experience the difference in the amount of stimulation in the information when in different states. Such experience would result in their learning to integrate their internal and external stimulation experience. For infants with CNS involvement or cocaine exposure, who are poor modulators, the range of total information processed would be restricted and the infant would not learn to modulate or control which information was processed. This would result in always attending to reduced stimulation if they are CNS-injured and increased stimulation if cocaine-exposed, producing less information about the world and perhaps greater individual stereopathy.

During the first year, information processing becomes much more investigative and sensory-specific. The infant who has experience with a whole range of input can now differentiate information coming from different sensory pathways at levels above the brainstem and midbrain areas. The infant is using this period to process enormous amounts of specific information about his or her world. For example, visual perception of forms, shapes, and complex patterns are all integrated at this time for use in skills such as visual recognition memory that then can be integrated into higher-level cortical categorization and cognition. However, we have found that these capabilities, for the most part, do not seem to be affected by CNS insult. Rather, our data indicate that tasks that tap into the organization of the arousal and self-regulatory components of attention may be better discriminators of CNS pathology than simple attention tasks alone. Thus, visual attention or discrimination tasks may be better for studying normal perceptual processes, but the arousal-modulating tasks may be better for differentiating individual differences if they indeed underlie the mechanisms for transitions in development.

Information-processing skills that emerge toward the end of the first year, such as those measured by focused attention and distractibility tasks, seem to have a large self-regulatory component. The ability to modulate arousal and attention as a neonate appears to be more necessary for these higher-level self-regulatory visual information-processing skills after a year than for the stimulus-specific processing skills during the first year. These higher-level skills relate to lack of modulation of early attention and are associated with early CNS insult. The infant who has poor modulation as a neonate tends to have poor autoregulation of information processing on higher-level tasks. Whether this is due to the restricted range of sensory input (i.e., all input at the more stimulating or at the less stimulating end of the range) or the lack of experience with integrating internal and external amounts of stimulation with state transitions, or some combination of both, is not known. But we do propose that it is not the initial brain injury that solely determines outcome after the first year of life, but rather the transaction between the consequences of the brain injury and the degree of recovery, and its changing effect on the infant's ability to attend to, control, or respond to his or her environment. Shifts in perceptual, cognitive, and self-regulatory development are found to be differentially related to early arousal/attention interactions as influenced by severity of CNS problems, with the degree of modulation of attention during the newborn period a major predictor of later functioning. More specifically, infants who have had early lower and mid-brainstem dysfunction as neonates (even if transient) that is manifest in poor autoregulatory control, have difficulty modulating their sensory input (as well as voluntary motor output), and hence may have difficulty in their ability to control, plan, and execute intentional acts as older children. Since such inhibitory control and executive function appear to be necessary for adequate cognitive performance, we would expect both individual differences and reductions in intellectual functioning associated with autoregulatory control deficits.

## NOTE

This research was supported in part by a National Institute of Child Health and Human Development Grant HD-21784 awarded to Judith M. Gardner, and a National Institute of Drug Abuse Grant DA-06644 awarded to Bernard Z. Karmel. The authors wish to express their sincere gratitude and appreciation to the medical and nursery staffs of the Neonatal Intensive Care Unit and Newborn Nursery of St. Vincent's Catholic Medical Center of New York for their cooperation. We are indebted to Robert L. Freedland for his overall contributions to all aspects of the studies from design to implementation as well as to his graphics expertise, and to Ivy Feldman for her overall contributions and helpfulness, especially for the focused-attention studies, which she designed, supervised, and analyzed. We thank Elise Conway, Amy Del Rosario, Sylvia Gregg, and our other research assistants past and present for their assistance in complex record keeping, subject recruitment and retainment, and data collection and entry.

Correspondence concerning this article should be addressed to Judith M. Gardner, NYS Institute for Basic Research in Developmental Disabilities, 1050 Forest Hill Road, Staten Island, New York 10314. Electronic mail may be sent to: bzksi@con2.com.

## REFERENCES


Aman, C.J., Roberts, R.J., & Pennington, B.F. (1998). A neuropsychological examination of the underlying deficit in attention deficit hyperactivity disorder. *Developmental Psychology, 34,* 956–969.

Azuma, S.D., & Chasnoff, I.J. (1993). Outcome of children prenatally exposed to cocaine and other drugs: a path analysis of three-year data. *Pediatrics, 92,* 396–402.

Bayley, N. (1969). *Bayley Scales of Infant Development.* New York: Psychological Corp.

Bayley, N. (1993). *Bayley Scales of Infant Development* (2nd ed.). San Antonio, TX: Psychological Corp.

Belser, R.C., Sudhalter, V., & Gardner, J.M. (2000, March). *Neonatal arousal modulated attention and linguistic outcome in at-risk children.* Paper presented at the 33rd annual Gatlinburg Conference on Research and Theory in Intellectual Disabilities, San Diego, CA.

Bender, L., (1938). A Visual Motor Gestalt Test and its clinical use. *American Orthopsychiatric Association Research Monograph, 3.*

Berlyne, D.E. (1958). Supplementary report: Complexity and orienting responses with longer exposures. *Journal of Experimental Psychology, 56,* 183.

Braddick, O., & Atkinson, J. (1988). Sensory selectivity, attentional control, and cross-channel integration in early visual development. In A. Yonas (Ed.), *Minnesota Symposia on Child Development: Vol. 20. Perceptual development in infancy* (pp. 105–143). Hillsdale: Erlbaum.

Brazelton, T.B., Nugent, J.K., & Lester, B.M (1987). Neonatal behavioral assessment scale. In J. Osofsy (Ed.), *Handbook of infant development,* (2nd ed., pp. 780–818). New York: Wiley.

Bronson, G.W. (1982) The structure, status, and characterics of the nervous system at birth. In P.M. Stratton (Ed.), *The psychobiology of the human newborn.* London: Wiley.

Calkins, S.D., Fox, N.A., & Marshall, T.R. (1996). Behavioral and physiological antecedents of inhibited and uninhibited behavior. *Child Development, 67,* 523–540.

Carpenter, M.B. (1993). *Core text of neuroanatomy* (4th ed.). Baltimore: Williams & Wilkins.

Chasnoff, I.J., Anson, A., Hatcher, R., Stenson, H., Iuakea, K., & Randolph, L.A. (1998). Prenatal exposure to cocaine and other drugs. Outcome at four to six years. In J.A. Harvey & B.E. Kosofsky (Eds.), *Cocaine: Effects on the developing brain. Annals of the New York Academy of Sciences: Vol. 846* (pp. 314–328). New York: NY Academy of Sciences.

Cleveland, W.S., & Devlin, S.J. (1988). Locally weighted regression: An approach to regression analysis by local fitting. *Journal of the American Statistical Association, 83,* 596–610.

Cox, C., Hack, M., Aram, D., & Borawski, E. (1992). Neonatal auditory brainstem response failure of very low birth weight infants: 8-year outcome. *Pediatric Research, 31,* 68–72.

Davis, E., Fennoy, I., Laraque, D., Kanem, N., Brown, G., & Mitchell, J. (1992). Autism and developmental abnormalities in children with perinatal cocaine exposure. *Journal of the National Medical Association, 84,* 315–319.

Dember,W.N., & Earl, R.W. (1957). Analysis of exploratory, manipulatory, and curiosity behavior. *Psychological Review, 64,* 563–574.

Denckla, M., & Rudel, R. (1976). Rapid automatized naming (R.A.N.): Dyslexia differentiated from other learning disabilities. *Neuropsychologia, 14,* 471–479.

Diamond, A. (1990). The development and neural bases of memory functions as indexed by the AB and delayed response tasks in human infants and infant monkeys. In A. Diamond (Ed.), *The development and neural bases of higher cognitive functions. Annals of the New York Academy of Sciences: Vol. 608* (pp. 267–317). New York: NY Academy of Sciences.

Diamond, A. (2000). Close interrelation of motor development and cognitive development and of the cerebellum and prefrontal cortex. *Child Development, 71,* 44–56.

Diamond, A., Prevor, M.B., Callender, G., & Druin, D.P. (1997). Prefrontal cortex cognitive deficits in children treated early and continuously for PKU. *Monographs of the Society for Research in Child Development, 62*(4, Serial No. 252).

Eakin, S., & Douglas, V. (1971). "Automatization" and oral reading problems in children. *Journal of Learning Disabilities, 4,* 31–38.

Eckerman, C.O., Hsu, H., Molitor, A. Leung, E.H.L., & Goldstein, R.F. (1999). Infant arousal in an en-face exchange with a new partner: Effects of prematurity and perinatal biological risk. *Developmental Psychology, 35,* 282–293.

Eckerman, C.O., Oehler, J.M., Hannan, T.E., & Molitor, A. (1995). The development prior to term age of very prematurely born newborns' responsiveness in en face exchanges. *Infant Behavior and Development, 17,* 55–70.

Ellingson, R.J. (1967). The study of brain electrical activity in infants. In L. Lipsitt & C.C. Spiker (Eds.), *Advances in child development and behavior: Vol. 3.* (pp. 53–97). New York: Academic Press.

Fagan, J.F. (1988). Screening infants for later mental retardation: From theory to practice. In P.M. Vietze & H.G. Vaughan (Eds.), *Early identification of infants with developmental disabilities* (pp. 71–95). Philadelphia: Grune & Stratton.

Fagan, J.F., & Singer, L.T. (1983). Infant recognition memory as a measure of intelligence. In L.P. Lippsitt (Ed.), *Advances in infancy research: Vol. 2* (pp. 31–78). Norwood, NJ: Ablex.

Fawer, C., Dubowitz, L.M.S., Levene, M.I., & Dubowitz, V. (1983). Auditory brainstem responses in neurologically abnormal infants. *Neuropediatrics, 14,* 88–92.

Feldman, I.J. (1999). *Attention and distractibility in high-risk infants.* Unpublished doctoral dissertation, St. John's University, New York.

Feldman, I.J., Freedland, R.L., Gardner, J.M., & Karmel, B.Z. (1996, April). *Distractibility in CNS impaired and non CNS imparied infants.* Poster session presented at the 10th biennial International Conference on Infant Studies, Providence, RI.

Feldman, I.J., Freedland, R.L., Gardner, J.M., Karmel, B.Z., & Gartenberg, S.C. (1999). Focused attention and distractibility in toddlers prenatally exposed to cocaine. In L.S. Harris (Ed.), *Problems of drug dependence 1998: Proceedings of the 60th annual scientific meeting* (NIDA Research Monograph No. 179, p. 98).Washington, DC: Department of Health and Human Services.

Feldman, I.J., Gardner, J.M., Karmel, B.Z., & Freedland, R.L. (1999, August). *Development of focused attention from 10 to 16 months in low- and high-risk infants.* Poster session presented at the IXth Conference on Developmental Psychology, Spetses, Greece.

Field, T.M. (1981). Infant arousal, attention, and affect during early interactions. In L.P. Lippsitt & C. Rovee-Collier (Eds.), *Advances in Infancy Research: Vol. 1* (pp. 51–100). Norwood, NJ: Ablex.

Field, T.M. (1982). Affective displays of high-risk infants during early interactions. In T. Field & A. Fogel (Eds.), *Emotion and early interaction* (pp. 101–125). Hillsdale: Erlbaum.

Field, T.M. (1995). Cocaine exposure and intervention in early development. In M. Lewis & M. Bendersky (Eds.), *Mothers, babies and cocaine: The role of toxins in development* (pp. 355–368). Hillsdale: Ehrlbaum.

Fox, N. (1994). The development of emotion regulation: Biological and behavioral considerations. *Monographs of the Society for Research in Child Development, 59*(No. 240).

Frank, D.A., Bresnahan, K., & Zuckerman, B.S. (1993). Maternal cocaine use: Impact on child health and development. *Advances in Pediatrics, 40,* 65–99.

Freedland, R.L., Karmel, B.Z., Gardner, J.M., & Lewkowicz, D.J. (1998). Prenatal cocaine exposure and stimulus behaviors during the first year of life. In J.A. Harvey & B.E. Kosofsky (Eds.), *Cocaine: Effects on the developing brain. Annals of the New York Academy of Sciences: Vol. 846* (pp. 386–390). New York: NY Academy of Sciences.

Gardner, J.M., & Karmel, B.Z. (1983). Attention and arousal in preterm and full-term neonates. In T. Field & A. Sostek (Eds.), *Infants born at risk: Behavior and development* (pp. 69–98). New York: Grune & Stratton.

Gardner, J.M., & Karmel, B.Z. (1984). Arousal effects on visual preferences in neonates. *Developmental Psychology, 20,* 374–377.

Gardner, J.M., & Karmel, B.Z. (1995). Development of arousal-modulated visual preferences in early infancy. *Developmental Psychology, 31,* 473–482.

Gardner, J.M., Karmel, B.Z., & Freedland, R.L. (2001). Determining functional integrity in neonates: A rapid neurobehavior assessment tool. In L.T. Singer & P.S. Zeskind (Eds.), *Biobehavioral assessment of the infant.* New York: Guilford Publications, 398–422.

Gardner, J.M., Karmel, B.Z., & Magnano, C.L. (1992). Arousal/visual preference interactions in high-risk neonates. *Developmental Psychology, 28,* 821–830.

Gardner, J.M., Lewkowicz, D.J., Rose, S.A., & Karmel, B.Z. (1986). Effects of visual and auditory stimulation on subsequent visual preferences in neonates. *International Journal of Behavioral Development, 9,* 251–263.

Gardner, J.M., & Turkewitz, G. (1982). The effect of arousal level on visual preferences in preterm infants. *Infant Behavior and Development, 5,* 369–385.

Geva, R. (1995). *Relationships between neonatal risk, cognitive development and inhibitory control at three years of age.* Unpublished doctoral dissertation, City University of New York.

Geva, R., Gardner, J.M., & Karmel, B.Z. (1998, April). Inhibitory control at 34 months of cocaine-exposed neonates. *Infant Behavior & Development, 21* (Special ICIS Issue), 431.

Geva, R., Gardner, J.M., & Karmel, B.Z. (1999). Feeding-based arousal effects on visual recognition memory in early infancy. *Developmental Psychology, 35,* 640–650.

Griffiths, R. (1984). *The abilities of young children. A comprehensive system of mental measurement for the first eight years of life.* Bucks: The Test Agency LTD.

Hack, M., Klein, N.K., & Taylor, H.G. (1995). Long-term developmental outcomes of low birth weight infants. *The Future of Children: Low Birth Weight, 5,* 176–196.

Hall, J.W., III (1992). *Handbook of auditory evoked responses.* Boston: Allyn & Bacon.

Harman, C., Posner, M.I., Rothbart, M.K., & Thomas-Thrapp, L. (1994). Development of orienting to locations and objects in human infants. *Canadian Journal of Experimental Psychology, 48,* 301–318.

Hoffmann, R. (1978). Developmental changes in human infant visual-evoked potentials to patterned stimuli recorded at different scalp locations. *Child Development, 49,* 110–118.

Hynd, G.W., & Willis, W.G. (1988). *Pediatric Neuropsychology.* Orlando: Grune & Stratton.

Jacobson, S.W., & Jacobson, J.L. (1995). Detecting effects of prenatal drug exposure. In M. Lewis, & M. Bendersky (Eds.), *Mothers, babies, and cocaine: The role of toxins in development* (pp. 111–128). Hillsdale: Erlbaum.

Jacobson, S.W., Jacobson, J.L., O'Neill, J.M., Padgett, R.J., Frankowski, J.J., & Bihun, J.T. (1992). Visual expectation and dimensions of infant information processing. *Child Development, 63,* 711–724.

Johnson, M.H. (1990). Cortical maturation and the development of attention in early infancy. *Journal of Cognitive Neuroscience, 2,* 81–95.

Johnson, M.H. (1996). From cortex to cognition: Cognitive neuroscience studies of infant attention and perception. In C. Rovee-Collier & L.P. Lipsitt (Eds.), *Advances in infancy research: Vol. 10* (pp. 161–217). Norwood, NJ: Ablex.

Johnson, M.H., Posner, M.I., & Rothbart, M.K. (1991). Components of visual orienting in early infancy: Contingency learning, anticipatory looking, and disengaging. *Journal of Cognitive Neuroscience, 3,* 335–344.

Kagan, J., Reznick, J.S., & Snidman, N. (1987). Physiology and psychology of behavioral inhibition. *Child Development, 58,* 1459–1473.

Karmel, B.Z., & Gardner, J.M. (1996). Prenatal cocaine exposure effects on arousal modulated attention during the neonatal period. *Developmental Psychobiology, 26,* 463–480.

Karmel, B.Z., Gardner, J.M., Brown, E., Zappulla, R., Magnano, C.L., Goodman, J., & Carr, L. (1988). Neurofunctional indicators of early brain insult. In P.M. Vietze & H.G. Vaughan (Eds.), *Early identification of infants with developmental disabilities.* Philadelphia: Saunders.

Karmel, B.Z., Gardner, J.M., & Freedland, R.L. (1996). Arousal-modulated attention at 4 months as a function of intrauterine cocaine exposure and CNS injury. *Journal of Pediatric Psychology, 21,* 821–832.

Karmel, B.Z., Gardner, J.M., & Freedland, R.L. (1998). Neonatal neurobehavioral assessment and Bayley I & II scores of CNS-injured and cocaine-exposed infants. In J.A. Harvey & B.E. Kosofsky (Eds.), *Cocaine: Effects on the developing brain. Annals of the New York Academy of Sciences: Vol. 846* (pp. 391–395). New York: NY Academy of Sciences.

Karmel, B.Z., Gardner, J.M., Lennon, E.M., Flory, M.J., Freedland, R.L., Phan, Ha T.T., & Miroschnichenko, I. (2002). Use of neonatal arousal-modulated attention and motor performance for evaluating current status and outcome in high-risk infants. Presented at the International Conference on Infant Studies, Toronto, Canada.

Karmel, B.Z., Gardner, J.M., & Lewkowicz, D.J. (1983, April). *Spatial-temporal interactions in neonates' visual preferences.* Paper presented at the Society for Research in Child Development, Detroit, MI.

Karmel, B.Z., Gardner, J.M., & Magnano, C.L. (1991). Attention and arousal in early infancy. In M.J. Weiss and P.R. Zelazo (Eds.), *Newborn attention: Biological constraints and the influence of experience.* Norwood, NJ: Ablex.

Karmel, B.Z., Gardner, J.M., Zappulla, R.A., Magnano, C.L., & Brown, E.G. (1988). Brainstem auditory evoked responses as indicators of early brain insult. *Journal of Electroencephalography and Clinical Neurophysiology, 71,* 429–442.

Karmel, B.Z., Kaye, H., & John, E.R. (1978). Developmental neurometrics: The use of quantitative analysis of brain electrical activity to probe mental function throughout the life-span. In A. Collins (Ed.), *XII Annual Minnesota Symposium in child psychology* (pp. 141–198). Minneapolis: University of Minnesota.

Karmel, B.Z., Lester, M.L., McCarvill, S.L., Brown, P., & Hofmann, M. (1977) Correlation of infants' brain and behavior response to temporal changes in visual stimulation. *Psychophysiology, 14,* 134–142.

Karmel, B.Z., & Maisel, E.B. (1975) A neuronal activity model for infant visual attention. In L. Cohen & P. Salapatek (Eds.), *Infant perception: From sensation to cognition: Vol. 1.* New York: Academic Press.

Kolb, B., & Whishaw, I.Q. (1990). *Fundamentals of human neuropsychology* (3rd ed.). New York: W.H. Freeman.

Koren, G., Nulman, J.R., Greenbaum, R., Loebstein, M., & Einarson, T. (1998). Long-term neurodevelopmental risks in children exposed in utero to cocaine. The Toronto adoption study. In J.A. Harvey & B.E. Kosofsky (Eds.), *Cocaine: Effects on the developing brain. Annals of the New York Academy of Sciences: Vol. 846* (pp. 296–305). New York: NY Academy of Sciences.

Korner, A.F. (1972) State as variable, as obstacle, and as mediator of stimulation in infant research. *Merrill-Palmer Quarterly, 18,* 77–94.

Lester, B.M. (1998). The maternal lifestyles study. In J.A. Harvey & B.E Kosofsky (Eds.), *Cocaine: Effects on the developing brain. Annals of the New York Academy of Sciences: Vol. 846* (pp. 296–305). New York: NY Academy of Sciences.

Lester, B.M., Corwin, M.J., Sepkoski, C., Seifer, R., Peucker, M., McLaughlin, S., & Golub, H.L. (1991). Neurobehavioral syndromes in cocaine-exposed newborn infants. *Child Development 62,* 694–705.

Levene, M.I., Wigglesworth, J.S., & Dubowitz, V. (1983). Hemorrhagic periventricular leukomalacia in the neonate: A real time ultrasound study. *Pediatrics, 71,* 794–797.

Lewis, T.L., Mondloch, C.J., Budreau, D.R., Maurer, D., Dannemiller, J.L., Stephens, B.R., & Kleiner, K.A. (1998, April). Face perception in young infants. *Infant Behavior & Development, 21* (Special ICIS Issue), 535.

Lezak, M.D. (1983). *Neuropsychological assessment* (2nd ed.). New York: Oxford University Press.

Lezak, M.D. (1989). Assessment of psychosocial dysfunctions resulting from head trauma. In *Assessment of the behavioral consequences of head trauma.*New York: Alan R. Liss, Inc.

Loader, C.R. (1999). *Local regression and likelihood.* New York: Springer.

Luria, A.R. (1973). The frontal lobes and the regulation of behavior. In K.H. Pribram & A.R. Luria (Eds.), *Psychophysiology of the Frontal Lobes.* New York: Academic Press.

Lyle, O.E., & Gottesman, I.I. (1977). Premorbid psychometric indicators of the gene for Huntington disease. *Journal of Consulting and Clinical Psychology, 45,* 1011–1022.

Lyle, O.E. & Quest, W. (1976). The Bender Gestalt: Use of clinical judgement versus recall scores in prediction of Huntington's disease. *Journal of Consulting and Clinical Psychology, 44,* 229–232.

Maurer, D., & Lewis, T.L. (1979). A physiologic explanation of infants' early visual development. *Canadian Journal of Psychology, 33,* 232–252.

Mayes, L.C., Granger, R.H., Frank, M.A., Schottenfeld, R., & Bornstein, M.H. (1993). Neurobehavioral profiles of neonates exposed to cocaine prenatally. *Pediatrics, 91,* 778–783.

Mayes, L.C., Grillon, C., Granger, R., & Schottenfeld, R. (1998). Regulation of arousal and attention in preschool children exposed to cocaine prenatally. In J.A. Harvey & B.E. Kosofsky (Eds.) *Cocaine: Effects on the developing brain. Annals of the New York Academy of Sciences: Vol. 846* (pp. 126–143). New York: NY Academy of Sciences.

Morton, J., & Johnson, M.H. (1991). CONSPEC and CONLERN: A two-process theory of infant face recognition. *Psychological Review, 98,* 164–181.

Murray, A.D. (1988). Newborn auditory brainstem evoked responses (ABRs): Longitudinal correlates in the first year. *Child Development, 59,* 1542–1554.

Olson, H.C., Grant, T.M., Martin, J.C., & Streissguth, A.P. (1995). A cohort study of prenatal cocaine exposure. Addressing methodological concerns. In M. Lewis & M. Bendersky (Eds.), *Mothers, babies, and cocaine: The role of toxins in development* (pp. 129–162). Hillsdale: Erlbaum.

Parmelee, A.H. (1975). Neurophysiological and behavioral organization of preterm infants in the first month of life. *Biological Psychiatry, 10,* 501–512.

Posner, M.I. (1988). Structures and functions of selective attention. In T. Boll & T. Bryant (Eds.), *Master lectures in clinical neuropsychology and brain function: Research, measurement, and practice* (pp. 171–202). Washington, DC: American Psychological Association.

Prechtl, H.F.R. (1974). The behavioral states of the newborn infant. *Brain Research, 76,* 185–212.

Richards, J.E. (2000). Localizing the development of covert attention in infants with scalp event-related potentials. *Developmental Psychology, 36,* 91–108..

Richards, J.E., & Hunter, S.K. (1998). Attention and eye movement in young infants: Neural control and development. In J.E. Richards (Ed.), *Cognitive neuroscience of attention: A developmental perspective* (pp. 131–162). Mahwah, NJ: Erlbaum.

Richards, J.E., & Hunter, S.K. (2002). Testing neural models of the development of infant visual attention. *Developmental Psychobiology, 40,* 226–36.

Riklan, M., & Diller, L. (1961). Visual motor performance before and after chemosurgery of the basal ganglia in Parkinsonism. *Journal of Nervous and Mental Disease, 132,* 307–314.

Rose, S.A. (1983). Differential rates of visual information processing in fullterm and preterm infants. *Child Development, 54,* 1189–1198.

Rose, S.A. (1988). Predicting cognitive development from infancy measures. In P.M. Vietze & H.G. Vaughan (Eds.), *Early identification of infants with developmental disabilities* (pp. 266–281). Philadelphia: Grune & Stratton.

Rose, S.A., Gottfried, A.W., & Bridger, W.H. (1979). Effects of haptic cues on visual recognition memory in fullterm and preterm infants. *Infant Behavior and Development, 2,* 55–67.

Rothbart, M.K., Posner, M.I., & Rosicky, J. (1994). Orienting in normal and pathological development. *Development and Psychopathology, 6,* 635–652.

Rothbart, M.K., Ziaie, H., & O'Boyle, C.G. (1992). Self-regulation and emotion in infancy. In N. Eisenberg & R.A. Fabes (Eds.), *Emotion and its regulation in early development* (New Directions in Child Development, No. 55, pp. 7–24). San Francisco: Jossey-Bass.

Ruff, H.A. (1988). The measurement of attention in high risk infants. In P.M. Vietze & H.G. Vaughan (Eds.), *Early identification of infants with developmental disabilities* (pp. 282–296). Philadelphia, PA: Grune & Stratton.

Ruff, H.A., & Lawson, K.R. (1991). Assessment of infants' attention during play with objects. In C.E. Schaefer, K. Gitlin & A. Sandgrund (Eds.), *Play diagnosis and assessment* (pp. 115–129). New York: John Wiley.

Ruff, H.A., & Rothbart, M.K. (1996). *Attention in early development. Themes and variations.* New York: Oxford University.

Satterfield, J.H., & Dawson, M.E. (1971). Electrodermal correlates of hyperactivity in children. *Psychophysiology,8,* 191–197.

Singer, L.T. (1999). Advances and redirections in understanding effects of fetal drug exposure. *Journal of Drug Issues, 29,* 253–262.

Singer, L.T., Arendt, R., Minnes, S., Farkas, K., & Salvator, A. (2000). Neurobehavioral outcomes of cocaine-exposed infants. *Neurotoxicological Teratology, 22,* 653–66.

Singer, L.T., Arendt, R., Minnes, S., Farkas, K., Salvator, A., Kirchner, H.L., & Kliegman, R. (2002). Cognitive and motor outcomes of cocaine-exposed infants. *JAMA, 287,* 1990–1.

Singer, L.T., Arendt, R., Song, L.Y., Warshawsky, E., & Kliegman, R. (1994). Direct and indirect interactions of cocaine with child birth outcomes. *Archives of Pediatrics and Adolescent Medicine, 148,* 959–964.

Singer, L.T., Garber, R., & Kliegman, R. (1991). Neurobehavioral sequelae of fetal cocaine exposure. *Journal of Pediatrics, 119,* 667–672.

Singer, L.T., Yamashita, T.S., Hawkins, S., Cairns, D., Baley, J., & Kliegman, R. (1994). Increased incidence of intraventricular hemorrhage and developmental delay in cocaine-exposed, very low birth weight infants. *Journal of Pediatrics, 124,* 765–771.

Sudhalter, V., Belser, R.C., Gardner, J.M., & Karmel, B.Z. (2001). Atypical language and neonatal CNS pathology in school-age children. Presented at the Xth European Conference on Developmental Psychology, Uppsala, Sweden.

Thoman, E.B. (1981) A biological perspective and a behavioral model for assessment of premature infants. In L.A. Bond & J.M. Joffee (Eds.), *Primary prevention of psychopathology: Vol. 6.* Hanover, NH: University Press of New England.

Turkewitz, G., Gardner, J.M., & Lewkowicz, D.J. (1984) Sensory/perceptual functioning during early infancy: The implications of a quantitative basis for responding. In G. Greenberg & E. Tobach (Eds.), *Behavioral evolution and integrative levels* (pp. 167–195). Hillsdale: Erlbaum.

Turkewitz, G., Lewkowicz, D.J., & Gardner, J.M. (1983). Determinants of infant perception. In J. Rosenblatt, R.A. Hinde, C. Beer, & M. Busnel (Eds.), *Advances in the study of behavior: Vol. 13* (pp. 39–62). New York: Academic Press.

Volpe, J.J. (1992). Effect of cocaine use on the fetus. *New England Journal of Medicine, 327,* 399–407.

Volpe, J.J. (2001). *Neorology of the newborn.* (4th ed.). Philadelphia: W.B. Saunders.

Wolf, M. (1984). Naming, reading, and the dyslexia: A longitudinal overview. *Annals of Dyslexia, 34,* 87–136.

Woodruff, D.S. (1978). Brain electrical activity and behavior relationships over the life span. In P. Baltes (Ed.), *Life span development: Vol. 1* (pp. 11–179). New York: Academic.

Zappulla, R.A., Malis, L.I., Greenblatt, E., & Karmel, B.Z. (1984). Utility of brainstem auditory evoked responses in diagnosis and surgical treatment of tumors of the cerebellopontine angle. In R.H. Nodar and C.Berber (Eds.), Evoked Potentials II. The 2nd International Evoked Potentials Symposium. Butterworth: Boston, MA, pp. 194–202.

Zeskind, P.S., & Marshall, T.R. (1991). Temporal organization in neonatal arousal: Systems, oscillations, and development. In M.J. Weiss and P.R. Zelazo (Eds.), *Newborn attention: Biological constraints and the influence of experience* (pp. 22–62). Norwood, NJ: Ablex.

Zuckerman , B., Frank, D.A., Hingson, R., Amaro, H., Levenson, S.M., Kayne, H., Parker, S., Vinci, R., Aboagye, K., Fried, L.E., Cabral, H., Timperi, R., & Bauchner, H. (1989). Effects of maternal marijuana and cocaine use on fetal growth. *The New England Journal of Medicine*, 762–768.

Chapter 6

# Visual Processing Strengths in Down Syndrome: A Case for Reading Instruction?

*Robert M. Hodapp and Tran M. Ly*

Throughout the twentieth century, researchers have pondered the issue of what causes mental retardation. So-called "defect theorists" have argued that a single, particularly deficited area of psychological functioning constitutes the primary cause of all mental retardation. Although different theorists have hypothesized different problems as that single defect, each focused on only one cause (e.g., Ellis, 1963; Luria, 1963; Zeaman & House, 1963). In contrast, developmentally oriented workers have contended that mental retardation involves a more global delay (Zigler, 1967, 1969). Particularly for those children and adults showing no clear organic cause for their mental retardation (the cultural–familial or "non-specific" form), developmentalists have historically believed that no single defect underlies delayed cognitive functioning.

To complicate the picture further, in recent years developmental approaches have been applied to children with various organic forms of mental retardation. Children with, for example, Down syndrome or fragile X syndrome are now known to develop somewhat differently from typical persons with mental retardation (Hodapp & Zigler, 1995). Such etiology-related differences arise in cognitive, linguistic, and adaptive profiles and trajectories, in maladaptive behavior, and even in how parents, siblings, families and others in the child's environment react to the child with mental retardation.

These studies generate new appreciations in two areas. First, we now understand that children with one versus another genetic etiology illustrate specific characteristics of human development. Children with Williams syndrome, for example, show extreme visuo-spatial deficits and relatively spared linguistic abilities (Bellugi, Wang, Jernigan, 1994; Udwin & Yule, 1990). Visuo-spatial abilities and language, then, may be two areas that are somewhat separable.

Such examples provide "experiments of nature" by which we gain complementary views of how children's development can (but may not always) proceed (Hodapp & Burack, 1990).

Second, as increasing numbers of etiology-related behaviors are discovered, we appreciate that mental retardation is not a single entity. Instead, children with one versus another type of mental retardation differ greatly in their behavioral outcomes. Children with Prader-Willi syndrome (caused by missing paternal information from chromosome 15) show life-threatening overeating and obesity (Dykens & Cassidy, 1996). Children with Smith-Magenis syndrome (caused by a deletion on chromosome 17) show significant sleep disturbances and self-abuse involving pulling out fingernails and toenails as well as putting objects into bodily orifices (Smith, Dykens, & Greenberg, 1998). Other genetic disorders also show characteristic behaviors that differ from those exhibited by most other persons with mental retardation.

As a result of these new appreciations, the mental retardation field of study may now be ready to abandon its one-size-fits-all perspective on behavioral functioning. As genetic and biomedical advances continue over the next few decades, a behavioral focus on samples that combine many different etiologies will become more and more indefensible. In the psychological realm as well, it now seems time to go beyond the field's search for one, single defect underlying all of mental retardation. Instead, we should begin the more difficult task of searching for specific deficits associated with mental retardation's many different etiologies.

This chapter constitutes an early attempt to employ this "multi-mental retardation" perspective to cognitive functioning. Instead of addressing cognitive functioning in all children with mental retardation, we limit our discussion to one etiological group, children with Down syndrome. But before addressing our main topic—visual processing and reading for persons with this syndrome—we provide some background. We first describe the issue of behavioral phenotypes of different genetic mental retardation disorders, then provide an overview of cognitive–linguistic profiles that characterize most persons with Down syndrome. In this way, we provide the rationale for examining behavior in different genetic disorders, as well as how this multi-mental retardation perspective operates in one particular etiological group.

## BEHAVIORAL PHENOTYPES AND "CHARACTERISTIC" COGNITIVE–LINGUISTIC PROFILES

As recently as thirty years ago, most mental retardation researchers felt that children's behaviors were totally unaffected by the child's cause of mental retardation. Witness, for example, Ellis' (1969) statement that "in spite of all the possible criticisms of ignoring etiology in behavioral research, it should be strongly emphasized that rarely have behavioral differences characterized different etiological groups" (p. 561). Similarly, Fisher and Zeaman (1970) concluded that "it does not appear to make any difference" how one becomes mentally retarded, because "the maturational results are the same" (p. 164). This perspective can be referred to as the "undifferentiated view" of behavior in persons with mental retardation.

### Differentiating mental retardation

In contrast to this undifferentiated view, a small group of researchers has argued for differentiating persons with mental retardation on the basis of etiology. Although such attempts began as early as the late nineteenth century (Burack, 1990), the move to differentiate by etiology has gained momentum over the past few decades. Due mainly to remarkable advances in human genetics over the past twenty years, approximately 750 different genetic disorders have now been associated with mental retardation, with new disorders discovered each year (Opitz, 1996). Although most such disorders account for only small portions of the mentally retarded population, a full third of all persons with mental retardation are currently diagnosed with one or another genetic disorder (Matalainen, Airaksinen, Mononen, Launiala, Kaariainen, 1995; Rutter, Simonoff, & Plomin, 1996). Not included in these percentages are the many persons with fetal alcohol syndrome, prematurity, and other organic but not genetic causes. Taken together, approximately 50 percent of persons with mental retardation may show some form of known organic etiology (Matalainen et al., 1995).

Alongside such genetic advances, many researchers have begun examining the so-called "behavioral phenotypes" of different genetic disorders (O'Brien & Yule, 1995). But what, exactly, is a behavioral phenotype? To date, even the term's definition has caused controversy. On one side are those who feel that the term should be applied to only certain behaviors. Flynt and Yule (1994) thus declare that "a behavioral phenotype should consist of a distinctive behavior that occurs in almost every case of a genetic or chromosomal disorder, and rarely (if at all) in other conditions" (p. 666). In contrast, Dykens (1995) argues that behavioral phenotypes are best characterized as "the heightened *probability* or *likelihood* [that] people with a given syndrome will exhibit certain behavioral sequelae relative to those without the syndrome" (p. 523, italics in original).

### Behavioral research and Down syndrome

Although we return later to these issues, the above debate might lead one to think that most researchers employ etiology-based research strategies. In fact, however, surveys consistently find that only about 10 percent to 15 percent of mental retardation behavioral studies group their subjects by etiology (Dykens, 1996; Hodapp & Dykens, 1994). Instead, most behavioral researchers in mental retardation continue to group their subjects on the basis of degree of intellectual impairment (mild, moderate, severe, and profound mental retardation). To most behavioral researchers, the thirty-year-old admonitions that "etiology doesn't matter" still apply.

But one exception does exist to this general inattention to etiology: Down syndrome. Studies of children with Down syndrome appear across the 140-year history of this disorder. Indeed, while the large majority of behavioral researchers pay little attention to etiology, small groups of British and American researchers have historically examined this particular syndrome. Part of Down syndrome's allure may stem from its relatively high prevalence rate (1/800 live births) and its diagnosis at birth, and part from its status as a "known" mental retardation disorder. Even the general public seems to know about Down syndrome, with individuals with the syndrome appearing on popular TV shows (e.g., *Life Goes On, ER*) and written popular books (*Count Us In*). Whatever the reasons for its research and public popularity, Down syndrome is the focus of almost half (46%) of all behavioral studies on any genetic mental retardation disorder (Hodapp, 1996). Before discussing the specific issues of visual strengths and reading, then, we first take advantage of this long-standing research tradition to present a basic overview of cognitive and linguistic functioning in this syndrome.

### *Cognitive–linguistic functioning in Down syndrome*

Although in some ways Down syndrome is well known, in others it is not. Specifically, people generally identify individuals with Down syndrome due to their characteristic physical appearance. Others may describe such persons as having sociable, outgoing personalities. But most members of the general public—and even many researchers—know little about the cognitive and linguistic aspects that typically characterize this syndrome.

A first important cognitive–linguistic characteristic involves the weak linguistic skills found in most persons with Down syndrome. Although some exceptions do occur (e.g., Rondal, 1995), most children and adults with the syndrome exhibit particularly poor linguistic abilities. Even within language, grammar is an area of special weakness, much below the person's overall levels of intellectual functioning (i.e., mental age; Fowler, 1990). Scarborough, Rescorla, Tager-Flusberg, Fowler, and Sudhalter (1991) find that, even at later ages, children with Down syndrome generally speak in short sentences, with few embedded clauses, relative clauses, and other sophisticated grammatical structures (see also Rondal, 1996).

A second issue relates to extremely high rates of speech problems among individuals with Down syndrome. Almost all parents (95%) report at least occasional difficulties in understanding their child with Down syndrome (Kumin, 1994), and such difficulties relate to deviant or impaired articulation, as well as to dysfluencies involving sound prolongations, pauses, and repetitions of sounds, syllables, parts of words, and whole words (Leddy, 1999). Even during adolescence, those with Down syndrome (compared to a mixed etiological group) correctly articulated fewer consonants on the Fristoe Test of Articulation (Rosin, Swift, Bless, & Vetter, 1988). These problems in articulation, in turn, lead children with the disorder to have speech that is much less intelligible.

Finally, children with Down syndrome appear to show a relative visual advantage in short-term memory. This advantage can be seen both when comparing different IQ subtests and when performing visual versus auditory short-term memory tasks. On the Kaufman Assessment Battery for Children (K-ABC; Kaufman & Kaufman, 1983), for example, one subtest (Hand Movements) involves repeating, in order, a series of hand movements shown by the examiner (visual short-term memory), whereas another (Number Recall) involves repeating, in order, a series of digits spoken by the examiner (auditory short-term memory). Several studies have now found that children with Down syndrome perform better on the visual than on the auditory test (Hodapp, Leckman, Dykens, Sparrow, Zelinsky, & Ort, 1992; Pueschel, Gallagher, Zartler, & Pezzullo, 1986). Studies of other IQ tests and of experimental measures also show the visual advantage for children with Down syndrome (for reviews, see Chapman, 1995; Hodapp, Evans, & Gray, 1999). Granted, the visual advantage is a relative one; that is, children with Down syndrome are still delayed several years (compared to their chronological ages) on visual short-term memory tasks. Still, children with the syndrome show higher visual versus auditory short-term memory performance. This relative visual advantage in children with Down syndrome has led to several studies of literacy and reading instruction in this population, the topic to which we now turn.

### *Reading in Down syndrome*

Spurred by both clinical and research findings of a visual advantage in children with Down syndrome, several researchers have begun to evaluate reading instruction. If these children show a visual advantage, might they be better able to acquire language (an otherwise deficited area) through literacy training?

Though preliminary, the findings are encouraging. From several case reports, it appears that individual children with Down syndrome can establish reading vocabularies of a fair number of "sight words," even during the preschool years (Buckley, 1995). In addition, Buckley, Bird, and Byrne (1996) report that as many as half of all children and adolescents with this syndrome may be able to read more than fifty words, with some reaching much higher levels (see also Johansson, 1993).

In an ongoing longitudinal study, Buckley and her colleagues have also begun to systematically compare readers versus nonreaders. Comparing seven readers with seven nonreaders over a four-year span, Laws, Buckley, Bird, MacDonald, and Broadley (1995) found that readers were ahead of nonreaders on tests of receptive vocabulary, receptive grammar, auditory memory, and visual memory. For many of these measures, an interaction also occurred between receiving reading instruction and developmental changes from time 1 to time 2. In effect, reading instruction seemed to allow children with Down syndrome an entryway to performing higher-level behaviors in other aspects of language (see also Buckley, 1999).

But just how far can children with Down syndrome develop in reading? Here several issues arise that are specific to reading per se. For example, one can distinguish between alphabetic readers, who sound out the letters, and logographic readers, who rely on visual memory for spelling and reading. Usually, children with Down syndrome appear to be logographic (essentially, "whole word") readers, at least initially. A question also arises as to just how good a reader one can become without understanding alphabetic principles; in typical children, for example, use of phonetic decoding is a strong predictor of both first- and third-grade reading skills (McGuinness, 1997).

A related issue concerns what reading researchers have called "phonological awareness." Phonological awareness concerns the ability to distinguish and manipulate phonemes within words—to hear the difference between the words "bat," "pat," and "hat." In contrast to more visual aspects of reading, then, phonological awareness might be considered a more auditory task that may also be related to one's ability to read. Tests of these skills typically include repeating polysyllabic words or non-words, producing and detecting rhyme and alliteration, blending a word from a series of spoken sounds, and segmenting a word using phoneme counting or deletion (Evans, 1994). Remember, too, that deviant or impaired articulation (as well as other speech dysfluencies) are present in most children with Down syndrome (Leddy, 1999).

At present, researchers debate the role of phonological awareness and learning to read. As Cossu, Rossini, and Marshall (1993) note, four possible connections have been hypothesized between phonological awareness and reading. First, phonological awareness may be a prerequisite to reading acquisition (Bradley & Bryant, 1993). Second, phonological awareness may be due to specific instruction in reading (Morais, Cary, Alegria, & Bertelson, 1979). Third, instruction in reading may lead to the development of phonological awareness, which in turn is critical for understanding the grapheme–phoneme correspondence (Morais, Alegria, & Content, 1987). Finally, phonological awareness may only be associated with reading, both of which develop via maturational processes (Liberman, Shankweiler, Liberman, Fowler, & Fisher, 1977).

Although this debate's resolution seems years away, studies are mixed in finding a relationship between phonological awareness and reading levels in children with Down syndrome. The most controversial study, by Cossu et al. (1993), did not find this relationship. In their study, ten Italian children with Down syndrome (mean age 11.4, mean IQ 44) were matched for reading ability with ten younger normal kids (mean age 7.3, mean IQ 111). Both groups were comparable in reading aloud regular and irregular words and pronouncing words and non-words (both groups were reading at approximately 7-year levels). On a variety of metalinguistic tasks that tap phonological awareness, the control group performed competently, whereas the children with Down syndrome performed poorly. Cossu et al. (1993) concluded that no causal connection exists between phonological awareness and reading. Evans (1994) also found that none of six children with Down syndrome were able to perform competently on both lower- and higher-level phonological awareness tasks, even though all showed some (whole word) reading (and four had reading ages of approximately 6.5 years).

Several points require discussion in relation to these two studies. First, among typically developing children, skills in phonological awareness do appear to promote reading acquisition, especially for the beginning reader. Especially in terms of lower-level skills such as rhyme or alliteration, phonological awareness precedes reading ability (Bradley & Bryant, 1993). In one study, for example, typical four- and five-year-olds could detect and produce rhyme and alliteration before reading instruction, suggesting that phonological awareness may be a prerequisite to learning to read. Specifically, Bradley and Bryant contend that some phonological awareness tasks such as rhyme and alliteration may precede reading, whereas other tasks such as phoneme deletion may be a result of learning to read. In asserting that children with Down syndrome are able to read in the absence of phonological awareness, then, both Cossu et al. (1993) and Evans (1994) are claiming a different sequence into reading for children with Down syndrome.

At the same time, other researchers criticize the tasks themselves. In commenting on the Cossu et al. (1993) study, Morton and Frith (1993) claim that failure to perform these particular phonological awareness tasks does not mean that children with Down syndrome lack the skill per se. Similarly, Bertelson (1993) believes that children who are unable to perform phonological awareness tasks but who can read may simply reflect these children's inability to understand task instructions. Others (e.g., Byrne, 1993) even suggest that other tasks allow better assessments of competence in phonological awareness.

An additional issue may involve the levels of reading involved. In both the Cossu et al. (1993) and Evans (1994) studies, children with Down syndrome were reading at very low levels. In Cossu et al. (1993), both the typically developing and Down syndrome groups were presumably functioning at about the 7-year level (average age of the typically developing children was 7.3 years). In Evans (1994), no child with Down syndrome had reached a 7-year reading age. It remains possible, then, that phonological awareness may be highly related to the ability to read, but that this relationship mainly occurs at higher levels of reading.

Such appears to be the case in published studies that examine the correlates of reading in Down syndrome across a wide range of reading abilities. Fowler, Doherty, and Boynton (1995) examined thirty-three adolescents and young adults with Down syndrome whose reading ranged from kindergarten to twelfth-grade levels. In addition to general cognitive measures such as the K-ABC and Peabody

Picture Vocabulary Test–Revised (PPVT–R; Dunn & Dunn, 1981), Fowler et al. (1995) also gave their subjects the Woodcock Reading Mastery Tests–Revised (Woodcock, 1987), as well as tests of phoneme awareness, verbal memory span, word retrieval, visual memory, and auditory comprehension of language.

In contrast to the findings of Cossu et al. (1993) and Evans (1994), Fowler et al. (1995) find clear relationships between reading abilities and several other measures. Even after accounting for general levels of ability (on the K-ABC and PPVT–R), these researchers found that phoneme awareness accounts for high percentages of the variance in reading tasks such as word recognition (49% of variance) and decoding (36%). In a similar study using non-word repetition as the phonemic task for children aged five to eighteen years, Laws (1998) finds a similar relationship between non-word repetition and reading ability (the K-ABC's reading–decoding subtest). Again, over 30 percent of the variance was accounted for by the phonemic test, even after controlling for the child's age and nonverbal abilities.

Moreover, the relationship between phonemic awareness and reading may constitute a "necessary but not sufficient" condition. As Fowler et al. (1995) note, "no person in our study achieved decoding skills beyond the first grade level without answering at least ten items correctly [from among 40] on our phoneme awareness measure" (p. 191). Similarly, "no one achieved beyond the third grade decoding level without responding correctly to at least half of the phonemic awareness items" (p. 192).

Finally, phoneme awareness may not be the only correlate of higher-level reading in children with Down syndrome. In the Fowler et al. (1995) study, both visual short-term memory and digit span (i.e., auditory short-term memory) correlated with higher levels of reading. And again (as with phonemic awareness), higher levels of both visual and auditory short-term memories were necessary, but not sufficient, for the achievement of higher levels of reading.

### *Down syndrome and etiology-based research and intervention*

In considering reading for children with Down syndrome, four partially or totally unresolved questions arise. These questions reflect both the possibilities of reading instruction in Down syndrome and the primitive state of the art in this and many etiology-related matters.

***Can children with Down syndrome be taught to read?*** To this, our initial question, we can tentatively answer "yes." It does appear that most children with Down syndrome can be taught some level of reading. Case studies as well as quasi-experimental longitudinal group comparisons converge on the finding that many—maybe even most—of these children can attain at least rudimentary reading abilities. Smaller percentages of the Down syndrome population can advance much further. At the very least, teaching reading to children with Down syndrome is not harmful and is probably beneficial.

If indeed reading instruction can be helpful to children with Down syndrome, a more widespread use of this intervention should be tried. To some extent, the change to reading instruction may already be happening, possibly as a byproduct of greater integration of these students into mainstreamed classrooms. In Great Britain, for example, the shift to mainstream placements was so rapid from 1990 to 1995 "that the majority [of children with Down syndrome] are in mainstream schools. Since they are all taught reading, this means it is no longer feasible to set up a local study to compare the cognitive development of children with Down's syndrome being taught to read with those who are not receiving literacy instruction" (Laws et al., 1995, p. 60).

***Which children with Down syndrome will benefit most?*** Reading instruction may help children with Down syndrome, but it remains unclear why some children read better than others. The ability to identify predictors of success would be particularly helpful for educators and others intervening with children with this syndrome.

In many ways, this issue of within-group heterogeneity is part of a larger issue of how one defines behavioral phenotypes. In contrast to the Flynt and Yule (1994) definition offered above, a more probabilistic view focuses on the way that genes predispose—but do not totally determine—whether one exhibits a certain behavior. To take an example outside of Down syndrome, Dykens and Kasari (1997) found that 80 percent of four- to nineteen-year-old children with Prader-Willi syndrome showed overeating behavior—many to very high degrees (Dykens & Kasari, 1997). But while extreme overeating occurred much more frequently in the group with Prader-Willi syndrome than in others with mental retardation, overeating was not noted as a salient problem in 20 percent of the Prader-Willi group. Similar patterns of "most but not all" occur for almost every etiology-specific behavior of genetic mental retardation disorders.

This more probabilistic perspective carries several important implications. First, what one considers as a "characteristic behavior" will not show itself in every child with any syndrome. Even if two people have the exact genetic disorder, they often differ behaviorally one from another. In this instance, most children with Down syndrome will benefit from reading instruction, but not every child will benefit from reading instruction or even show the syndrome's characteristic advantage in visual versus auditory processing. Within-group heterogeneity thus provides an important lesson against stereotyping.

Equally important, within-group heterogeneity presents both limits and opportunities for researchers and interventionists. On one hand, heterogeneity forces researchers to be more cautious when specifying characteristic behaviors of one versus another etiology; such heterogeneity also implies that etiology-specific interventions will not be successful for every person with a particular syndrome. At the same time, such heterogeneity presents opportunities. Why, for instance, do certain children with Down syndrome—but not others—learn to read at relatively high levels? Are these the children who have been introduced to reading earlier, or who have been taught using a specific approach, or who have certain other genetic, neurological, or cognitive characteristics? Although so far unanswered, such questions may present the most interesting future challenges to etiology-based behavioral researchers (Dykens, 1995; Fidler & Hodapp, 1998).

***Does reading constitute an entryway into other aspects of language for these children?*** In the few studies that have included children with Down syndrome who vary from low to high levels in their reading abilities, the child's ability to read seems highly correlated with abilities in other areas. Some of these skills are similar to correlates of reading found in typically developing children. Here the relations with both phonemic awareness and (auditory) short-term memory come to mind. But other relations—for example to visual memory—may be specific to the Down syndrome group (Fowler et al., 1995).

Beyond the presence of interrelated skills is the question of the degree to which teaching one skill advances related skills. Can one, for example, use reading instruction to aid the child's articulation, or visual or auditory short-term memory? Laws et al. (1995) even offer the possibility that gains in reading may lead to gains in linguistic grammar and vocabulary. But much more support seems necessary before we truly understand how achievements in a single skill relate to skill levels in other areas.

***To what extent is the visual advantage or benefits of reading instruction unique to individuals with Down syndrome?*** A second tenet of the Flynt and Yule (1994) definition involves the uniqueness of etiology-related behaviors, the idea that a specific phenotypic behavior should occur " ... rarely (if at all) in other conditions" (p. 666). To date, certain behaviors do seem to be found in one and only one etiological group. For most other behaviors, however, two or more etiological groups share a particular behavior. By employing the K-ABC test, for example, several research groups have found that boys with fragile X syndrome generally display a weakness in sequential (i.e., step-by-step) versus simultaneous (i.e., Gestalt) processing skills (Dykens, Hodapp, & Leckman, 1987 Kemper, Hagerman, Altshul-Stark, 1988 Powell, Houghton, & Douglas, 1997). But this sequential deficit is also found in most children with Prader-Willi syndrome (Dykens, Hodapp, Walsh, & Nash, 1992). As such, both represent behavioral phenotypes—each group shows a cognitive profile that differs from profiles shown by most children with mental retardation—but that profile is not unique to only one syndrome.

Is the visual advantage in Down syndrome totally specific to Down syndrome or partially specific to Down syndrome plus a few other disorders (Hodapp, 1997)? Theoretically, total specificity implies that a single pathway may operate from genetic disorder to brain functioning to behavior. In contrast, partial specificity implies that two or more specific genetic disorders lead to similar neurological outcomes—it is as if two (or more) paths come together somewhere in the complicated, many-stepped path from gene to brain. At present, we cannot say for certain which is the case in either the Down syndrome visual advantage or in the likely propensity of children with the syndrome to benefit from literacy training.

Practically as well, if the visual advantage and benefits from reading are partially—and not totally—specific, then a single strategy of intervention may prove effective for children with both Down syndrome and with one or more additional genetic mental retardation disorders. Such appears the case for children with fragile X syndrome and with Prader-Willi syndrome. Since both show relative strengths in simultaneous processing, then for both groups one should be able to capitalize on these strengths in educational programs (Hodapp & Fidler, 1999). Indeed, the *KSOS: Kaufman Sequential or Simultaneous?* (Kaufman, Kaufman, & Goldsmith, 1984) presents a variety of different intervention approaches to employ when one teaches a "simultaneous learner" as opposed to a "sequential learner." Such may also be the case in visually based learning strategies for both Down syndrome and one or more additional groups.

As these discussions illustrate, we have many more questions than answers when it comes to the specific issues involved in reading instruction in Down syndrome. But just as clinical work is informed by knowledge of etiology (Dykens & Hodapp, 1997), so too should educational interventions be informed by etiology-related strengths and weaknesses (Hodapp & Fidler, 1999). Reading in Down syndrome seems among the most promising of these etiology-related interventions. Indeed, if these children can use their visual strengths to learn to read—and if reading helps them develop in other areas—then mental retardation professionals have made a major advance. We may be well on our way to going beyond the one-size-fits-all research and intervention approach, to the benefit of children with many different types of mental retardation.

## NOTE

We would like to thank Elisabeth Dykens and Alison Imbens-Bailey for their helpful comments on earlier drafts of this chapter.

## REFERENCES

Bellugi, U., Wang, P., & Jernigan, T. (1994). Williams Syndrome: An unusual neuropsychological profile. In S.H. Broman & J. Grafman (Eds.), *Atypical cognitive deficits in developmental disorders* (pp. 23–56). Hillsdale, NJ: Erlbaum.

Bertelson, P. (1993). Reading acquisition and phonemic awareness testing: How conclusive are data from Down's syndrome? Remarks on Cossu, Rossini, and Marshall (1993). *Cognition, 48,* 281–283.

Bradley, L., & Bryant, P.E. (1993). Categorising sounds and learning to read—a causal connection. *Nature, 301,* 419–421.

Buckley, S. (1995). Teaching children with Down syndrome to read and write. In L. Nadel & D. Rosenthal (Eds.), *Down syndrome: Living and learning in the community* (pp. 158–169). New York: Wiley-Liss.

Buckley, S. (1999). Promoting the cognitive development of children with Down syndrome: The practical implications of recent psychological research. In J.A. Rondal, J. Perera, & L. Nadel (Eds.), *Down's syndrome: A review of current knowledge* (pp. 99–110). London: Whurr Publishers Ltd.

Buckley, S., Bird, G., & Byrne, A. (1996). The practical and theoretical significance of teaching literacy skills to children with Down's syndrome. In *Down's syndrome: Psychological, psychbiological, and socio-educational approaches* (pp. 119–128). London: Whurr Publishers Ltd.

Burack, J.A. (1990). Differentiating mental retardation: The two-group approach and beyond. In R.M. Hodapp, J.A. Burack, & E. Zigler (Eds.), *Issues in the developmental approach to mental retardation* (pp. 27–48). New York: Cambridge University Press.

Byrne, B. (1993). Learning to read in the absence of phonemic awareness? (A comment on Cossu, Rossini, and Marshall (1993).) *Cognition, 48,* 285–288.

Chapman, R.S. (1995). Language development in children and adolescents with Down syndrome. In P. Fletcher & B. MacWhinney (Eds.), *Handbook of child language* (pp. 641–663). Oxford: Blackwell Publishers.

Cossu, G., Rossini, F., & Marshall, J.C. (1993). When reading is acquired but phonemic awareness is not: A study of literacy in Down's syndrome. *Cognition, 46,* 129–138.

Dunn, L., & Dunn, L. (1981). *Peabody Picture Vocabulary Test–Revised.* Circle Pines, MN: American Guidance Service.

Dykens, E.M. (1995). Measuring behavioral phenotypes: Provocations from the "new genetics." *American Journal on Mental Retardation, 99,* 522–532.

Dykens, E.M. (1996). DNA meets DSM: The growing importance of genetic syndromes in dual diagnosis. *Mental Retardation, 34,* 125–127.

Dykens, E.M., & Cassidy, S.B. (1996). Prader-Willi syndrome: Genetic, behavioral, and treatment issues. *Child and Adolescent Psychiatry Clinics, 5,* 913–927.

Dykens, E.M., & Hodapp, R.M. (1997). Treatment issues in genetic mental retardation syndromes. *Professional Psychology: Research and Practice, 28,* 263–270.

Dykens, E.M., Hodapp, R.M., & Leckman, J.F. (1987). Strengths and weaknesses in intellectual functioning of males with fragile X syndrome. *American Journal of Mental Deficiency, 92,* 234–236.

Dykens, E.M., Hodapp, R.M., Walsh, K.K., & Nash, L. (1992). Profiles, correlates, and trajectories of intelligence in Prader-Willi syndrome. *Journal of the Academy of Child and Adolescent Psychiatry, 31,* 1125–1130.

Dykens, E.M. & Kasari, C. (1997). Maladaptive behavior in children with Prader-Willi syndrome, Down syndrome, and nonspecific mental retardation. *American Journal on Mental Retardation, 102,* 228–237.

Ellis, N.R. (1963). The stimulus trace and behavioral inadequacy. In N.R. Ellis (Ed.), *Handbook of mental deficiency, psychological theory and research* (pp. 134–158). New York: McGraw-Hill.

Ellis, N.R. (1969). A behavioral research strategy in mental retardation: Defense and critique. *American Journal of Mental Deficiency, 73,* 557–566.

Evans, R. (1994). Phonological awareness in children with Down's syndrome. *Down's Syndrome: Research and Practice, 2,* 102–105.

Fidler, D.J., & Hodapp, R.M. (1998). The importance of typologies for science and service in mental retardation. *Mental Retardation, 36,* 489–495.

Fisher, M.A., & Zeaman, D. (1970). Growth and decline of retardate intelligence. *International Review of Research in Mental Retardation, 4,* 151–191.

Flynt, J., & Yule, W. (1994). Behavioural phenotypes. In M. Rutter, E. Taylor, & L. Hersov (Eds.), *Child and adolescent psychiatry: Modern approaches* (3rd ed., pp. 666–687). London: Blackwell Scientific.

Fowler, A.E. (1990). Language abilities in children with Down Syndrome: Evidence for a specific syntactic delay. In D. Cicchetti & M. Beeghly (Eds.), *Children with Down Syndrome: A developmental approach* (pp. 302–328). New York: Cambridge University Press.

Fowler, A.E., Doherty, B.J., & Boynton, L. (1995). The basis of reading skill in young adults with Down syndrome. In L. Nadel & D. Rosenthal (Eds.), *Down syndrome: Living and learning in the community* (pp. 182–196). New York: Wiley-Liss.

Hodapp, R.M. (1996). Down syndrome: Developmental, psychiatric, and management issues. *Child and Adolescent Psychiatric Clinics of North America, 5,* 881–894.

Hodapp, R.M. (1997). Direct and indirect behavioral effects of different genetic disorders of mental retardation. *American Journal on Mental Retardation, 102,* 67–79.

Hodapp, R.M., & Burack, J.A. (1990). What mental retardation tells us about typical development: The examples of sequences, rates, and cross-domain relations. *Developmental Psychopathology, 2,* 213–225.

Hodapp, R.M., & Dykens, E.M. (1994). Mental retardation's two cultures of behavioral research. *American Journal on Mental Retardation, 98,* 675–687.

Hodapp, R.M., Evans, D., & Gray, F.L. (1999). Intellectual development in children with Down syndrome. In J.A. Rondal, J. Perera, & L. Nadel (Eds.), *Down's syndrome: A review of current knowledge* (pp. 124–132). London: Whurr Publishers Ltd.

Hodapp, R.M., & Fidler, D.J. (1999). Special education and genetics: Connections for the 21st century. *Journal of Special Education, 33,* 130–137.

Hodapp, R.M., Leckman, J.F., Dykens, E.M., Sparrow, S.S., Zelinsky, D., & Ort, S.I. (1992). K-ABC profiles in children with fragile X syndrome, Down syndrome, and non-specific mental retardation. *American Journal on Mental Retardation, 97,* 39–46.

Hodapp, R.M., & Zigler, E. (1995). Past, present, and future issues in the developmental approach to mental retardation and developmental disabilities. In D. Cicchetti & D. Cohen (Eds.), *Manual of developmental psychopathology: Vol. 2. Risk, disorder, and adaptation* (pp. 299–331). New York: Wiley.

Johansson, I. (1993). Teaching prereading skills to disabled children. *Journal of Intellectual Disability Research, 37,* 413–417.

Kaufman, A.S., & Kaufman, N.L. (1983). *Kaufman Assessment Battery for Children (K-ABC).* Circle Pines, MN: American Guidance Service.

Kaufman, A.S., Kaufman, N.L., & Goldman, B.Z. (1984). *K-SOS: Kaufman Sequential or Simultaneous?* Circle Pines, MN: American Guidance Service.

Kemper, M.B., Hagerman, R.J., & Altshul-Stark, D. (1988). Cognitive profiles of boys with fragile X syndrome. *American Journal of Medical Genetics, 30,* 191–200.

Kumin, L. (1994). Intelligibility of speech in children with Down syndrome in natural settings: Parents' perspective. *Perceptual and Motor Skills, 78,* 307–313.

Laws, G. (1998). The use of non-word repetition as a test of phonological memory in children with Down syndrome. *Journal of Child Psychology and Psychiatry, 39,* 1119–1130.

Laws, G., Buckley, S., Bird, G., MacDonald, J., & Broadley, I. (1995). The influence of reading instruction on language and memory development in children with Down's syndrome. *Down's Syndrome: Research and Practice, 3,* 59–64.

Leddy, M. (1999). The biological bases of speech in people with Down syndrome. In J.F. Miller, M. Leddy, & L.A. Leavitt (Eds.), *Improving the communication of people with Down syndrome* (pp., 61–80). Baltimore: Paul H. Brookes.

Liberman, I.Y., Shankweiler, D., Liberman, A.M., Fowler, C., & Fisher, W.F. (1977). Phonetic segmentation and recoding in the beginning reader. In A.S. Reber & D.L. Scarborough (Eds.), *Toward a psychology of reading* (pp. 207–223). Hillsdale, NJ: Erlbaum.

Luria, A.R. (1963). Psychological studies of mental deficiency in the Soviet Union. In N.R. Ellis (Ed.), *Handbook of mental deficiency* (pp. 353–387). New York: McGraw-Hill.

Matalainen, R., Airaksinen, E., Mononen, T., Launiala, K., & Kaariainen, R. (1995). A population-based study on the causes of severe and profound mental retardation. *Acta Pediatrica, 84,* 261–266.

McGuinness, D. (1997). Decoding strategies as predictors of reading skill: A follow-up study. *Annals of Dyslexia, 47,* 117–150.

Morais, J., Alegria, J., & Content, A. (1987). The relationship between segmental analysis and alphabetic literacy: An interactive view. *Cahiers de Psychologie Cognitive, 7,* 415–438.

Morais, J., Cary., L., Alegria, J., & Bertelson, P. (1979). Does awareness of speech as a sequence of phones arise spontaneously? *Cognition, 7,* 323–331.

Morton, J., & Frith, U. (1993). What lesson for dyslexia from Down's syndrome? (Comments on Cossu, Rossini, and Marshall 1993.) *Cognition, 48,* 289–296.

O'Brien, G., & Yule, W. (Eds.). (1995). *Behavioural phenotypes.* London: MacKeith Press.

Opitz, J.M. (1996, March). *Historiography of the causal analysis of mental retardation.* Speech to the 29th annual Gatlinburg Conference on Research and Theory in Mental Retardation and Developmental Disabilities, Gatlinburg, TN.

Powell, L., Houghton, S., & Douglas, G. (1997). Comparison of etiology-specific cognitive functioning profiles for individuals with Fragile X and individuals with Down syndrome. *Journal of Special Education, 31,* 362–376.

Pueschel, S.R., Gallagher, P.L., Zartler, A.S., & Pezzullo, J.C. (1986). Cognitive and learning profiles in children with Down syndrome. *Research in Developmental Disabilities, 8,* 21–37.

Rondal, J. (1995). *Exceptional language development in Down syndrome.* Cambridge: Cambridge University Press.

Rondal, J. (1996). Oral language in Down's syndrome. In J.A. Rondal, J. Perera, L. Nadel, & A. Comblain (Eds.), *Down's syndrome: Psychological, psychobiological, and socio-educational perspectives* (pp. 99–117). London: Whurr Publishers Ltd.

Rosin, M.M., Swift, E., Bless, D., & Vetter, D.K. (1988). Communication profiles of adolescents with Down syndrome. *Journal of Childhood Communication Disorders, 12,* 49–64.

Rutter, M., Simonoff, E., & Plomin, R. (1996). Genetic influences on mild mental retardation: Concepts, findings, and research implications. *Journal of Biosocial Science, 28,* 509–526.

Scarborough, H.S., Rescorla, L., Tager-Flusberg, H., Fowler, A.E., & Sudhalter, V. (1991). The relation of utterance length to grammatical complexity in normal and language-disordered groups. *Applied Psycholinguistics, 12,* 23–45.

Smith, A.C.M., Dykens, E.M., & Greenberg, F. (1998). Behavioral phenotype of Smith-Magenis syndrome. *American Journal of Medical Genetics, 28,* 179–185.

Udwin, O., & Yule, W. (1990). Expressive language of children with Williams Syndrome. *American Journal of Medical Genetics* (Supplement), *6,* 108–114.

Woodcock, R. (1987). *Woodcock Reading Mastery Tests–Revised.* Circle Pines, MN: American Guidance Service.

Zeaman, D., & House, B. (1963). The role of attention in retardate discriminant learning. In N.R. Ellis (Ed.), *Handbook of mental deficiency, psychological theory and research* (pp. 159–223). New York: McGraw-Hill.

Zigler, E. (1967). Familial mental retardation: A continuing dilemma. *Science, 155,* 292–298.

Zigler, E. (1969). Developmental versus difference theories of retardation and the problem of motivation. *American Journal of Mental Deficiency, 73,* 536–556.

Chapter 7

# Visual Variability Discrimination

*Michael E. Young and Edward A. Wasserman*

We examined college students' discrimination of complex visual displays that involved different degrees of variability. The participants' accuracy and reaction-time scores on a two-alternative forced-choice discrimination of multi-picture displays disclosed that people can and do use entropy to classify different levels of visual display variability. Stimulus control by the abstract dimension of entropy (1) was a function of the specific entropy values to be discriminated and (2) may serve as the substrate of adaptive conceptual behavior in both human and nonhuman animals.

In 1785, the English poet William Cowper famously observed that, "variety's the very spice of life." But, just what is variety? How might we detect it? Why is it useful for us to do so? And are humans the only creatures who are able to discriminate variety? In this chapter we will attempt to deal with these and related questions.

## Defining variability

Variety is tantamount to the amount of information in a collection of events. Whether the events in that collection are simultaneously or successively presented, the more different kinds of events, the greater the variety. Specifically, the information–theoretic concept of "entropy" (Shannon & Weaver, 1949) is commonly enlisted to scale variety or randomness.

## Detecting variability

Detecting variety is a behavioral process that surely involves three cognitive activities: discriminating, comparing, and deciding. Detecting variety in sets of

**Figure 7.1**
Examples of 16-Icon Same and Different Arrays Used in Young and Wasserman (1997) and Current Study. These arrays consisted of 16 icons that were chosen from one of two sets of 16 or 24 icons to create Same and Different arrays, with these icons randomly located in a 5 3 5 grid.

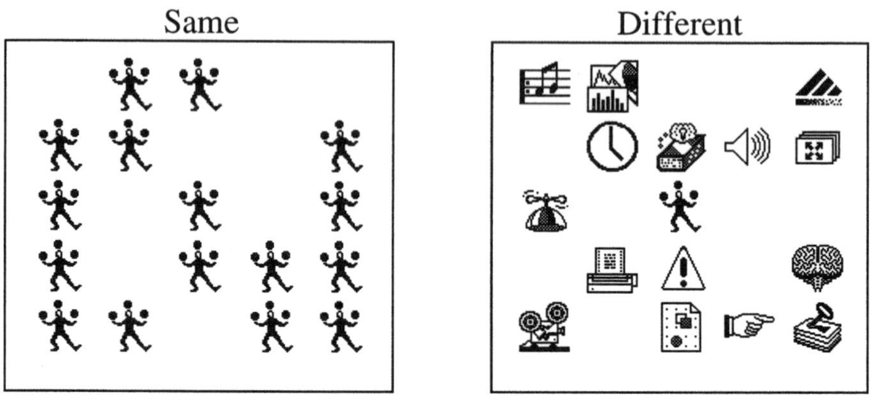


complex displays, like those pictured in Figure 7.1, entails our discriminating two or more of the sixteen items that constitute the display. We must then compare those items with one another to appreciate their relation to one another. Finally, we must decide on the precise degree of variety in the display.

## Utility of variability

Early research on categorization focused on people's ability to classify items as a function of their proximity to the central tendency of a category. Many theories of categorization refer to that central tendency as the category's prototype (Medin & Smith, 1984). All novel instances are assumed to be compared with stored prototypes; category membership is determined by the category of the prototype that is most similar to the novel instance. Categorizing similar things together provides an important cognitive economy, obviating the need to treat each new item as something unique. The utility of detecting item variability is not as readily apparent, but, it is much more important than one might initially surmise.

Posner and Keele (1968, 1970) and Fried and Holyoak (1984) demonstrated that people extract knowledge about the variability, as well as the central tendency, of categories. Highly variable categories are harder to learn, but people are more likely to classify highly distorted novel patterns into a category if the category's training instances are highly variable. This behavior is optimal from a statistical perspective; the likelihood of a novel instance belonging to a category is higher when the trained category comprises more diverse items. Thus, if an instance is equidistant from the prototypes of two categories (one with low variability and one with high variability), the statistically optimal decision is to place that instance within the more variable category. Ashby and Gott (1988) have demonstrated that category variability must be considered to achieve optimal classification.

In addition to the role of variability in improving similarity-based categorization, everyday experience suggests that variability can also be the sole basis of people's decisions. There are situations in which variety is preferred and situations in which uniformity is preferred. For example, when visiting a new city, people often sample a broad range of restaurants. Variety in dining is considered desirable; by sampling a variety of restaurants, the chances are higher of identifying an exceptional establishment. During the travel to that city, however, one often seeks out a McDonalds or a Perkins. These restaurants strive for national (and international) uniformity across their chain, guaranteeing a certain consistency with which we have all become familiar.

The tension between the advantages of diversity and uniformity are prevalent in today's business world. Many corporations have standardized on one computer platform and on one suite of applications. This uniformity eases the transfer of files from machine to machine, simplifies the technical support of those machines, and facilitates the transfer of skills from one position to another. This homogeneity, however, has drawbacks in that it decreases the likelihood of an employee always having the best tool for a particular task. The one-size-fits-all mentality can also undermine an organization's ability to compete most efficiently in a highly flexible consumer market: although a toolbox full of wrenches can help in a variety of circumstances, this tool is of little use when one needs a screwdriver.

Unlike the preferred uniformity among computer platforms, human diversity in the workforce is now considered a desirable commodity. A diverse psychology department (in skills and backgrounds) improves the ability of that unit to meet the needs of its heterogeneous students. This diversity, however, can create difficulty in building a department that works in unison toward the attainment of common goals. The push for conformity among a diverse population recognizes the potential problems with too much variety in an organization.

It should now be clear that the detection of variability is useful in our everyday lives and that it is the basis of important decisions that people make in a wealth of practical situations, so humans' ability to detect the variety that is pictured in the displays of Figure 7.1 should come as no surprise to the discriminating reader. What may be surprising is that pigeons, too, detect pictorial variety. We have reported (Young & Wasserman, 1997) that pigeons' discrimination of variety almost perfectly accords with the measured entropy in the depicted stimuli. Because this pigeon research motivated the experimental design used in the current study, we will first provide an overview of the work on the pigeon's discrimination of variety in visual displays. We will then present a metric for quantifying variability—entropy—and demonstrate that people's variability discrimination is largely a function of entropy, too.

### Variability and same–different discrimination

Human beings and other animals are constantly confronted with an extraordinarily complex array of external stimuli. Yet, sense is somehow made of these varied stimuli. One way of reducing the demands on an organism's sensory and information-processing systems is for it to treat similar stimuli as members of a single class; by doing so, substantial cognitive economy can be achieved, thus freeing its adaptive machinery to deal with other competing exigencies of survival. Conceptualization also permits an organism to treat novel stimuli as members of a particular class and to generalize knowledge about that class to these new members; then, an organism need not be bound to respond to only those stimuli with which it has had prior experience, thereby enhancing its ability to cope with a continually changing world.

Although theorists have often extolled these adaptive virtues of conceptualization, we remain far from understanding exactly how organisms process stimuli to partition the world into classes of related objects and events. Indeed, given early writings on the subject, we should wonder whether nonhuman animals are even capable of conceptual behavior. A century ago, C. Lloyd Morgan denied animals the ability to behave conceptually. To do so, he said, requires that we "neglect all that is variable and focus the attention on the uniform relation. [Then] we have reached a conception, and this conception is not concrete, particular, and individual, but abstract, general, and of universal application" (1896, p. 263). Morgan believed that only adult humans and not even children are capable of conceptualization.

Several recent lines of research are radically changing that initial opinion (Wasserman, 1995). One of those new lines of research concerns the acquisition of the same–different concept by animals. The classification of two or more items as the same as or different from one another requires a level of abstract conceptualization previously thought to be unique to human beings.

Evidence of this ability in nonhuman primates is now substantial (e.g., Premack, 1976; Wright, Santiago, Urcuioli, & Sands, 1983); however, early evidence of same–different conceptualization in the pigeon was promising, but weak. Indeed, early attempts to train pigeons to discriminate same from different stimuli led to some rather pessimistic conclusions:

> Because [a Same–different concept] transcends the items themselves there is a possibility that it may be a difficult concept to learn. There is the possibility that monkeys can abstract the task and learn the Same–different concept whereas pigeons cannot. (Wright et al., 1983, p. 316)

Most of the prior research on abstract conceptualization in animals has involved the classification of just two visual items as the same as or different from one another (e.g., Edwards, Jagielo, & Zentall, 1983; Santiago & Wright, 1984; Wright, Santiago, Sands, Kendrick, & Cook, 1985; Wright et al., 1983). Two items represent the smallest number necessary for a same–different discrimination. But it is also possible, and perhaps even easier, to make same–different judgments with displays involving more than two items.

Wasserman, Hugart, and Kirkpatrick–Steger (1995) reported an experiment in which pigeons were taught to discriminate displays involving sixteen items, rather than two items, that were either all the same as one another or all different from one another. The displays were 4 × 4 arrays of sixteen computer icons that were chosen from a set of 16 possible icons; the Same displays involved a single randomly selected icon repeated sixteen times, whereas the Different displays involved all 16 icons in one of sixteen different spatial configurations. Pigeons received food pellet reinforcement for pecking one of two buttons ("same") in the presence of Same displays and for pecking the other button ("different") in the presence of Different displays. This same–different discrimination was readily acquired and it prompted strong transfer to novel 16-item displays; accuracy on trained displays averaged 83 percent correct after approximately 13,000 training trials, and 71 percent correct on new displays that were created from a set of 16 untrained icons.

Young and Wasserman (1997) elaborated on the basic design of Wasserman et al. (1995) by randomly locating the 16 items within a 5 × 5 array of possible locations (for examples of the Same and Different displays used in their study, see Figure 7.1), effectively making all of the icon configurations in training and testing trial-unique. Acquisition of the same–different discrimination was very rapid and it prompted strong transfer to novel 16-item displays; accuracy on training displays averaged 93 percent correct after 4,800 trials and 79 percent correct on new testing displays that were created from untrained icons.

The stronger transfer to novel displays that was observed with stimuli that were even more complex than those used in most prior research raised the intriguing possibility that the larger number of items actually aided the pigeon in discriminating Same from Different displays and in acquiring an abstract same–different concept. Young, Wasserman, and Garner (1997) showed that this possibility is a reality. In tests of displays comprising fewer items—both after initial training with 16-item displays (Experiment 1) and during acquisition (Experiment 2)—the pigeon's ability to discriminate Same from Different displays deteriorated as the number of items was reduced. Why?

The results of several additional experiments pointed to display variability as the controlling dimension in the pigeon's discrimination of Same from Different displays. After discriminative same–different training with 16-icon displays, pigeons in Experiments 2, 3, and 4 of Young and Wasserman (1997) exhibited clear sensitivity to display variability; discriminative performance with arrays involving intermediate degrees of variability was systematically graded. As an example, the arrays used in Experiment 3 of that study are shown in Figure 7.2; the corresponding performance profile is shown in Figure 7.3. The mD/nS notation designates a display comprising *m* Different icons and *n* Same icons; as the number of different icons in a display increases (and the number of same icons in a display decreases), the variability of the display increases. Ar-

**Figure 7.2**
Mixture Arrays Used in Experiment 3 of Young and Wasserman (1997). The mD/nS notation indicates the presence of *m* different icons and *n* same icons.

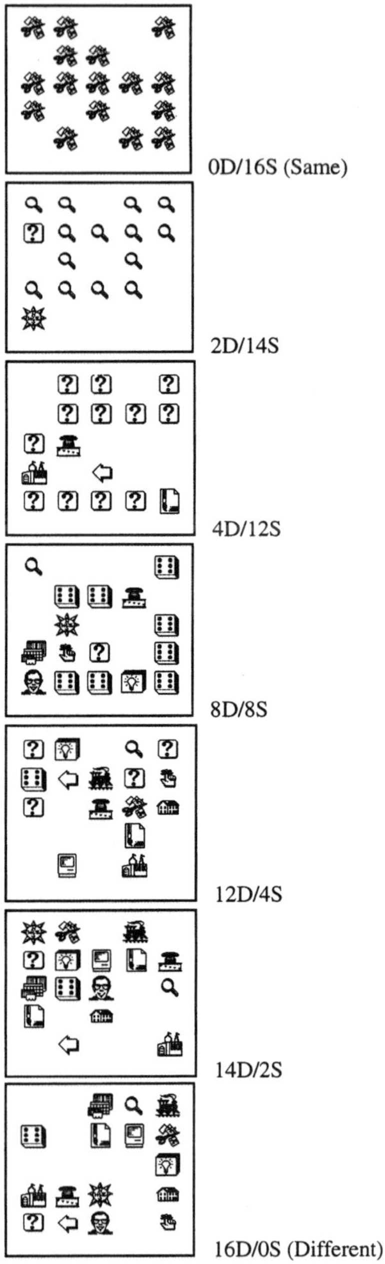

**Figure 7.3**
Response Pattern Observed in Experiment 3 of Young and Wasserman (1997) for Mixture Arrays Illustrated in Figure 7.2

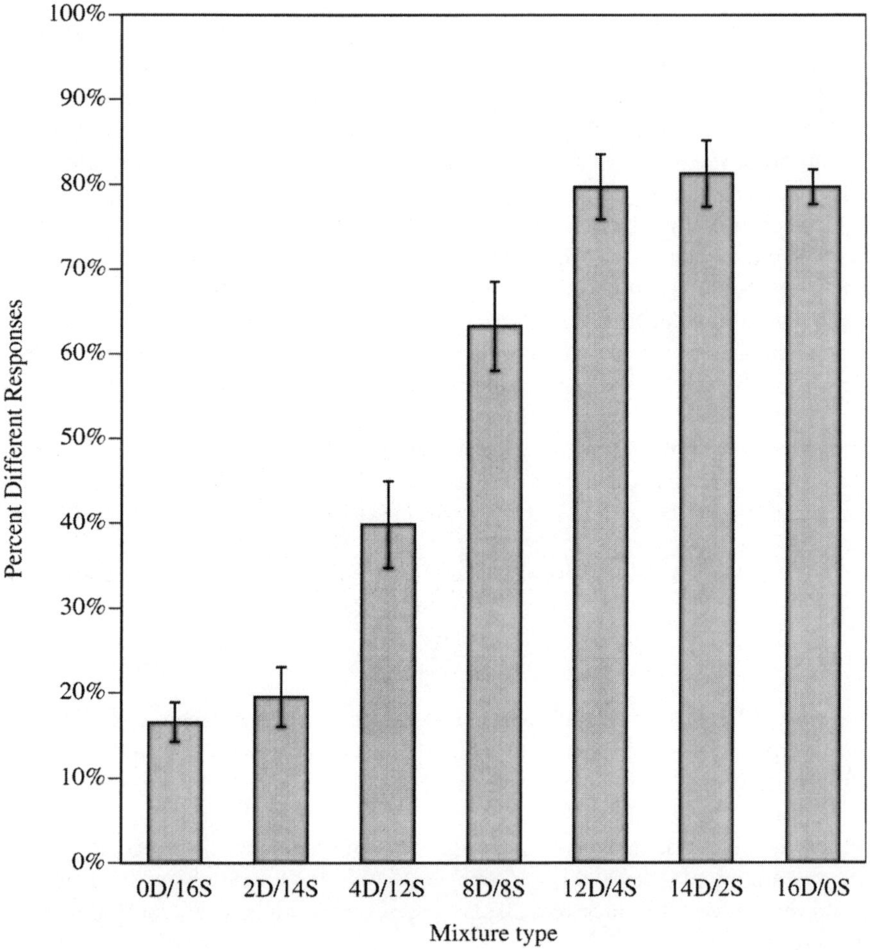


rays with high degrees of variability were very likely to be reported as different, whereas those with decreasing degrees of variability were increasingly likely to be reported as same.

## Quantifying variability

Variability in a continuous variable is often quantified using variance. A 16-icon display, however, represents a frequency distribution of a *categorical* variable—icon type. For each display, there are 16 possible icon types, each of which may have a frequency ranging from 0 (not present) to 16 (the only type present). When a pigeon responds to an array with an intermediate degree of variability, it must determine whether the frequency histogram represented by that array is more similar to those represented by the Same arrays or to those represented by the Different arrays. Intuitively, we recognize that the Same arrays exhibit the lowest possible variability and that the Different arrays exhibit the highest possible variability. But, intuition is a very poor way to quantify the variability of the wide range of possible variability levels. Information theory (Shannon & Weaver, 1949) offers a single metric—entropy—that elegantly accomplishes that end.

Entropy measures the amount of variety or diversity in a categorical variable by a weighted average of the number of bits of information that are required to predict each of the categories of the variable. Rare- or low-frequency categories carry a great deal of information (i.e., they are very important), whereas common categories carry very little information (i.e., they are less important). Prediction of the category of a variable that is observed to have only one value is easy and requires no information: entropy is zero. When all of the observed categories have a frequency of one, entropy is maximal. Our Same and Different arrays represent these two endpoints of the entropy dimension: the Same arrays have no entropy and the Different arrays have maximal entropy for sixteen observed categories.

To quantify entropy, we used the following equation (Shannon & Weaver, 1949):

$$H(A) = -\sum_{a \varepsilon A} p_a \log_2 p_a \qquad (1)$$

where $H(A)$ is the entropy of categorical variable $A$, $a$ is a category of $A$, and $p_a$ is the proportion of observed values within that category. When a display has 16 identical icons, there is only one category with a probability of occurrence of 1.0. Because $\log_2$ (1.0) = 0.0, the entropy of the Same displays is 0.0. The Different displays consist of one occurrence of each of 16 icons or categories, yielding an entropy of $-.0625 \times \log_2(.0625) \times 16$, or 4.0. During testing, the pigeons could thus have responded to a display based on whether its entropy was closer to 0.0 or to 4.0.

When Young and Wasserman (1997) fit their pigeons' performance to the entropy measure, the fit was remarkably good (Figure 7.4); it accounted for 95 percent of the variance in mean choice behavior across three separate experiments (each involving four pigeons) in which entropy was changed in three entirely different ways. When entropy was compared to many other possible predictors (the number of icon types in a display, the frequency of the most common icon type, the frequency of the least common icon type, and the mean frequency across the icon types), it was clearly superior and eclipsed any explanatory power offered by these other plausible predictors.

Interpreting same–different conceptualization as entropy detection also explained the pigeons' superior discrimination of 16-item Same from Different

**Figure 7.4**
Scaled Logit Transforms of the Percentage of Different Responses Across Experiments 2, 3, and 4 as a Function of Stimulus Display Entropy (As Reported by Young & Wasserman, 1997)

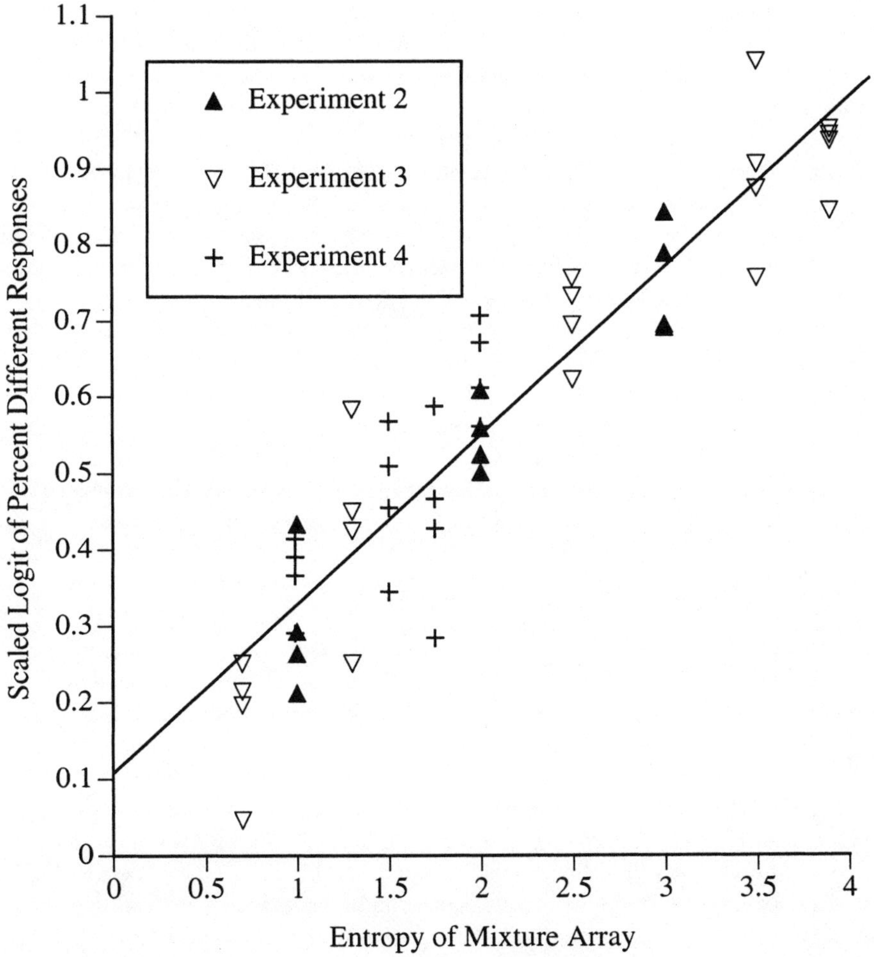


arrays compared to its inferior discrimination of arrays comprising fewer items. When pigeons are trained to discriminate 16-icon Same from Different arrays, they learn to make one response to displays with an entropy of 0.0 and another response to displays with an entropy of 4.0. During testing, a bird would then be expected to distribute its responses to novel arrays as a function of their entropy; displays with entropies closer to 0.0 should be more likely to be reported as same, whereas those with entropies closer to 4.0 should be more likely to be reported as different. The entropy of Same displays is 0.0 regardless of the number of items within the display; there is only one category in the display, no matter how many items are present. Thus, 2-item Same displays should readily be reported as "same" after initial training with 16-item Same and Different displays. The entropy of Different displays, however, is reduced as the number of display items is decreased because the display comprises fewer and fewer categories. Thus, 2-item Different displays (which have an entropy of 1.0) should be judged to be more similar to 16-item Same displays (which have an entropy of 0.0) than to 16-item Different displays (which have an entropy of 4.0). These predictions were confirmed in Experiment 1 of Young et al. (1997). After initial training with 16-item Same and Different displays, the pigeons' responses to Same displays were unchanged across a range of smaller display sizes, whereas Different displays were increasingly likely to be reported as same as the number of items was reduced; indeed, 2-item Different displays were consistently and otherwise inexplicably reported as same.

The smaller entropy difference between 2-item Same and Different displays (0.0 vs. 1.0) and the larger entropy difference between 16-item Same and Different displays (0.0 vs. 4.0) also led us to expect that acquisition of the same–different discrimination would be more difficult with fewer items. As noted earlier, in Experiment 2 of Young et al. (1997), we found just such an effect of item number, representing important and sometimes unexpected confirmations of the pigeons' use of entropy in their classification of complex visual displays.

## Humans' sensitivity to variability

Because we were unaware of any empirical evidence on the matter, in the current experiment we sought to determine whether humans too would be sensitive to the variability in a collection of visual items. Our goal was to quantify this sensitivity and ascertain whether this sensitivity would be similar to that observed in pigeons.

Prior work on the formation of basic-level categories (Ashby & Gott, 1988; Fried & Holyoak, 1984; Posner & Keele, 1968; 1970) demonstrated that people extract knowledge about the variability and central tendency of categories. Each of these prior studies showed that knowledge of variability is critical to the extraction of optimal, or even good, discriminative performance in a categorization task. The only quantitative examinations of the effects of variability, however, have been in the service of categorization tasks in which continuous features were varied (e.g., height, width, angle). Continuous-feature variability is easily quantified by variance, but categorical-feature variability (e.g., icon type, has-feathers, hair style) must be quantified using entropy or an analogous measure. Unlike the prior studies documenting the effects of variability on a central-tendency discrimination task, the categorization task that we used in the current study defines category membership solely as a function of a display's variability, not its central tendency (which was the same across trials with our Same and Different arrays).

Although the Same and Different arrays of Figure 7.1 obviously differ in their variability (representing the two endpoints of a variability continuum), we were interested in whether people's sensitivity to variability would be a function of entropy. Although pigeons demonstrated sensitivity to the absolute entropy of a display, we suspected that people might be preferentially sensitive to the relative entropy of a display. Absolute entropy is a function of both the variability and the number of items in a display; although the absolute entropy of Same arrays, 0.0, remains constant as the number of items increases, the absolute entropy of Different arrays increases as the number of items increases. Sensitivity to absolute entropy explained the pigeons' asymmetrical response pattern as the number of items in a display was reduced; this manipulation affected the pigeons' discriminative performance on Different trials, but not on Same trials.

In contrast to absolute entropy, relative entropy does not vary with the number of items in a display; it scales the variability in a collection of items relative to the maximum variability that is possible given that number of items. Formally, relative entropy is the absolute entropy of a display divided by the maximum possible absolute entropy, given that number of items; relative entropy is 0.0 when variability is minimal and 1.0 when variability is maximal. For example, if a display contains four items that are identical, then that display has an absolute entropy of 0.0. Given that the maximum absolute entropy for four items is 2.0 when all of the items are different (by Equation 1), four-item Same arrays have a relative entropy of 0.0 / 2.0 = 0.0. In contrast, if a display contains four items that are all different, then that display has an absolute entropy of 2.0. Given that the maximum absolute entropy for four items is also 2.0, this four-item Different array has a relative entropy of 2.0 / 2.0 = 1.0. We considered it most improbable that people would be increasingly likely to classify Different arrays as "same" when the number of items was reduced from sixteen to two (as required by a dependence on absolute entropy); rather, we believed that people would be sensitive to relative entropy, classifying all of the Same arrays as same and all of the Different displays as different, independent of the number of items in the display.

We conducted an experiment designed to disclose people's classification of display variability. Recent research by Young and Wasserman (2001) revealed that when people were trained to discriminate Same from Different (i.e., using the same general training procedure used with pigeons), tests involving arrays of intermediate variability revealed that most people (80%) tended to categorically discriminate Same arrays from arrays with any level of variability (i.e., participants were highly likely to choose "different" for all of the mixtures). A small proportion of the participants (20%) in this study, however, clearly responded to mixture arrays as a function of their entropy. To encourage participants to respond in a less categorical fashion, in the present study we assessed the degree of dimensional control that might be exerted by entropy by training the participants to discriminate 16-icon displays with an entropy less than 2.0

from 16-icon displays with an entropy greater than 2.0. Subsequent testing involved 2-, 4-, 8-, and 12-icon Same and Different arrays to distinguish the participants who rely on absolute entropy from those who rely on relative entropy.

## METHOD

### Participants

A total of sixty students enrolled in an introductory psychology course at The University of Iowa served as voluntary participants. They received one course credit for their participation.

### Procedure

#### *Visual stimuli*

The participants were trained and tested with arrays of icons that were chosen from a set of 24 icon types. The training arrays had the following eight levels of entropy: 0.0, 0.5, 1.0, 1.5, 2.5, 3.0, 3.5, and 4.0, all of which comprised 16 icons. Dimensional control by entropy predicts that participants will discriminate displays with entropy scores less than 2.0 from displays with entropy scores greater than 2.0 and that the discrimination of displays with more extreme entropies (e.g., 0.0 and 4.0) will surpass that of the displays with more intermediate entropies (e.g., 1.5 and 2.5).

We also included testing with Same and Different arrays comprising fewer icons (either 2, 4, 8, or 12 icons) to separate absolute entropy from relative entropy. Table 7.1 shows the array compositions and their absolute entropy, relative entropy, and number of icon types.

In producing arrays with entropies of 1.0 or 3.0, we used three different methods. This maneuver insured that a subset of the training arrays would have the same entropy, but that they would differ with respect to the number of icon types, thus allowing us to differentiate control by entropy from control by the number of icon types (a possible confounding variable). The icon distributions for the three types of Entropy 1.0 and the three types of Entropy 3.0 training arrays are also shown in Table 7.1 (marked by an asterisk) and exemplified in Figure 7.5. In Method 1 (top row of Figure 7.5), each type of icon appeared an equal number of times. In Method 2 (middle row of Figure 7.5), one of the types of icons appeared three times and one other was allowed to appear more than once, whereas the others could each appear only once. In Method 3 (bottom row of Figure 7.5), one of the types of icons was allowed to appear more than once, whereas the others could each appear only once. We adjusted the number of types of icons used in each method to achieve the appropriate entropy level.

**Table 7.1**
Statistical Attributes of the Mixture Arrays Used

| Icon Distribution | Absolute Entropy | Relative Entropy | Number of Icon Types |
|---|---|---|---|
| 16A | 0.0 | 0.0 | 1 |
| 14A2B | 0.5 | 0.1 | 2 |
| *12A3B1C | 1.0 | 0.2 | 3 |
| *13A1B1C1D | 1.0 | 0.2 | 4 |
| *8A8B | 1.0 | 0.2 | 2 |
| 11A1B1C1D1E1F | 1.5 | 0.4 | 6 |
| 4A4B4C4D | 2.0 | 0.5 | 4 |
| 8A1B1C1D1E1F1G1H1I | 2.5 | 0.6 | 9 |
| *6A1B1C1D1E1F1G1H1I1J1K | 3.0 | 0.8 | 11 |
| *2A2B2C2D2E2F2G2H | 3.0 | 0.8 | 8 |
| *5A3B1C1D1E1F1G1H1I1J | 3.0 | 0.8 | 10 |
| 4A1B1C1D1E1F1G1H1I1J1K1L1M | 3.5 | 0.9 | 13 |
| 1A1B1C1D1E1F1G1H1I1J1K1L1M1N1O1P | 4.0 | 1.0 | 16 |
| Same Arrays | | | |
| 2 icons | 0.0 | 0.0 | 1 |
| 4 icons | 0.0 | 0.0 | 1 |
| 8 icons | 0.0 | 0.0 | 1 |
| 12 icons | 0.0 | 0.0 | 1 |
| Different Arrays | | | |
| 2 icons | 1.0 | 1.0 | 2 |
| 4 icons | 2.0 | 1.0 | 4 |
| 8 icons | 3.0 | 1.0 | 8 |
| 12 icons | 3.6 | 1.0 | 12 |

*Note:* A letter designates a randomly chosen icon type that was different from those designated by the other letters within the string. The number preceding a letter indicates that the icon type occurred that number of times. The training arrays are denoted by an asterisk.

### *Training*

The participants entered the experimental room and were seated at one of four identically configured Macintosh PowerPC 7100/80 computers. The experimenter read the participants a set of general instructions, handed each of the participants a set of written instructions, and asked the participants to put on a pair of headphones connected to the computer (the headphones were used to present auditory feedback). In addition to the text explaining the mechanics

**Figure 7.5**
Examples of the Entropy 1.0 vs. Entropy 3.0 Training Arrays. The three methods that were used for achieving these levels of entropy are discussed in the text and are outlined in Table 7.1.

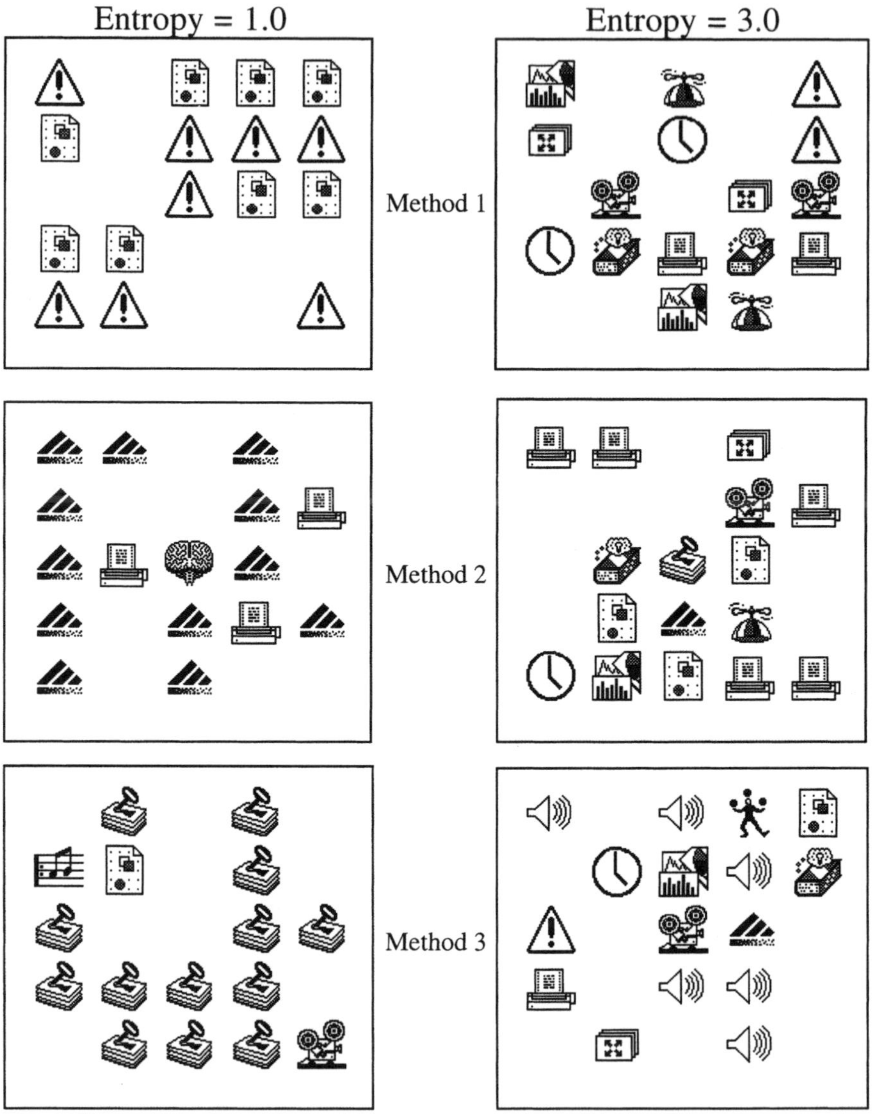

of the experiment and exhorting the participants to make their choices as quickly as possible while still being accurate, the participants read the following introductory paragraph:

> You will be observing a series of displays and attempting to learn which response is correct for each display. You will make your best guess as to which of *two* responses is correct. You will then be provided feedback in the form of an auditory tone (if correct) or a flash of the screen (if incorrect). This information will assist you in improving your chances of being correct for subsequent displays. Your goal is to accurately predict the correct response for each display by the end of training.

No information was provided that could have directed the participant toward any particular aspect of the displays. At no point in the instructions were the words "same" or "different" used. Once each participant indicated an understanding of the procedure, the experimenter started each of the programs.

The training program comprised 144 trials: 6 randomized blocks of 24 trials each, 3 of each of the entropy levels, 0.0, 0.5, 1.0, 1.5, 2.5, 3.0, 3.5, and 4.0. The Entropy 1.0 arrays and Entropy 3.0 arrays were produced using three different methods; each of the arrays comprising levels of entropy other than 1.0 and 3.0 were produced using a single method (see Table 7.1). For half of the participants, the "1" key was correct for low-variability arrays and the "3" key was correct for high-variability arrays; for the other half of the participants, the key assignments were reversed.

### *Testing*

After 144 training trials, the testing period began. The session continued without a noticeable change, but the testing arrays were randomly interspersed among the training arrays at a relatively low rate. During the testing phase, 3 randomized blocks of 80 trials were given consisting of 9 of each of the trained entropy levels [0.0, 0.5, 1.0, 1.5, 2.5, 3.0, 3.5, and 4.0], with the Entropy 1.0 and Entropy 3.0 arrays again generated using three different methods: 4 Same arrays comprising fewer than 16 icons (either 2, 4, 8, or 12 icons), and 4 Different arrays comprising fewer than 16 icons (either 2, 4, 8, or 12 icons). Differential feedback continued for trials involving training arrays. Nondifferential feedback was given on testing trials; the participants received "correct" feedback regardless of their response.

### *Questionnaire*

After the training and testing periods, the participants were asked to complete a questionnaire. We requested demographic information (age, gender, handedness, grade-point average (GPA), and Ameican College Test (ACT) score) and asked three questions: (a) Describe any characteristics of the displays that you used in determining which of the two responses was correct, (b) Did you notice that some of your choices were correct no matter how you responded? (yes/no options were provided), and (c) If so, then please describe the situations under which your choices were always correct.

## RESULTS

A large proportion of the participants mastered the discrimination within the 144 allotted trials. We chose an inclusion criterion of 75 percent correct during the final 24-trial block of training; any participant failing to meet this criterion was dropped from subsequent analyses. Of the 60 participants, 44 (or 73%) met criterion. Among those who were deemed to have mastered the discrimination, learning was rapid and strong; during the final two blocks of training, these participants averaged 88 percent correct. All subsequent analyses excluded the 144-trial training period.

We noticed that there were substantial individual differences in discriminative responding. These differences were seen exclusively on the Different displays comprising fewer icons (no other differences were readily apparent). To separate the different response profiles, we performed a cluster analysis of each participant's responses on the Different arrays and focused on the largest two-cluster division; 20 participants were in the Absolute Entropy cluster (demonstrating a sensitivity to the number of icons in the display) and 24 participants were in the Relative Entropy cluster (demonstrating an insensitivity to the number of icons in the display). We used only the Different arrays in our cluster analysis.

As shown in Figure 7.6, the difference between the clusters of participants was substantial. The participants in the smaller cluster (Absolute Entropy, $n$ = 20) were more likely to choose "same" for the Different arrays as the number of icons was reduced, but they consistently chose "same" for the Same arrays regardless of icon number. The Absolute Entropy participants also exhibited strong sensitivity to the full range of variability on the Mixture displays; as the Mixture was changed from mostly Same to mostly Different, reports changed from mostly "same" responses to mostly "different" responses. This behavior very closely parallels that observed in pigeons (Young et al., 1997; Young & Wasserman, 1997).

The participants in the larger cluster (Relative Entropy, $n$ = 24) exhibited a very different response profile (Figure 7.6). These participants were largely unaffected by the number of icons in the Same and Different displays, but they were highly sensitive to the full range of variability that was present in the Mixture displays.

To confirm these observations, we first examined the participants' performance on Same and Different testing arrays. A $2 \times 2 \times 5$ repeated-measures Analysis of Variance (ANOVA) of percentage of different responses as a function of Cluster (Absolute Entropy vs. Relative Entropy), Array Type (Same vs.

**Figure 7.6**
Mean Percentage of Different Responses for the Same, Different, and Mixture Arrays. The response patterns are separately portrayed for the Absolute Entropy and Relative Entropy participant clusters.

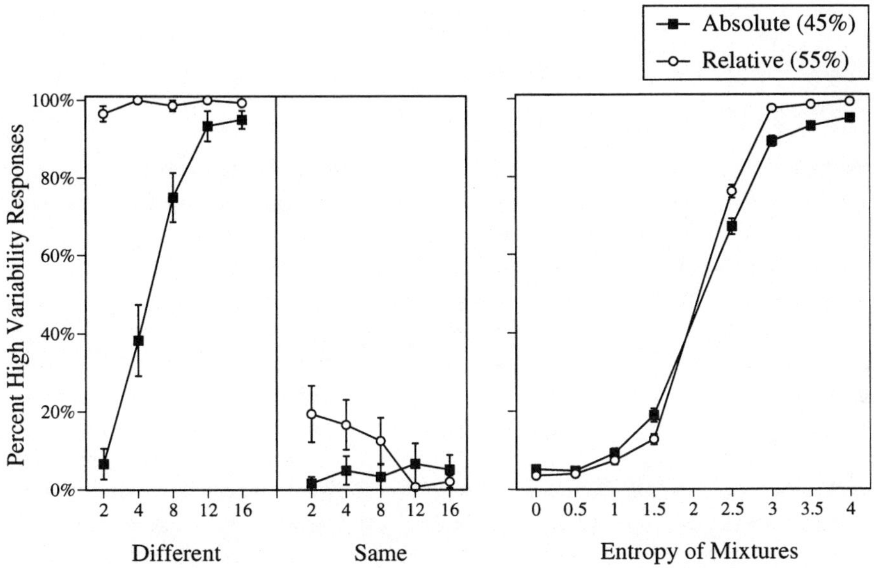


Different), and Number of Icons (2, 4, 8, 12, and 16) revealed that all of the main effects and interactions were statistically significant (all *ps* $< .0001$). Most importantly, the three-way Cluster × Array Type × Number of Icons interaction was significant, $F(4,170) = 20.33$, $p < .0001$, as evidenced by the differential effects of icon number in Same and Different displays for the two clusters of participants (left panel of Figure 7.6).

In a separate analysis, we considered the participants' performance on the Mixture arrays during the testing phase. A 2 × 8 repeated-measures ANOVA of the percentage of different responses as a function of Cluster (Absolute vs. Relative) and Mixture Entropy (0.0, 0.5, 1.0, 1.5, 2.5, 3.0, 3.5, 4.0) revealed no main effect for Cluster, $F < 1$, $p > .10$, a significant main effect for Mixture Entropy, $F(7,295) = 709.66$, $p < .0001$, and a significant Cluster × Mixture Entropy interaction, $F(7,295) = 3.39$, $p < .01$. Planned orthogonal tests ($\alpha = .05$) comparing the Absolute Entropy cluster of participants to the Relative Entropy cluster of participants at each of the tested entropy values revealed that the Absolute Entropy cluster exhibited slightly weaker learning than the Relative Entropy cluster; the Relative Entropy cluster produced slightly more low-variability responses for the Entropy 1.5 arrays (*M* difference = 7.3%) and slightly more high-variability responses for the Entropy 2.5 arrays (*M* difference = 9.2%) and the Entropy 3.0 arrays (*M* difference = 8.5%). An identical analysis of the training data revealed no significant effects involving

Cluster, although the Cluster × Mixture Entropy interaction approached significance.

In examining reaction times for the Same and Different arrays with fewer than 16 icons, the reaction times (RTs) grew progressively longer as the number of icons was reduced (left panel of Figure 7.7), with the effect of icon number being larger for the Relative Entropy cluster than for the Absolute Entropy cluster. Different arrays produced longer RTs than Same arrays; this difference was larger for the Absolute Entropy cluster (Different $M$ = 680 ms, Same $M$ = 508 ms) than for the Relative Entropy cluster (Different $M$ = 692 ms, Same $M$ = 595 ms).

RTs for the Mixture arrays (during the testing phase) exhibited a systematic pattern with the longest RTs occurring in the middle of the entropy range (Entropy 2.5 produced the longest mean RT, $M$ = 902 ms) and progressively shorter RTs as the entropy of an array approached the endpoints of the entropy scale (0.0 and 4.0), as shown in the right panel of Figure 7.7. There was a decided asymmetry in the relationship between entropy and RT; the RTs for arrays with entropies greater than 2.0 (the midpoint of the 0.0 to 4.0 entropy range) were longer than those with entropies less than 2.0. This disparity is also evident at the endpoints, where the Same arrays produced much shorter RTs ($M$ = 557 ms) than the Different arrays ($M$ = 655 ms). The two clusters of participants evidenced similar RT trends across changes in entropy; the only difference between the clusters was that the Relative Entropy cluster had longer RTs in the middle of the distribution than did the Absolute Entropy cluster.

To confirm these observations, we first examined the participants' RTs on the Same and Different testing arrays. A 2 × 2 × 5 repeated-measures ANOVA of average log(RT) (averaged across trials) as a function of Cluster (Absolute vs. Relative), Array Type (Same vs. Different), and Number of Icons (2, 4, 8, 12, and 16) revealed no effect for Cluster, $F = 1.57$, $p > .10$, a significant main effect for Array Type, $F(1,42) = 53.99$, $p < .0001$, and a significant main effect for Number of Icons, $F(4,168) = 45.96$, $p < .0001$. The effect of icon number differed across array types, with Different arrays being more sensitive to changes in icon number than Same arrays, as confirmed by a significant Array Type × Number of Icons interaction, $F(4,168) = 3.51$, $p < .01$. The effect of icon number also differed across the two clusters of participants; the Relative Entropy cluster was more sensitive to changes in icon number than was the Absolute Entropy cluster, as confirmed by a significant Cluster × Number of Icons interaction, $F(4,168) = 3.09$, $p < .05$. Finally, the difference between the Different and Same array RTs was larger for the Absolute Entropy cluster of participants ($M$ difference = 172 ms) than for the Relative Entropy cluster of participants ($M$ difference = 97 ms), as confirmed by a significant Cluster × Array Type interaction, $F(1,42) = 5.50$, $p < .05$. The three-way interaction was not statistically significant ($p > .10$).

We next considered the participants' RTs for the Mixture testing arrays. A 2 × 8 repeated-measures ANOVA of log(RT) as a function of Cluster (Absolute

**Figure 7.7**
Mean Reaction Times for Same, Different, and Mixture Arrays. The RTs are separately portrayed for the Absolute Entropy and Relative Entropy participant clusters.

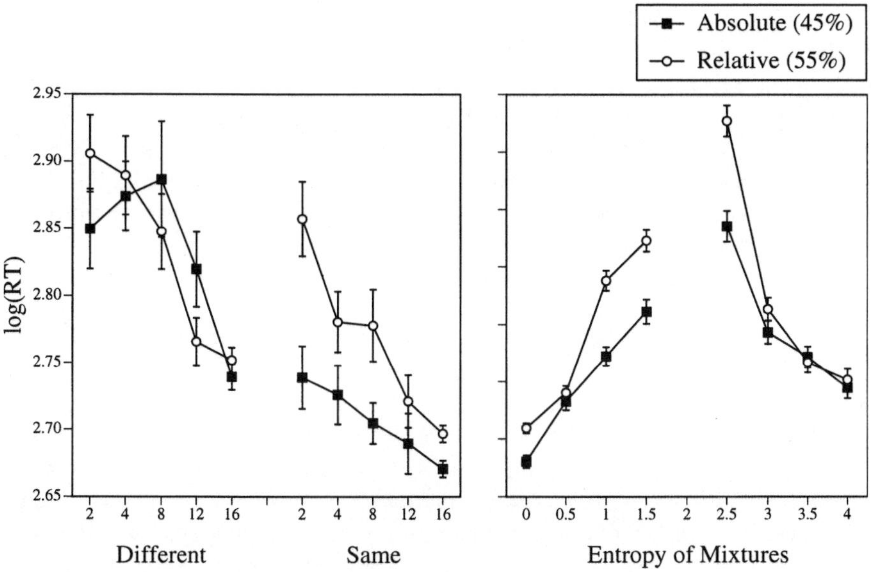


vs. Relative) and Mixture Entropy (0.0, 0.5, 1.0, 1.5, 2.5, 3.0, 3.5, and 4.0) revealed no significant main effect for Cluster, $F(1,42) = 2.10$, $p > .10$, a significant main effect for Mixture Entropy, $F(7,294) = 48.54$, $p < .0001$, and a significant Cluster $\times$ Mixture Entropy interaction, $F(7,294) = 2.73$, $p < .01$. Planned comparisons ($\alpha = .05$) of the Cluster differences revealed that the Relative Entropy cluster of participants had longer RTs than the Absolute Entropy cluster of participants for arrays with an entropy of 1.0 (*M* difference = 96 ms), 1.5 (*M* difference = 114 ms), and 2.5 (*M* difference = 180 ms).

In our final set of analyses, we examined the demographic data to determine (a) the participants' reported strategies, (b) whether the participants noticed that some trials were always correct, and (c) if there were any participant characteristics that could predict his or her cluster (Absolute Entropy vs. Relative Entropy). Of the 44 participants who learned the task, we classified 39 as having reported a variability strategy, 2 as having reported an individual icon strategy, and 3 as having reported "other" (e.g., "if on the right of screen, choose key three"). A total of 39 percent of the participants reported noticing that some of the trials were always correct on at least one trial, although their retrospective reports were not accurate as to which trials were always correct and the participants typically noticed the nondifferential feedback on only a few trials.

Neither age, gender, handedness, GPA, ACT score, nor whether the individual noticed that responding on some trials was always correct was predictive of whether a participant would learn the task nor, among the learners, his or her cluster (all *ps* $> .10$). Those participants who reported a strategy based on variability were more likely to be among the learners ($M = 91\%$) than those who reported another strategy ($M = 29\%$), $F(2,57) = 18.75$, $p < .0001$. Because classification as a "learner" required a high level of accuracy (75%) by the end of the training phase, it is possible for someone to have eventually learned the task, but not to have been classified as a learner.

Among the learners, there were differences in the strategies reported by the Relative Entropy and Absolute Entropy clusters, $\chi^2(2,41) = 8.66$, $p < .05$. Most notably, participants in the Relative Entropy cluster were more likely to have reported a variability strategy (100%) than were those in the Absolute Entropy cluster (75%).

## DISCUSSION

The participants who were trained to discriminate arrays with an entropy less than 2.0 from arrays with an entropy greater than 2.0 were highly likely to learn the discrimination and to use entropy as the basis of their discriminative performance. As observed in studies involving perceptual dimensions like luminosity, the accuracy of our participants' responses was highest for values distant from the categorical boundary (in this case, an entropy of 2.0) and lowest for values close to that boundary. The participants appear to have relied on entropy as the basis of their discriminative performance, with the majority (55%) using relative entropy and the minority (45%) using absolute entropy.

The observed RTs corroborate the accuracy data; the participants had more difficulty (i.e., longer RTs) discriminating arrays with entropies close to the category boundary. Interestingly, we observed that displays with low entropy generated shorter RTs than those with high entropy. There was a notable exception; when the number of icons in a display was reduced, RTs tended to increase even when the absolute entropy was constant (in the Same arrays) or decreased (in the Different arrays). In the absence of any evidence to the contrary, we believe this finding is the result of a novelty gradient in which RTs increase for novel displays.

The consistently shorter RTs for lower entropy arrays suggest the operation of a basic cognitive process whose duration of action is a direct function of display entropy. This process functions in concert with other mechanisms known to affect RTs. For example, RTs increase as response uncertainty increases near categorical boundaries (Ashby, Boynton, & Lee, 1994) and with stimulus novelty (Nosofsky, 1991; cf. Ashby et al., 1994). These three factors—longer RTs for higher entropy arrays, longer RTs at category boundaries, and longer RTs for more novel items—appear jointly to determine the RTs observed in the current experiment.

In our attempt to identify the effective discriminative dimension for the two clusters of participants, we considered absolute entropy, relative entropy, and the number of icon types as predictors of participant behavior. Because the relationships between the predictors and the dependent variable tended to be sigmoidal (e.g., see Figure 7.6), we standardized the data (Heinemann, Avin, Sullivan, & Chase, 1969) and fit the standardized scores using the following equation:

$$\frac{1}{1+e^{-\mathit{gradient} \bullet (\text{predictor} + \mathit{boundary})}} \qquad (2)$$

Each predictor was first standardized to lie on a scale of 0.0 and 1.0 to insure valid comparisons of the best-fitting parameters across predictors. The resulting function was fit to the data using nonlinear regression (specifically, the Gauss–Newton iterative procedure). Because iterative procedures can fail to converge for some sets of initial parameter values, we maximized the likelihood of convergence by initializing the parameters to plausible values after visual inspection of the data.

The values of the best–fitting parameter scores reflect the participants' classificatory behavior (see Figure 7.8 for an example). When the gradient of the best-fitting sigmoid was high, the participants exhibited a sharp distinction between the categories associated with the two keys; when the gradient of the best-fitting sigmoid was low, the participants exhibited a gradual distinction between the categories. The boundary parameter represents the point of maximum inflection for the sigmoid; thus, a boundary of 0.5 would indicate that the point of maximum inflection was at the middle value within the range of scaled predictor values [0.0, 1.0] and any deviation from 0.5 would indicate that the categorical boundary is shifted away from this theoretical midpoint. The degree of control is an index of how strongly discriminative behavior was under the control of the predictor; this aspect was factored out by the standardization procedure (Heinemann et al., 1969) before applying Equation 2.

Because a range of entropy values was used in the training phase as well as in the testing phase, we fit both phases using the three predictors. Table 7.2 summarizes the quality of the best-fitting sigmoid for each predictor as indexed by root mean square error (RMSE) and the parameter values for that best-fitting function.

### Training phase

Because all of the displays that were used during training involved 16 icons, relative entropy and absolute entropy were perfectly correlated (and, when absolute entropy was scaled to fall between 0.0 and 1.0, the two were identical). Thus, the training data allow the comparison of entropy and the number of icon types as predictors of the percentage of different responses, but they do not allow us to distinguish relative entropy from absolute entropy. The RMSE for entropy was less than half of that for the number of icon types ($p < .01$); this result was true in both the Absolute Entropy and the Relative Entropy clusters. The Absolute Entropy cluster of participants exhibited a weaker degree of control by entropy (0.694) than the Relative Entropy cluster of participants

**Figure 7.8**
A Sigmoid Fit of Hypothetical Data. The figure shows the three parameter values that describe a group's response pattern.

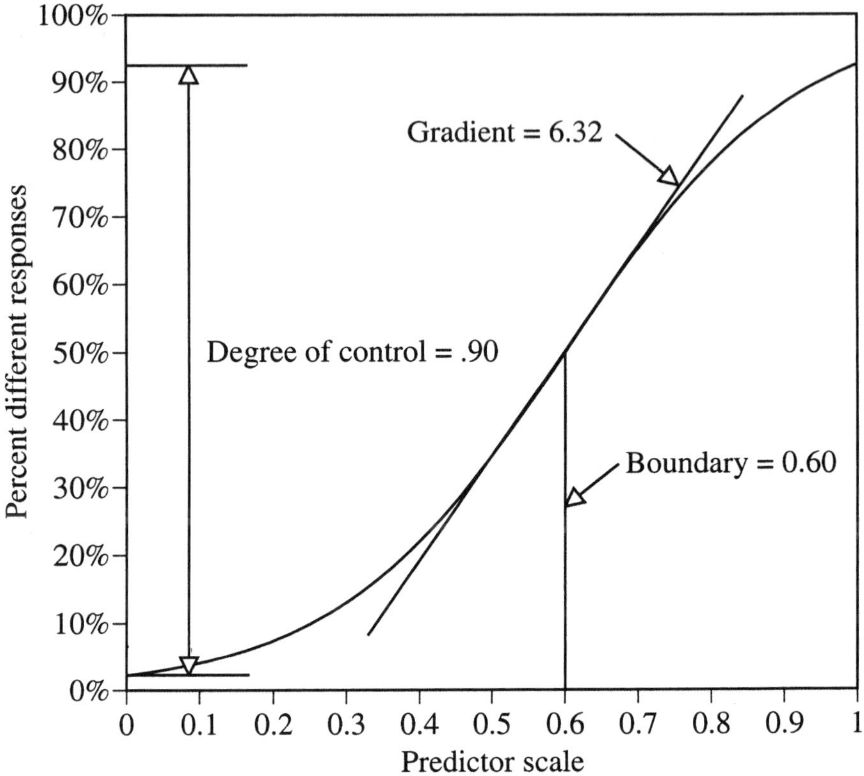


(0.791), but both produced similar gradients and boundaries. As expected given the training procedure, each cluster's boundary (0.524 and 0.511) was very close to the midpoint of the predictor scale.

### Testing phase

Within the Absolute Entropy cluster of participants, the RMSE for absolute entropy was much lower than that for relative entropy and the number of icon types ($p < .001$); within the Relative Entropy cluster of participants, the RMSE for relative entropy was much lower than that for absolute entropy and the number of icon types ($p < .0001$). The learning advantage for the Relative Entropy cluster of participants that was in evidence during training persisted into the testing phase; the Relative Entropy cluster of participants had a higher degree of control and a sharper distinction (manifested in the higher gradient of the best-fitting sigmoid for that cluster) between the low- and high-variability

**Table 7.2**
Fits for Each of the Alternative Predictors of the Percentage of Different Responses for Both the Absolute Entropy and Relative Entropy Clusters of Participants

| | Absolute Entropy | | | Relative Entropy | | |
|---|---|---|---|---|---|---|
| Parameter | Absolute Entropy | Relative Entropy | Number of icon types | Absolute Entropy | Relative Entropy | Number of icon types |
| | | | Training | | | |
| Degree of control | 0.694 | 0.694 | .694 | 0.791 | 0.791 | 0.791 |
| Gradient | 8.688 | 8.688 | 11.312 | 8.340 | 8.340 | 10.239 |
| Boundary | .524 | .524 | .390 | .511 | .511 | .393 |
| RMSE | .051 | .051 | .116 | .043 | .043 | .110 |
| | | | Testing | | | |
| Degree of control | 0.933 | 0.933 | 0.933 | 0.987 | 0.987 | 0.987 |
| Gradient | 9.555 | 5.358 | 10.195 | 8.140 | 12.132 | 6.900 |
| Boundary | .532 | .561 | .368 | .390 | .519 | .233 |
| RMSE | .039 | .231 | .105 | .223 | .071 | .293 |

*Note:* During training, absolute entropy and relative entropy were perfectly correlated producing the identical fits. The parameter values for the curves with RMSEs that are statistically lower than the other curves within that cluster are underscored.

categories than that found in the Absolute Entropy cluster of participants. The boundaries were once again very near the midpoint of the entropy scale and were very similar in both clusters.

## IMPLICATIONS

We were interested in quantifying people's sensitivity to visual display variability and in determining whether their discrimination of these displays would parallel that of the pigeons studied in our earlier research (Young et al., 1997; Young & Wasserman, 1997). By training people to discriminate low variability (Entropy < 2.0) from high variability (Entropy > 2.0) pictorial arrays, participants used display entropy to solve this variability discrimination problem (right side of Figure 7.6). Subsequent testing involving Same-and-Different displays composed of fewer icons revealed that the participants differed in their use of absolute (45%) vs. relative (55%) entropy (left side of Figure 7.6).

To our knowledge, this work represents the first demonstration in humans that discriminative behavior can be based primarily, if not solely, on the variability of a set of complex visual items. We further have strong evidence that people discriminated the displays using the same dimension—entropy—that pigeons use. The two species differed, however, in their use of absolute vs. relative entropy as measures of display variability; a majority of the people who learned the required discrimination (55%) used relative entropy as the basis of their discriminative behavior, whereas none of Young and Wasserman's (1997) pigeons did so. Still, 45 percent of the people responded very similarly to our pigeons.

Unfortunately, we were unsuccessful at identifying factors that might influence the use of relative entropy rather than absolute entropy; none of the demographic variables predicted a participant's cluster. Computationally, relative entropy is more difficult to calculate than absolute entropy because an observer must determine both the absolute entropy of the array and the maximum possible absolute entropy of the array; this greater effort was nicely revealed in the longer RTs observed for the Relative Entropy cluster (Figure 7.7). Those participants, however, willing to put forth the extra effort benefited in that their final performance was more strongly under the control of variability (as revealed by the higher degree of control in the Relative Entropy cluster; see Table 7.2). Unfortunately, it is not yet clear why participants relying on relative entropy would apparently find the discrimination both more difficult (evidenced by the longer RTs) and easier (evidenced by the higher degree of control) nor why they chose this strategy over the computationally simpler use of absolute entropy.

The appearance of individual differences offers an opportunity to better understand the cognitive system and its processing of these complex arrays. Although we did not require participants to adopt one or the other strategy, we consider it quite likely that all people could be trained to use either absolute or relative entropy under the proper set of conditions. The question then arises: Which strategy would be more difficult for people to adopt and what conditions would be sufficient to motivate each strategy? Given that pigeons (Young & Wasserman, 1997) and baboons (Wasserman, Fagot, & Young, unpublished data) strongly prefer absolute entropy, we anticipate that even those participants who preferentially used relative entropy in our task would find it relatively easy to use absolute entropy. Conversely, those participants who preferentially used absolute entropy in our tasks might find it more difficult to use relative entropy.

By training participants using displays involving different numbers of items, we could require participants to either use absolute or relative entropy in order to solve the discrimination. We might also be successful at producing a bias towards relative entropy by emphasizing the irrelevance of item number in the instructions that precede the task or allowing unlimited time for a response. These manipulations might provide further insight into the human visual variability processing system.

## Nonspecific preference for variety

We have succeeded in showing that human and nonhuman animals can discriminate different levels of visual display variability. This ability is quite systematic, conforming to a mathematically precise measure of variability—entropy. But, just why do organisms possess this ability? Does it endow them with some adaptive advantage?

Previously, we noted situations in which variety is desired and situations in which uniformity is desired. Low variability assures consistency, whereas high variability increases the chances of finding creative solutions or tools. Hence, situational demands may dictate when consistency or variety is preferred.

In addition to these factors, personal preferences may influence one's choice of consistency or variety. William Cowper's declaration that variety is the "very spice of life" suggests that variability in everyday life is something to be sought (although in our food, spices are generally preferred in measured quantities!). Too little variety breeds boredom and creates a predictable environment that fails to stimulate learning; too much variety, however, can make life frenetic and create an unpredictable environment that precludes one's ability to leverage learning. Still, there are individuals who generally prefer the ordinary (people who are strongly rooted in tradition) and others who prefer the exotic (people who will try anything for a good time).

There exists some experimental evidence of people's preferences for variability in visual stimuli. For example, Munsinger and Kessen (1966) examined adults' preferences for various letter strings. The participants were asked to state their preference for pairs of letter strings that possessed one of eight levels of variability, ranging from redundant strings of letters (low variability) through random words to random letters (high variability). Intermediate levels of variability were preferred over both low and high levels. Munsinger and Kessen (1966) also found that the participants had a more difficult time recalling stimulus strings as their variability was increased. Relatedly, our participants exhibited faster and more accurate discrimination for the lower entropy displays in the present study. Munsinger and Kessen (1966) speculated that participants' preferences for intermediate levels of variability was determined by their ability to process variability: variability at or near one's optimal processing capacity is highly preferred.

Recall that entropy, as a measure of variability, is an information–theoretic concept: entropy measures the amount of information in a multi-element stimulus. Thus, it is not surprising that computing the amount of entropy, or information, in a stimulus would provide utility to an organism. Our goal was not to determine the amount of entropy that is preferred by a participant, but rather to document that humans, as well as pigeons, can discriminate complex collections of items as a function of their entropy. It seems likely, however, that humans cannot only discriminate low from high variability, but that personal preferences and external circumstances dictate which are favored and when.

### Pigeons and people

Our results suggest that both people and pigeons use entropy to categorize display variability. The two species differed, however, in one fundamental way: a significant number of college student participants (and no pigeons) relied on relative entropy rather than on absolute entropy. Given the documented difficulty that pigeons have with discriminating displays involving two identical items from those involving two different items (a distinction between minimal and maximal relative entropy), it is likely that pigeons are inclined to use absolute entropy, and that they find it very difficult to compute entropy relative to some reference point (e.g., the maximal entropy possible given the number of items present, as is required to compute relative entropy).

We expect that future investigations into the discrimination of variability by humans and nonhuman animals will continue to elucidate important cognitive mechanisms that underlie a wide range of adaptive behaviors such as abstract conceptualization. That conceptualization is no longer the private province of human beings is now clear (Wasserman, 1993). What is not clear is the extent to which the effective mechanisms of conceptualization in humans and animals are alike.

## NOTE

We would like to thank Nicole Hill and April Logsdon for their assistance in conducting the experiment. The data reported was presented at the 1998 meeting of the Psychonomic Society. Correspondence about this paper should be addressed to Michael E. Young, Department of Psychology, Southern Illinois University, Carbondale, IL, 62901-6502. Electronic mail may be sent to meyoung@siu.edu.

## REFERENCES

Ashby, F. G., Boynton, G., & Lee, W. W. (1994). Categorization response time with multidimensional stimuli. *Perception & Psychophysics, 55,* 11–27.

Ashby, F. G., & Gott, R. E. (1988). Decision rules in the perception and categorization of multidimensional stimuli. *Journal of Experimental Psychology: Learning, Memory and Cognition, 14,* 33–53.

Edwards, C. A., Jagielo, J. A., & Zentall, T. R. (1983). "Same/different" symbol use by pigeons. *Animal Learning & Behavior, 11,* 349–355.

Fried, L. S., & Holyoak, F. J. (1984). Induction of category distributions: A framework for classification learning. *Journal of Experimental Psychology: Learning, Memory and Cognition, 10,* 234–257.

Heinemann, E. G., Avin, E., Sullivan, M. A., & Chase, S. (1969). Analysis of stimulus generalization with a psychophysical method. *Journal of Experimental Psychology, 80,* 215–224.

Medin, D. L., & Smith, E. E. (1984). Concepts and concept formation. *Annual Review of Psychology, 35,* 113–138.

Morgan, C. L. (1896). *An introduction to comparative psychology.* London: Walter Scott, Ltd.

Munsinger, H., & Kessen, W. (1966). Preference and recall of stimulus variability. *Journal of Experimental Psychology, 72,* 311–312.

Nosofsky, R. M. (1991). Typicality in logically defined categories: Exemplar-similarity versus rule instantiation. *Memory & Cognition, 19,* 131–150.

Posner, M. I., & Keele, S. W. (1968). On the genesis of abstract ideas. *Journal of Experimental Psychology, 77,* 353–363.

Posner, M. I., & Keele, S. W. (1970). Retention of abstract ideas. *Journal of Experimental Psychology, 83,* 304–308.

Premack, D. (1976). *Intelligence in ape and man.* Hillsdale, NJ: Erlbaum.

Santiago, H. C., & Wright, A. A. (1984). Pigeon memory: Same/different concept learning, serial probe recognition acquisition, and probe delay effects on the serial-position function. *Journal of Experimental Psychology: Animal Behavior Processes, 10,* 498–512.

Shannon, C. E., & Weaver, W. (1949). *The mathematical theory of communication.* Urbana: University of Illinois Press.

Wasserman, E. A. (1993). Comparative cognition: Beginning the second century of the study of animal intelligence. *Psychological Bulletin, 113,* 211–228.

Wasserman, E. A. (1995). The conceptual abilities of pigeons. *American Scientist, 83,* 246–255.

Wasserman, E. A., Hugart, J. A., & Kirkpatrick-Steger, K. (1995). Pigeons show same–different conceptualization after training with complex visual stimuli. *Journal of Experimental Psychology: Animal Behavior Processes, 21,* 248–252.

Wright, A. A., Santiago, H. C., Sands, S. F., Kendrick, D. F., & Cook, R. G. (1985). Memory processing of serial lists by pigeons, monkeys, and people. *Science, 229,* 287–289.

Wright, A. A., Santiago, H. C., Urcuioli, P. J., & Sands, S. F. (1983). Monkey and pigeon acquisition of same/different concept using pictorial stimuli. In M. L. Commons, R. J. Herrnstein, & A. R. Wagner (Eds.), *Quantitative analyses of behavior, Vol. 4.* (pp. 295–317). Cambridge, MA: Ballinger.

Young, M. E., & Wasserman, E. A. (1997). Entropy detection by pigeons: Response to mixed visual displays after same–different discrimination training. *Journal of Experimental Psychology: Animal Behavior Processes, 23,* 157–170.

Young, M. E., Wasserman, E. A., & Garner, K. L. (1997). Effects of number of items on the pigeon's discrimination of same from different visual displays. *Journal of Experimental Psychology: Animal Behavior Processes, 23,* 491–501.

Young, M. E., & Wasserman, E. A. (2001). Entropy and variability discrimination. *Journal of Experimental Psychology: Learning, Memory, and Cognition, 27,* 278–293.

Part II

# Animal Visual Processing

Chapter 8

# The Multiplicity of Visual Search Strategies in Pigeons

*Jeffrey S. Katz and Robert G. Cook*

It has been a little over a hundred years since Thorndike (1898) began the experimental analysis of animal intelligence. During that time a variety of approaches have tried to understand the structure and mechanisms underlying the animal mind. Among the many facets of intelligence that have attracted experimental attention has been whether animals engage in rule-governed behavior, and if so, of what kinds. Human behavior is often expressed as rule-based (e.g., Anderson, 1993; Haberlandt, 1997; Jones, Ritter, & Wood, 2000; Lee, 1998; Mackay, Stromer, & Serna, 1998; Newell & Simon, 1972; Nisbett, 1993; Sidman, Rauzin, Lazar, Cunningham, Tailby, & Carrigan, 1982; Sidman & Tailby, 1982). For instance, we easily answer questions about things with which we have no direct experience, such as novel math problems. This competence reflects our developed ability to learn simplifying rules or general principles that abstract the relations among a set of elements. The great benefit of this cognitive ability is that it releases behavior from the direct control of the stimulus and its history of reinforcement, and permits flexible and adaptive solutions to novel problems. As a result, we often engage in behaviors that are unbounded by our experience with specific stimuli.

Over the last three decades, there has been growing evidence that animals can also use both object-based and rule-based concepts to guide behavior with novel stimuli in a variety of situations. This suggests that the ability to use concepts and rules is a widespread cognitive process in the animal kingdom, and may reflect a type of cognitive primitive that may have emerged early in the evolution of information processing by nervous systems. In birds, the class of animal of direct concern in this chapter, one of the most common methods for studying such concepts has been with photographs of realistic natural objects in go/no-go and choice-discrimination procedures. It has been found, for instance, that object concepts, such as those formed by pictures of trees, cars, cats, flowers, birds, mammals, fish, oak leaves, and humans, are easily acquired by pigeons (Bhatt, Wasserman, Reynolds, & Knauss, 1988; Cerella, 1979; Cook, Wright, & Kendrick, 1990; Herrnstein & de Villiers, 1980; Herrnstein, Loveland, & Cable, 1976). More recently, research from our lab has suggested that pigeons may have an unappreciated capacity for learning and using abstract rule-like concepts, such as those involved with same–different discriminations (Cook, Cavoto, & Cavoto, 1995; Cook, Katz, & Cavoto, 1997, 1998; Cook, Katz, & Kelly, 1999; Cook & Wixted, 1997).

Humans also engage in stimulus–specific strategies in which the stimulus or specific context governs what we do, such as when we use the same rote procedure for studying facts for exams, doing statistics problems, or swinging a golf club. Likewise, pigeons also provide abundant evidence that they also utilize stimulus–specific strategies, being capable of memorizing large numbers of exemplars (Vaughn & Green, 1984) and stimulus relations (Carter & Werner, 1978; Wright, 1997) when solving discriminations. As expected in such circumstances, the birds show little transfer to new situations, because their learning is tied closely to the experiences they have received.

As the evidence in both cases is compelling, it appears that learning in pigeons is a mixture of both relational and item–specific behaviors. This duality raises a host of interesting questions to explore. How are these two distinct forms of learning to be reconciled? Do they represent contrasting learning mechanisms that are implemented in different areas of the brain or have evolved separately? Or do they in some way reflect the operation of one learning mechanism? What and why do certain conditions promote the use of one over the other? In this chapter, we begin to explore this mixture of discrimination behaviors within the specific confines of the visual search behavior of pigeons.

Visual search tasks typically have the subject find a target item (e.g., red square) among a set of distractor items (e.g., blue squares) and indicate its presence or absence. In the experiments described later, we employed a visual search task for pigeons in which they had to locate and peck at an odd target region within a larger stimulus context. These experiments grew out of a research program looking at how pigeons perceive and process visual textures (Cook, 1992a, 1992b, 1992c, 1993a, 1993b, 2000; Cook, Cavoto, & Cavoto, 1996; Cook et al., 1998). Texture stimuli are multidimensional displays in which global regions are formed from the grouping of repeated local elements. Figure 8.1 shows an example of a texture display in which the odd target differs in color (top) and one where it differs in shape (bottom). The pigeons' task is to visually search for and peck at the odd target region, which varies randomly from trial to trial in its location and its dimensional difference (i.e., color or shape) from the surrounding distractor elements. Locating the target successfully leads to food reinforcement, while pecking at the region of surrounding distractor elements leads to a brief time out.

**Figure 8.1**
Examples of Color and Shape Texture Displays Used in Experiments. The color and shape of the elements used to construct the textures were randomly selected for baseline displays. The location of the target region was also randomly selected.

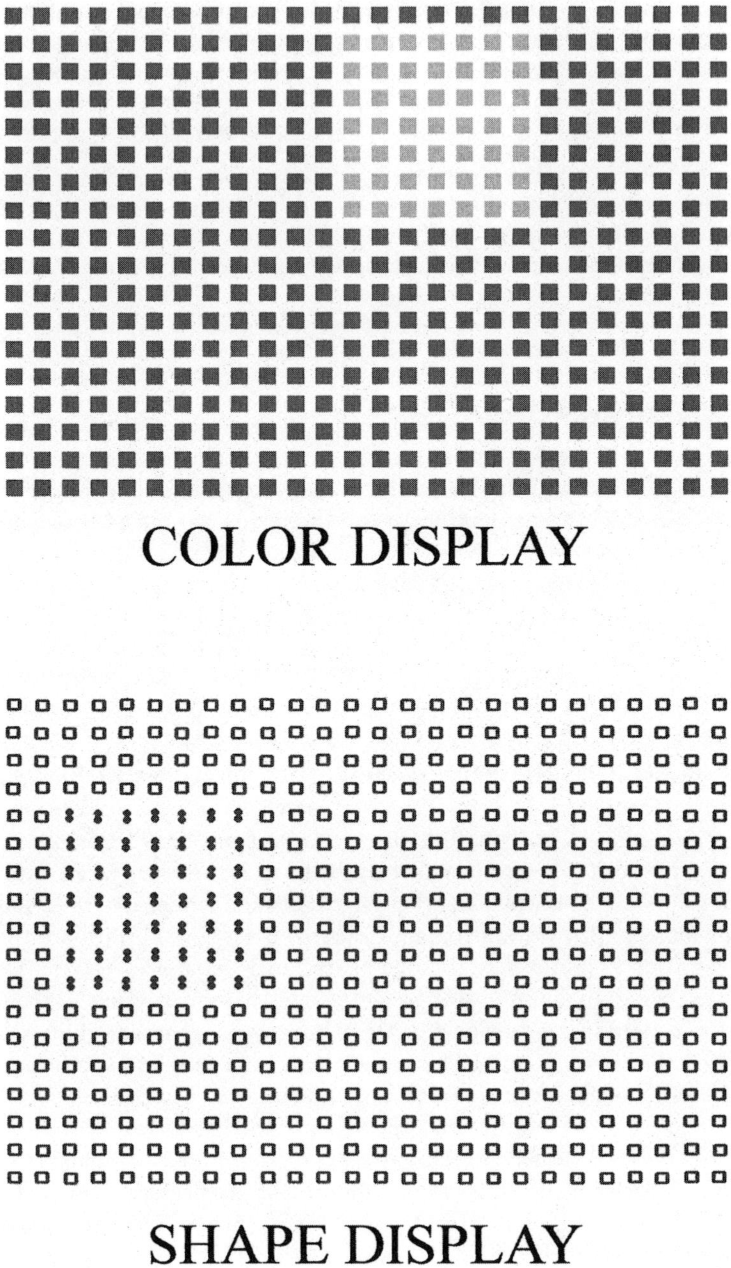

We have used this "target localization" task extensively to investigate visual cognition in pigeons (Cook, 1992a, 1992b, 1992c, 1993a, 1993b, 2000; Cook et al., 1996; Cook, Cavoto, Katz, & Cavoto, 1997; Cook et al., 1998; Katz & Cook, 2000). Our overarching goal has been to better understand the mechanisms of perception and discrimination behavior in these highly visual, but distantly related, creatures, and determine the extent to which their underlying psychological mechanisms are the same or different from our own. In the course of these investigations, we have found that pigeons can apparently use both relational (Cook, 1992a) and item–specific strategies (Cook et al., 1996; Katz & Cook, 2000) in searching for the target, creating an opportunity to examine the interaction between these different modes of behavior and when they are implemented in their solution to this discrimination task. The chapter reviews some of the recent evidence concerning this question and provides some new data about this general issue.

## AVIAN SEARCH STRATEGIES

How exactly do the pigeons locate the target in such visual search tasks? Previous research has suggested there is a variety of different strategies that are employed. In the present chapter we use the term "strategies" as a shorthand for describing recurring patterns of behavior that depend on the presence or absence of certain cues. Although the pigeons actively selecting strategically among these is a possibility, it is not one we intend by our use of the term.

One of the earliest strategies revealed to be used by birds involved learning to search for specific target features (P.M. Blough, 1989, 1991, 1992, 1996; P.M. Blough & Lacourse, 1994; Bond, 1983; Bond & Riley, 1991; Langley, 1996; Langley, Riley, Bond, & Goel, 1996; Pietrewicz & Kamil, 1979; Plaisted, 1997; Plaisted & Mackintosh, 1995; Reid & Shettleworth, 1992; Vreven & P.M. Blough, 1998). In these experiments, the birds typically search for targets (colors, shapes, or their combination) that are mutually exclusive from the distractors in which they are embedded. The performance of the birds is then compared in two basic conditions. In one condition, a particular target item is repeated over successive trials or increased in testing frequency across a session. Performance in these high-repetition conditions is then compared to a session in which the target items are mixed at random over an equivalent baseline period. During high-repetition conditions, search performance improves, as measured by a decrease in reaction time or an increase in accuracy. This improved search efficiency indicates the birds were using an "item-specific target approach rule." That is, the birds had learned to look for and approach a particular target during its repetition. It has been suggested that this effect is representative of the birds learning to form a "search image" or focusing attention on particular features of the target (e.g., D.S. Blough & P.M. Blough, 1997; Bond, 1983; Langley, 1996; Pietrewicz & Kamil, 1979; Reid & Shettleworth, 1992).

In the example just mentioned, the birds could learn what to search for on each trial because of the limited size of the search set, but in some tasks the exact nature of the target is not easily known; other strategies must be used. One such tactic is an "odd-item search strategy" (D.S. Blough, 1989; Cook, 1992a). In our texture-based target-localization task, this odd-item strategy involves comparing the perception of a stimulus with a generalized pattern or Gestalt of a texture's regional properties (e.g., a small odd area embedded within a rectangular area; Cook, 1992b, 1993a, 1993b; Cook et al., 1998). By relying on target oddity per se, rather than absolute identity of the specific elements of the display, this strategy can be quite general, allowing the birds to easily transfer to novel-display test items they have not been tested with before (D.S. Blough, 1989; Cook, 1992a). Perhaps not too surprisingly, given the structure of the task, it appears that pigeons can solve search tasks in part by learning to approach both the general and specific properties of the target.

Katz and Cook (2000) recently extended this type of work by looking at the effects of distractor repetition on visual search behavior. The results of their research suggest that additional factors tied to the processing of the distractors also can influence the success with which a target is localized. Because the Katz and Cook's stimulus repetition procedure is critical to the new experiments described in this chapter, its procedures and logic are described next in some detail. Six highly experienced pigeons were involved in experiments using the texture-based visual search task. At the beginning of the experiments, a typical daily session involved 180 randomly chosen color or shape trials. These color and shape trials could be constructed from any combination of twenty colors and thirty-four shape features. Importantly, because these features were always randomly selected to be targets or distractors on every trial, it was impossible for the pigeons to utilize any pre-existing item-specific rules to locate the target.

Then we began a series of test sessions in which specific features of the target or distractor elements were repeated within a session to examine how item-specific information became incorporated into their ongoing search behavior. Each session consisted of three phases, beginning with a baseline phase in which the birds saw a set of randomly chosen color and shape trials. Starting on trial 100 of a session, this was followed by a sixty-trial repetition phase. During this repetition phase, a randomly selected shape or color feature was repeated across trials for either the target or distractor region. The identity of the repeated feature changed on a daily basis, although testing was conducted in blocks of either color or shape sessions. For the distractor repetition condition, the color feature of the distractors was repeated from trial to trial across the repetition phase. For the target repetition condition, the color feature of the targets was repeated from trial to trial across the repetition phase. Figure 8.2 illustrates one possible example of a distractor (bottom panel) and target repetition (top panel) sequence for color displays (although not shown in the diagram, the value of the irrelevant shape features used to construct these regions varied randomly across trials for both conditions). Identical manipulations were also

**Figure 8.2**
Top and Bottom Panels Respectively Illustrate for Color Displays Target and Distractor Repetition. During Target Repetition, the color feature of the target is repeated across trials. During Distractor Repetition, the color feature of the distractor is repeated across trials. The location of the target is randomly selected every trial. Note that the shape features are not represented but are also selected at random on every trial.

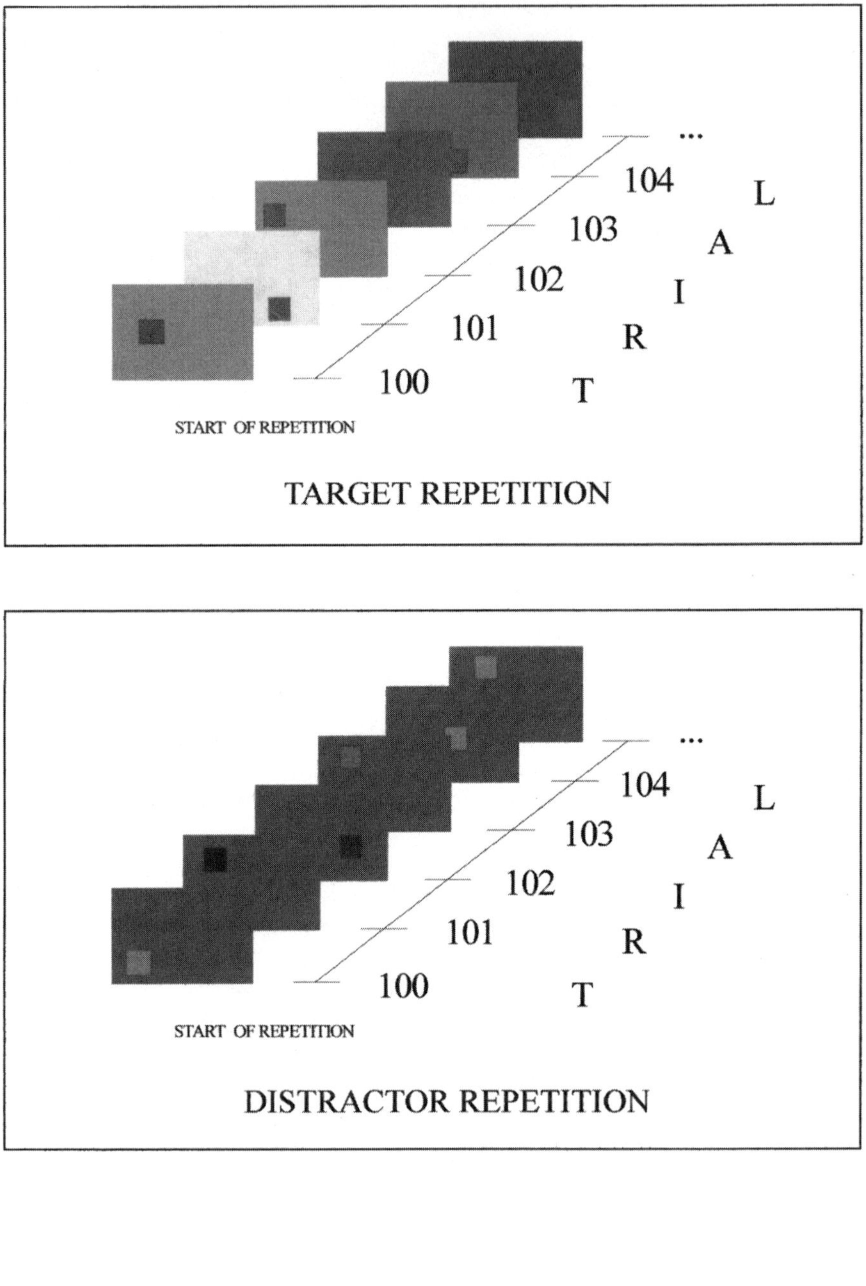

conducted for testing the repetition of shape displays. Following the repetition phase, the pigeons returned to baseline testing with both color and shape displays for the remainder of the session.

It is important to note, again, that during the baseline trials the pigeons must utilize a generalized relational strategy that does not depend on the specific elements of the displays to solve the task because of the large number of displays being tested (>35,000) and the unknowable nature of the target's identity on any trial. Thus, if we found that search performance improved during either the distractor or target repetition condition, it would indicate that both relational and item-specific strategies could contribute to search performance. If improvement occurs during target repetition, the pigeons are capable of learning and using an item-specific target approach strategy. If improvement occurs during distractor repetition, the pigeons are learning and using a distractor avoidance strategy that in turn facilitates locating the unknown targets.

The stimulus repetition experiments revealed the following results. Overall, the pigeons showed a monotonic increase in localization accuracy over the period of trials in which a specific distractor feature was repeated. This effect, however, was restricted to trials in which the dimensional difference was from the shape dimension. Figure 8.3 shows mean target localization accuracy for distractor and target repetition for color (top) and shape (bottom) displays across trial blocks. Because distractor repetition improved search performance, the pigeons used a "distractor avoidance strategy" that involved remembering and using the identity of the repeated distractor feature from trial to trial.

We think the reason this repetition effect was stronger for shape displays is because of their lower discriminability in relation to the color displays. We have found that color displays are discriminated more accurately and faster than shape displays. Thus, the generally more conspicuous color targets were found by means of a different strategy, such as the odd-item strategy. A similar absence of stimulus-repetition effects has also been found in other search experiments with birds in which highly discriminable displays are presented (Bond, 1983; Bond & Riley, 1991; Langley, 1996; Langley et al., 1996; Reid & Shettleworth, 1992).

The absence of a target repetition effect surprised us at first, because the past research with pigeons had shown that repeating target features consistently improved search (e.g., D.S. Blough & P.M. Blough, 1997; Bond, 1983; Langley et al., 1996; Pietrewicz & Kamil, 1979; Reid & Shettleworth, 1992). We believe that the basis for this disparity lies in a key difference between our odd-item task and the fixed-item procedures of the other studies. In our odd-item task, distractors and targets are randomly selected from a single pool of elements, whereas in a fixed-item search task, the distractor and target elements are selected from separate sets of elements. This mutually exclusive mapping in fixed-item search tasks helps allow the animals to more easily learn item-specific search rules for locating the target; when target frequency is increased, birds can effectively limit their search to the frequent target.

**Figure 8.3**
Mean Target Localization and Standard Errors of Mean Across Successive Trial Blocks Within a Session for Color (Top Panel) and Shape (Bottom Panel) Displays (from Katz & Cook, 2000). For baseline (open symbols), only color (top panel) and shape (bottom panel) performance is represented. The repetition phase is denoted by filled symbols. The first twenty trials of each session served as warm-ups whereby the pigeons were required to successfully find the target and are, thus, not depicted. *Some information from the American Psychological Association.*

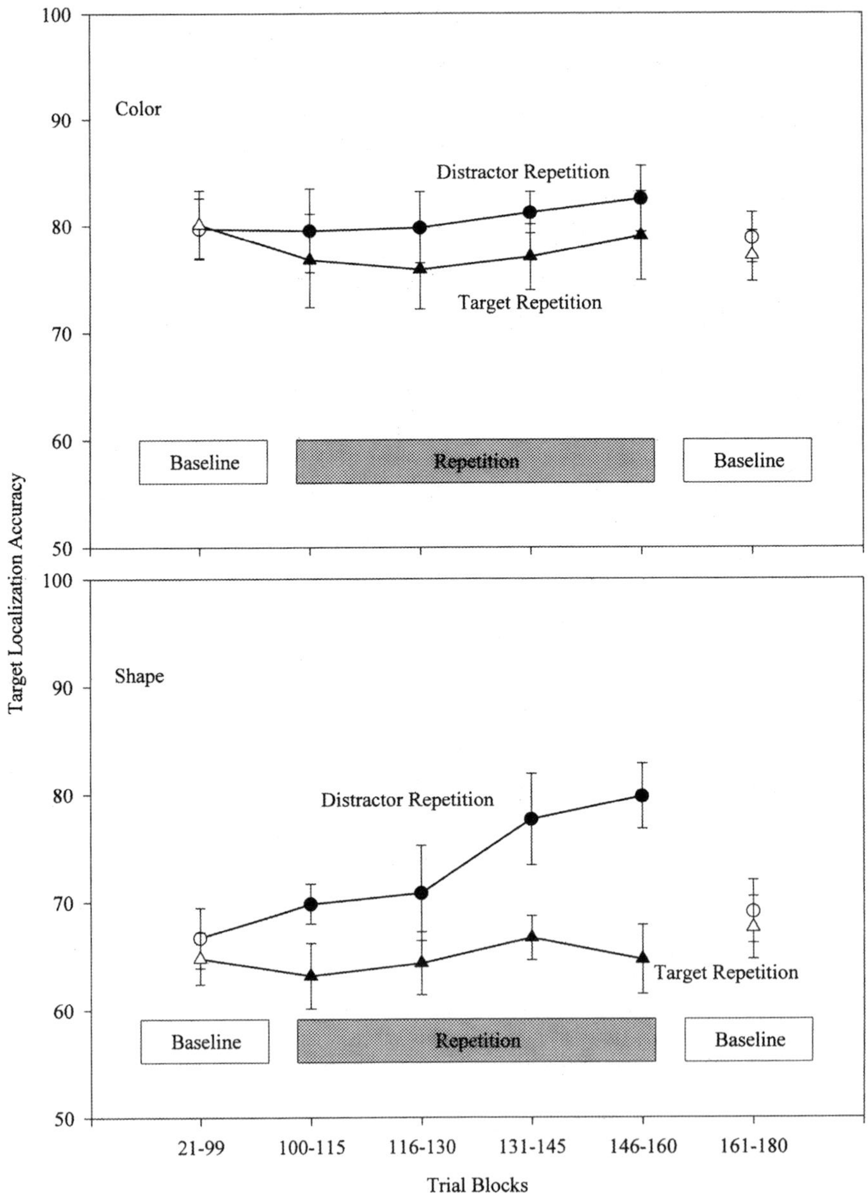

The absence of mutually exclusive distractor and target sets in our oddity task results in search performance being primarily controlled by different strategies than the item-specific target approach rule. Differences in strategy usage based on distractor and target element mapping have also been reported in cats and humans (De Weerd, Vandenbussche, & Orban, 1992; Schneider & Shiffrin, 1977; Shiffrin & Schneider, 1977). The results in Figure 8.3, from Katz and Cook (2000) indicate a distractor-avoidance strategy. It is of interest to note that this item-specific strategy may also represent a form of generalized distractor avoidance as well. That is, the pigeons may have used distractor avoidance as a key part of locating the target even prior to these repetition tests. This strategy can be considered to be generalizable because it could be applied to any element used to construct a texture display. Hence, it is possible that such a general-distractor avoidance strategy might have served as the progenitor for an item-specific distractor-avoidance strategy detected here.

In summary, birds appear to use a variety of strategies to solve visual search tasks. These strategies can be general (oddity and distractor avoidance) rules that can be applied to any array of stimulus elements; They can also be item-specific rules that can be target approach or distractor avoidance-based. In general, perhaps the most impressive aspect of the pigeon's ability to solve visual search tasks is the multiplicity of strategies that are used by the pigeon. The following experiments further explore some of the factors influencing these search strategies.

## NEW EXPERIMENTS ON EXAMINING REPETITION EFFECTS

This part describes two further experiments conducted to follow up on Katz and Cook's results. In these experiments, we further explore the different strategies used by pigeons in solving the odd-item visual search task. The first experiment explores identity repetition and examines how the number of repeated features affects performance. We present evidence that search performance improves with increasing element specificity. The second experiment explores repetition of target location. While our previous experiments all explore feature repetition of identity (what), this experiment extends feature repetition by using location (where). We present evidence that pigeons can also use a location-based strategy.

## EFFECTS OF NUMBER OF REPEATED FEATURES

This experiment further examined the effect of repeating specific display features in our target localization task. The primary issue of interest was the effect of increasing the number of repeated features associated with the distractors. In our earlier experiments, we repeated only the relevant dimensional feature of the distractor. What would be the effect of repeating both the color and shape features of the distractors? Would increasing the distractor element's specificity aid in search? For example, the pigeons might learn not just to "avoid circles" but to "avoid red circles." What would be the effect of repeating distractor and target features at the same time? Would the pigeons benefit from this pairing more than from just repeating the distractor features?

Seven conditions were tested. The same basic experimental procedure from Katz and Cook (2000) was used, although only shape displays were tested for repetition effects. Table 8.1 summarizes the features repeated across these different test conditions. The "R" represents which features were repeated during the repetition phase. The "*" represents which features were randomly selected on every trial of the repetition phase. The following conditions were tested: Distractor Repetition: repetition of the distractor shape feature; Target Repetition: repetition of the target shape feature; Relevant Repetition: repetition of both the relevant distractor and target shape features; Irrelevant Repetition: repetition of the irrelevant color feature; Distractor+Irrelevant Repetition: a combination of the distractor and irrelevant conditions; Target+Irrelevant Repetition: a combination of the target and irrelevant conditions; and Complete Repetition: repetition of all features of the display.

Seven pigeons were tested. Six were the same as tested by Katz and Cook (2000), along with one additional pigeon with similar experience. Pigeons were tested in a computerized operant chamber, where stimuli could be presented on a high-resolution color CRT equipped with a touchscreen that was used to record the birds' pecking behavior directed to these stimuli. All stimulus displays were presented in color on a computer monitor built in to the front panel of the chamber. Each texture display was 20 centimeters wide, 13 centimeters high, and consisted of 468 elements arranged in a 26 x 18 array. Individual elements ranged from 3 to 6 millimeters in size, based on their respective shape. Forty-nine of the 468 elements were arranged in a contrasting 7 x 7 target region. This target region was randomly placed in one of 240 possible target positions on each trial. These stimuli were composed of pairs of elements drawn from a pool of 740 elements (pairwise combinations of 20 colors and 37 shapes, see http://www.pigeon.psy.tufts.edu/jep/blink/blink_elements.htm for some examples of these values). Altogether there were 40,700 different trials that could be used.

Each trial began with a peck to a randomly located white circular ready signal (2 cm in diameter) that was immediately followed by presentation of a texture display. The pigeon's task was to locate the randomly located odd target in the display. A trial was considered correct if five pecks were directed to the target region prior to five pecks being directed to the distractor region. Correct target localization responses were rewarded with 2-second access to mixed grain, while incorrect responses were punished with a 10-second time-out in a darkened chamber.

**Table 8.1**
Randomly Selected and Repeated Features for the Seven Repetition Conditions Involving Shape Displays

| | Distractor | | Target | |
|---|---|---|---|---|
| Condition | Color | Shape | Color | Shape |
| Distractor | * | R | * | * |
| Target | * | * | * | R |
| Relevant | * | R | * | R |
| Irrelevant | R | * | R | * |
| Distractor+Irrelevant | R | R | R | * |
| Target+Irrelevant | R | * | R | R |
| Complete | R | R | R | R |

*Note.* "*" represents features selected at random on every trial. "R" represents features that were repeated across trials.

Sixty test sessions were conducted. Each session lasted 180 trials. In trials 1 to 100 and trials 161 to 180, pigeons were tested with a random mixture of color and shape baseline displays. Trials 101 to 160 were used to test the repetition conditions. The seven repetition conditions described previously were tested in two groupings. The first grouping tested the Complete, Relevant, Distractor, Target, and Irrelevant Repetition conditions in six randomized blocks of five sessions. The second grouping tested the Distractor, Target, Irrelevant, Distractor+Irrelevant, and Target+Irrelevant conditions, also in six randomized blocks of five sessions.

Figure 8.4 shows mean target localization accuracy for each condition across blocks of 15 trials during the repetition phase of each session. Filled symbols represent repetition trials. Unfilled symbols represent shape performance from baseline phases that preceded or followed repetition (trials 21–100 and 161–180 respectively). The top left panel shows the results from all seven conditions and the remaining panels show the individual conditions. The legend in the top left panel ranks the different conditions in descending order of target localization accuracy based on trial block 146–160, capturing the relative degree of improvement caused by the repeated items within a session.

Generally, we found that increasing the number of features repeated across trials improved performance. That is, as the specificity of the distractors or targets in the displays was increased, so did the degree of improvement observed across trials. This is reflected in the numerical ordering of the conditions in Figure 8.4 (top left panel). Repeated measures analyses of variance comparing

**Figure 8.4**
Mean Target Localization and Standard Errors of the Mean Across Successive Trial Blocks Within a Session for Seven Different Identity Repetition Conditions. For baseline (open symbols), only shape performance is represented. The upper left panel combines all seven conditions. Within the legend, the test conditions were arranged in descending order of target localization accuracy based on trial block 146–160. The first twenty trials of each session served as warm-ups whereby the pigeons were required to successfully find the target and are, thus, not depicted.

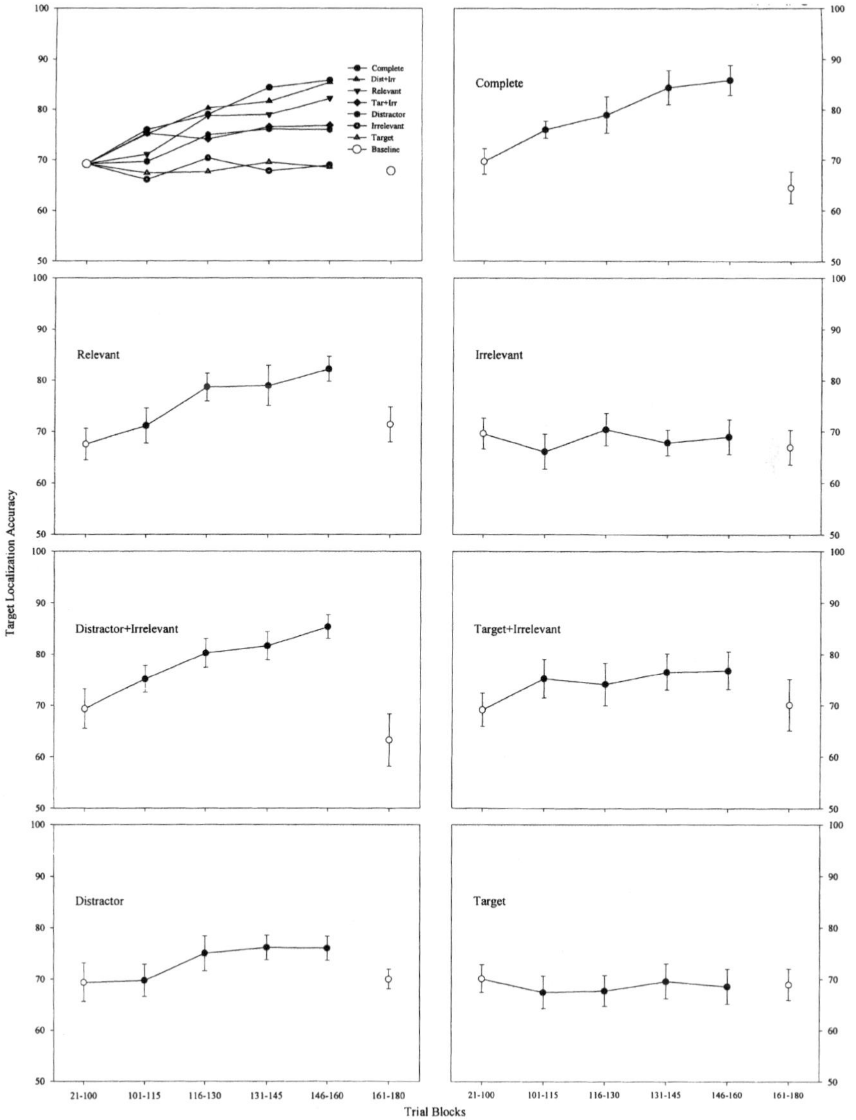

blocks of trials (21–100, 101–115, 116–130, 131–145, 146–160) across a session found significant improvements for the Complete, $F(4, 24) = 15.4, p < .000005$; Distractor+Irrelevant, $F(4, 24) = 15, p < .000005$; Relevant, $F(4, 24) = 6, p < .005$; Distractor, $F(4, 24) = 8.9, p < .0005$; and Target+Irrelevant, $F(4, 24) = 2.2, p < .098$. Target+Irrelevant Repetition only approached significance because learning was completed during the first trial block of the repetition phase. Indeed, the first trial block (101–115) of repetition was significantly different from the baseline trial block (21–100), $t(6) = 2.6, p < .05$. Neither the Target nor Irrelevant Repetition conditions produced any increase in accuracy. The increase in performance with Distractor Repetition and corresponding failure to see any improvement in Target Repetition replicates the findings reported in Katz and Cook (2000).

Overall, the results indicate that target localization accuracy improves with increasing element specificity. This was true for both distractor and target elements. As can be seen in Figure 8.4, repeating the irrelevant feature in conjunction with the relevant distractor or target feature improved search for Distractor+Irrelevant and Target+Irrelevant respectively in comparison to Distractor and Target Repetition. There was also an effect of repeating the relevant distractor and target feature at the same time during Relevant Repetition. As can be seen in Figure 8.4, Relevant Repetition improved search accuracy beyond that of Distractor Repetition or Target Repetition. The latter finding is likely related to an item-specific "conditional expectancy strategy" first described by D.S. Blough (1993).

The conditional expectancy strategy is a form of within-trial priming involving conditions in which both distractor and target features are simultaneously repeated over trials. In D.S. Blough's experiments, pigeons were highly trained in an odd-item visual search task to discriminate four items that occurred equally often as targets or distractors in eight different displays during baseline sessions. In the test sessions, pigeons experienced four of these eight different displays frequently and four displays infrequently. The pigeons learned to localize the target items in frequent pairings faster than those occurring in infrequent pairings. Such results indicate that within a trial, a distractor rapidly primes the identity of a highly expected target item. Overall, D.S. Blough concluded such results indicated that pigeons learned conditional expectancies over sessions, such that distractor A (e.g., circle) predicts target B (e.g., triangle), for instance. This type of distractor-mediated item-specific search strategy likely accounts for the Relevant Repetition effect found in the current experiment.

Our results provide data consistent with two possible extensions of the conditional expectancy strategy. One extension is the possibility of extra-dimensional priming, as opposed to intra-dimensional priming. In intra-dimensional priming, priming occurs within the same dimension (e.g., yellow distractors prime blue targets). In extra-dimensional priming, priming occurs across two different dimensions such as color and shape (e.g., yellow distractors prime circle targets). Consider the increase in accuracy during the Target+Irrelevant condition. It is possible that the repeated irrelevant color feature in the distractor region predicted the relevant shape feature of the target. For instance, in a shape display, blue distractors predict triangle or blue triangle targets; hence, the increase in search performance during Target+Irrelevant Repetition could have been due to a conditional expectancy strategy and not an item-specific target approach rule. The other extension of the conditional expectancy strategy is the faster rate at which the distractor-target pairings were learned. In D.S. Blough's experiments (1993), the expectation effect materialized over sessions. In the present experiment, the conditional priming effect seems to have occurred quite rapidly within a session. This disparity is probably due to procedural differences. In the current experiment, the pigeons were required to learn a single distractor-target pairing, whereas in D.S. Blough's, the pigeons simultaneously learned four distractor-target pairings that each occurred at a lesser frequency (23.5% of the trials in a session).

## EFFECTS OF REPEATING TARGET LOCATION

The previous experiments had looked at the effects of repeated visual features; the next experiment examined the effect of repeating the target's spatial location across trials. The physical location of items in visual space is a property that is critical in human models of vision. Research with humans in a visual search task has shown that learning the frequent locations of where a target might occur facilitates search for a target in those locations (Shaw & Shaw, 1977). Similarly, learning where a target will be located has also been shown to improve search performance in pigeons (D.S. Blough, 1993; P.M. Blough & Lacourse, 1994). D.S. Blough (1993) found that pigeons could learn to search for targets in a specific area of visual search array based on a specific distractor. P.M. Blough and Lacourse (1994) found that pigeon search performance improves with brief (3-trial) repetition of target location.

We tested whether our six pigeons would use location information to improve their search performance in our repetition procedure. That is, the only feature of the textures that remained constant was target location; all other properties varied as in the baseline trials. Both one-location and two-location repetition conditions were tested. During one-location repetition, the target region was presented at a single location across the repetition phase. During two-location repetition, the target region was presented equally often at two locations across the repetition phase.

For one-location repetition, 26 test sessions were conducted: a randomly selected target location was held constant in trials 101 to 160. For two-location repetition, 42 sessions were conducted: two pseudorandomly selected target locations were held constant in trials 31 to 150. The target locations did not overlap. During the 120-trial repetition phase, the two locations were presented in randomized, two-trial blocks so that each location occurred 60 times. For both conditions, baseline and repetition trials were simply randomly generated shape and color displays.

Figure 8.5 shows mean target localization accuracy across trial blocks for one- (circles) and two-location (triangles) repetition. Filled symbols represent repetition trials and the open symbols represent color and shape baseline trials. Note that the blocking of the trials for one-location repetition (11–100, 101–115, 116–130, 131–145, 146–160, 161–180) and two-location repetition (11–20, 21–60, 61–90, 91–120, 121–150, 151–180) are slightly different. Separate, repeated-measures ANOVAs for the one-location and two-location repetition condition revealed a significant increase in accuracy across blocks of trials for one-location, $F(4, 20) = 10.3$, $p < .001$, but not for the two-location condition. Further, the one-location repetition improved search for both color and shape trials.

These results have several implications. First, repeating one target location improved search more than repeating two target locations, with the latter producing little benefit. This contrasts with what we found when we repeated two visual features across a session in a similar manner (Experiment 4, Katz & Cook, 2000). In that study, we found that pigeons could easily learn to avoid two types of shape distractors within a block of repetition. Why did the birds have more trouble with learning two locations? One possibility is that the two-location condition required the pigeons to still analyze the display's color and shape properties, whereas the one location did not. In the latter, the pigeons could go directly to the specific location without examining the specific features of the displays. In the former condition, however, the pigeons had to analyze whether there is a dimensional difference at a particular location in order to determine which of the two locations actually contained the target. This additional requirement may have made it more difficult for the birds to detect the repetition per se because of its greater similarity to the target search processes required on the baseline displays.

Another implication of the results is that the location of a target may be more important than the identity of a target in our localization task. Recall that repeating the identity of a target feature failed to produce any change in accuracy in the previously described experiment, or in Katz and Cook (2000). In the current experiment, however, repeating the target's location while changing its identity did produce an increase in search accuracy. These results are slightly different from that found by P.M. Blough and Lacourse (1994). They generally found that repetition of target identity and location were equally effective on pigeon search performance. It is difficult to draw any strong conclusions concerning these differences because of numerous procedural differences between tasks. Nevertheless, the above findings further indicate that pigeons can learn to use repetition of a target's spatial location to improve their search of a display.

**Figure 8.5**
Mean Target Localization Across Successive Trial Blocks Within a Session for One- (Circle Symbols) and Two- (Triangle Symbols) Target Location Repetition

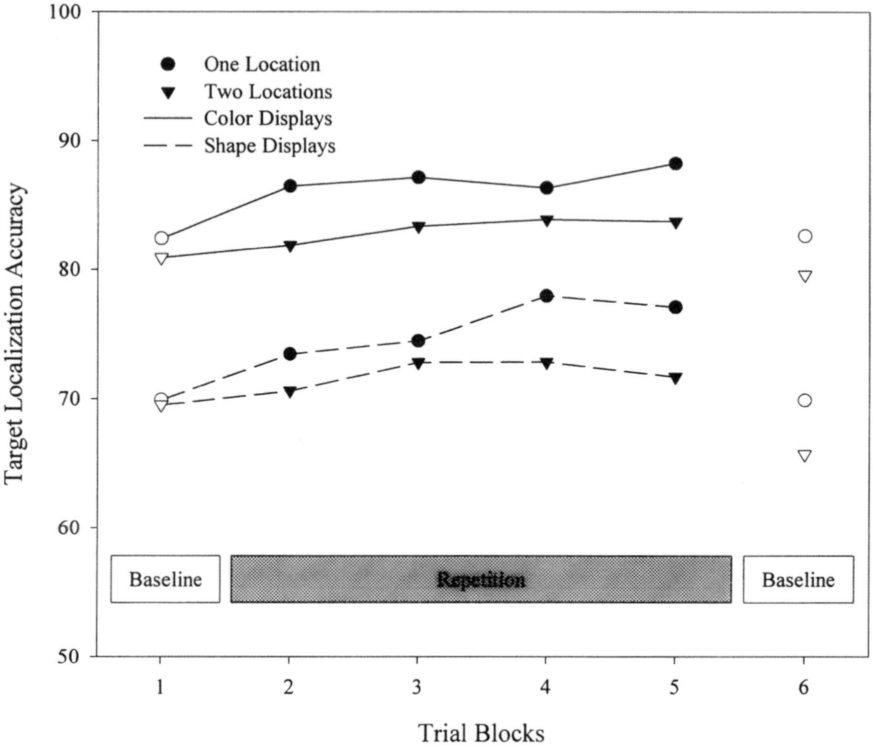


## CONCLUDING REMARKS

Two new findings are most important from the above experiments. First, pigeon search accuracy improves with increasing element specificity. Second, pigeon search accuracy improves with repetition of target location information. These findings add to the growing list of stimulus properties that pigeons can simultaneously use to solve visual search problems. Together, these results and the other past findings that have been reported support the theory of pigeons' flexibility to employ different search strategies. The pigeons seem sensitive to two broad classes of display attributes that support different strategies, one that is relational in nature—that these properties are derived by comparing information from across different portions of the display—and the other that is item-specific—that these properties are derived from the absolute properties of the display's features.

## Relational strategies

The birds seem to rely on at least two different relational strategies in solving our target localization task. The first is the "odd-item strategy," which functions by comparing the global perception of a texture stimulus with a generalized representation of the texture's regional properties (see also D.S. Blough, 1989). This process is most evident when the target is highly discriminable in relation to the distractors. When the target region is more cryptic and difficult to locate, other strategies are incorporated to find the target. These strategies require more effort to search and likely involve processing specific features within a display. The one we have found the most potential evidence for is a relational "distractor avoidance strategy." It functions by allowing the pigeons to better avoid distractors until the target is located. The latter strategy is an interesting mixture of specific and general properties. Although specific features of the display are processed within a trial, we have placed it in the class of relational strategies in part because it may be employed as a general strategy across independent trials. That is, it can be used to process any display from trial to trial regardless of its specific features. It is a combination of these two generalized strategies that contributes to the pigeon's ability to discriminate novel displays (Cook, 1992a) and maintain the successful discrimination of as many as 70,000 different texture displays.

## Item-specifics strategies

Despite this remarkable flexibility, the birds remain sensitive to item-specific information in the displays; then, when the opportunity occurs, as in the current repetition procedures, the pigeons can readily lock on to specific features of the display to improve their performance. The most prominent in our situation is the use of the "item-specific distractor avoidance strategy" in which birds learn within a session to avoid specific distractors (Katz & Cook, 2000). This strategy is evidenced by the pigeons' ability to remember and avoid specific distractor features, with the foregoing experiments adding that increasing this element specificity further increases this repetition effect. This process is probably a more concrete version of the general distractor avoidance strategy discussed previously. A closely related item-specific approach involves the "conditional expectancy strategy" (D.S. Blough, 1993). This strategy is also distractor-mediated, and occurs when a specific distractor can prime or alter the birds' search for a particular target or possible location. In the present experiments, we have found new evidence for this factor in that repetition of both the distractor and target elements produced better search than either repetition of the distractors or target alone. Also, this effect may be used both intra- and extra-dimensionally. Finally, the last item-specific strategy is the well-studied "item-specific target approach strategy." This strategy relies on learning and attending to the specific properties of the target (P.M. Blough, 1989, 1991, 1992,

1996; P.M. Blough & Lacourse, 1994; Bond, 1983; Bond & Riley, 1991; Langley, 1996; Langley et al., 1996; Pietrewicz & Kamil, 1979; Plaisted, 1997; Plaisted & Mackintosh, 1995; Reid & Shettleworth, 1992; Vreven & P.M. Blough, 1998). These target properties can be either identity- or location-based.

Taken as a whole, the evidence reviewed in this chapter suggests that pigeons solve visual search tasks by using virtually every piece of information available to them—simultaneously using a combination of both relational and item-specific search strategies that can involve properties of both the target and the distractors. One classic issue in the study of animal cognition has been to what degree discrimination learning is controlled by absolute and relational factors (e.g., Kohler, 1947; Lashley, 1942; Pavlov, 1927; Spence 1936). This debate was often cast as being mutually exclusive alternatives. This apparently is not the case. What seems abundantly clear is that animals, such as pigeons and primates, are quite capable of being controlled by both factors. The twist of the current work is that it suggests that both relational and absolute factors can operate concurrently in the same steady-state discrimination.

The direction for future research in this area lies in determining why and when animals apply these different solutions in solving problems. Several factors have already become evident; for example, the frequency of repeating particular features strongly promotes item-specific learning in the presence of relational strategies, as shown in the current experiments and by Katz and Cook, (2000). Likewise, using large numbers of exemplars, which reduces the frequency of a feature or image's appearance, also strongly promotes relational learning (Cook, Katz, & Cavoto, 1997; Santiago & Wright, 1984; Wasserman, Hugart, & Kirkpatrick-Steger, 1995; Wright, Cook, Rivera, Sands, & Delius, 1988). Whether the features making up the target and distractor elements of a trial are mutually exclusive or can change roles from trial to trial also seems to contribute to the type of strategy that predominantly guides search behavior (De Weerd et al., 1992; Katz, & Cook, 2000). Perceptual discriminability also influences whether relational or item-specific strategies are used (Bond, 1983; Bond & Riley, 1991; Katz & Cook, 2000; Langley, 1996; Langley et al., 1996; Reid & Shettleworth, 1992).

A final question of interest centers around whether the animals actively select among these strategies, much like Krechevsky (1932) suggested some years ago. We often explicitly select one type of strategy over another in solving problems, puzzles, and games. Can animals similarly engage in the active selection of their problem-solving approach? The evidence here is very limited. For our pigeons, we would suggest that the parallel nature of the strategies detected here do not involve any active selection. Rather, the learning that takes place is more implicit in nature, and is a direct function of contextual variables. If this is so, perhaps one benefit of language has been to permit humans to become strategically involved in deciding what behaviors to engage in next. Regardless, the continued analysis of different visual discriminations, such as the present search task studies, will help us to understand the patterns of behavior, and their underlying mechanisms, as an animal learns about different environmental events and their relations. Such discoveries will provide us with an understanding of the differences and similarities in the mechanisms of intelligence between human and nonhuman species.

## NOTE

This research was supported by a National Science Foundation Grant to Robert G. Cook. Preparation of this article was supported by a National Institute of Mental Health Individual National Research Service Award MH12511–01 to Jeffrey S. Katz. We thank Anthony Wright for comments on this article and Brian Cavoto for assistance in conducting these experiments.

Correspondence concerning this article should be addressed to either author Jeffrey S. Katz, Department of Psychology, 226 Thach Hall, Auburn University, Alabama, 36849, or Robert G. Cook, Department of Psychology, Paige Hall, Tufts University, Medford, Massachusetts 02155. Electronic mail may be sent via Internet to katzjef@auburn.edu or rcook1@emerald.tufts.edu.

## REFERENCES

Anderson, J.R. (1993). *Rules of the mind.* Hillsdale, NJ: Erlbaum.

Bhatt, R.S., Wasserman, E.A., Reynolds, W.F., & Knauss, K.S. (1988). Conceptual behavior in pigeons: Categorization of both familiar and novel examples from four classes of natural and artificial stimuli. *Journal of Experimental Psychology: Animal Behavior Processes, 14,* 219–234.

Blough, D.S. (1989). Odd-item search in pigeons: Display size and transfer effects. *Journal of Experimental Psychology: Animal Behavior Processes, 15,* 14–22.

Blough, D.S. (1993). Effects on search speed of probability of target-distractor combinations. *Journal of Experimental Psychology: Animal Behavior Processes, 19,* 231–243.

Blough, D.S. & Blough, P.M. (1997). Form perception and attention in pigeons. *Animal Learning & Behavior, 25,* 1–20.

Blough, P.M. (1989). Attentional priming and visual search in pigeons. *Journal of Experimental Psychology: Animal Behavior Processes, 15,* 358–365.

Blough, P.M. (1991). Selective attention and search images in pigeons. *Journal of Experimental Psychology: Animal Behavior Processes, 17,* 292–298.

Blough, P.M. (1992). Detectability and choice during visual search: Joint effects of sequential priming and discriminability. *Animal Learning & Behavior, 20,* 293–300.

Blough, P.M. (1996). Priming during multiple-target search: The cumulative effects of relative target frequency. *Animal Learning & Behavior, 24,* 394–400.

Blough, P.M., & Lacourse, D.M. (1994). Sequential priming in visual search: Contributions of stimulus-driven facilitation and learned expectancies. *Animal Learning & Behavior, 22,* 275–281.

Bond, A.B. (1983). Visual search and selection of natural stimuli in the pigeon: The attention threshold hypothesis. *Journal of Experimental Psychology: Animal Behavior Processes, 9,* 292–306.

Bond, A.B. & Riley, D. (1991). Searching image in the pigeon: A test of three hypothetical mechanisms. *Ethology, 87,* 203–224.

Carter, D.E., & Werner, T.J. (1978). Complex learning and information processing by pigeons: A critical analysis. *Journal of the Experimental Analysis of Behavior, 29,* 565–601.

Cerella, J. (1979). Visual classes and natural categories in the pigeon. *Journal of Experimental Psychology: Human Perception and Performance, 5,* 68–77.

Cook, R.G. (1992a). The acquisition and transfer of texture visual discriminations by pigeons. *Journal of Experimental Psychology: Animal Behavior Processes, 18,* 341–353.

Cook, R.G. (1992b). Dimensional organization and texture discrimination in pigeons. *Journal of Experimental Psychology: Animal Behavior Processes, 18,* 354–363.

Cook, R.G. (1992c). The visual perception and processing of textures by pigeons. In W.K. Honig, & G. Fetterman (Eds.), *Cognitive aspects of stimulus control* (pp. 279–299). Hillsdale: Erlbaum.

Cook, R.G. (1993a). Gestalt contributions to visual texture discriminations by pigeons. In T. Zentall (Ed.), *Animal cognition: A tribute to Donald A. Riley* (pp. 251–269). Hillsdale: Erlbaum.

Cook, R.G. (1993b). The experimental analysis of cognition in animals. *Psychological Science, 4,* 174–178.

Cook, R.G. (2000). The comparative psychology of avian visual cognition. *Current Direction in Psychological Science, 9,* 83–89.

Cook, R.G., Cavoto, K.K., & Cavoto, B.R. (1995). Same/Different texture discrimination and concept learning in pigeons. *Journal of Experimental Psychology: Animal Behavior Processes, 21,* 253–260.

Cook, R.G., Cavoto, K.K., & Cavoto, B.R. (1996). Mechanisms of multidimensional grouping, fusion, and search in avian texture discrimination. *Animal Learning & Behavior, 24,* 150–167.

Cook, R.G., Cavoto, B.R., Katz, J.S., & Cavoto, K.K. (1997). Pigeon perception and discrimination of rapidly changing texture stimuli. *Journal of Experimental Psychology: Animal Behavior Processes, 23,* 390–400.

Cook, R.G., Katz, J.S., & Cavoto, B.R. (1997). Pigeon same–different concept learning with multiple stimulus classes. *Journal of Experimental Psychology: Animal Behavior Processes, 23,* 417–433.

Cook, R.G., Katz, J.S, & Cavoto, B.R. (1998). Processes of visual cognition in the pigeon. In S. Soraci & B. McIlvane (Eds.), *Perspectives on fundamental processes in intellectual functioning: Vol. 1. A survey of research approaches.* Greenwich, CT: Ablex.

Cook, R.G., Katz, J.S., & Kelly, D.M. (1999). Pictorial same–different categorical learning and discrimination in pigeons. Special issue of *Current Psychology of Cognition, 18,* 805–843.

Cook, R.G., Wright, A.A., & Kendrick, D.F. (1990). Visual categorization by pigeons. In M.L. Commons, R.J. Herrnstein, S.M. Kosslyn, and D.B. Mumford (Eds.), *Quantitative Analyses of Behavior, Vol. 8.* Hillsdale: Erlbaum.

Cook, R.G., & Wixted, J.T. (1997). Same–different discrimination in pigeons: Testing competing models of discrimination and stimulus integration. *Journal of Experimental Psychology: Animal Behavior Processes, 23,* 401–416.

De Weerd, P., Vandenbussche, E., & Orban, G.A. (1992). Texture segregation in the cat: A parametric study. *Vision Research, 32,* 305–322.

Haberlandt, K. (1997). *Cognitive psychology.* Needham Heights: Allyn & Bacon.

Herrnstein, R.J., & de Villiers, P.A. (1980). Fish as a natural category for people and pigeons. In G.H. Bower (Ed.), *The psychology of learning and motivation: Vol. 14* (pp. 59–95). New York: Academic Press.

Herrnstein, R.J., Loveland, D.H., & Cable, C. (1976). Natural concepts in pigeons. *Journal of Experimental Psychology: Animal Behavior Processes, 2,* 285–302.

Jones, G., Ritter, F.E., & Wood, D.J. (2000). Using a cognitive architecture to examine what develops. *Psychological Science, 11,* 93–100.

Katz, J.S., & Cook, R.G. (2000). Stimulus repetition effects on texture-based visual search by pigeons. *Journal of Experimental Psychology: Animal Behavior Processes, 26,* 220–236.

Kohler, W. (1947). *Gestalt Psychology: An introduction to new concepts in modern psychology.* New York: Liveright.

Krechevsky, I. (1932). "Hypotheses" in rats. *Psychological Review, 39,* 516–532.

Langley, C.M. (1996). Search images: Selective attention to specific visual features of prey. *Journal of Experimental Psychology: Animal Behavior Processes, 22,* 152–163.

Langley, C.M., Riley, D.A., Bond, A.B., & Goel, N. (1996). Visual search for natural grains in pigeons (*Columba livia*): Search images and selective attention. *Journal of Experimental Psychology: Animal Behavior Processes, 22,* 139–151.

Lashley, K.S. (1942). An examination of the "continuity theory" as applied to discrimination learning. *Journal of General Psychology, 26,* 241–265.

Lee, A.Y. (1998). Transfer as a measure of intellectual functioning. In S. Soraci & B. McIlvane (Eds.), *Perspectives on fundamental processes in intellectual functioning: Vol. 1. A survey of research approaches.* Greenwich, CT: Ablex.

Mackay, H.A., Stromer, R., & Serna, R.W. (1998). Emergent behavior and intellectual functioning: Stimulus classes, generalization, and transfer. In S. Soraci & B. McIlvane (Eds.), *Perspectives on fundamental processes in intellectual functioning: Vol. 1. A survey of research approaches.* Greenwich, CT: Ablex.

Newell, A., & Simon, H.A. (1972). *Human problem solving.* Englewood Cliffs: Prentice-Hall.

Nisbett, R.E. (1993). *Rules for reasoning.* Hillsdale: Erlbaum.

Pavlov, I.P. (1927). *Conditional reflexes.* London: Oxford University Press.

Pietrewicz, A.T., & Kamil, A.C. (1979, June 22). Search image formation in the blue jay (*Cyanocitta cristata*). *Science, 204,* 1332–1333.

Plaisted, K. (1997). The effect of interstimulus interval on the discrimination of cryptic targets. *Journal of Experimental Psychology: Animal Behavior Processes, 23,* 248–259.

Plaisted, K.C., & Mackintosh, N.J. (1995). Visual search for cryptic stimuli in pigeons: Implications for the search image and search rate hypotheses. *Animal Behaviour, 50,* 1219–1232.

Reid, P.J., & Shettleworth, S.J. (1992). Detection of cryptic prey: Search image or search rate? *Journal of Experimental Psychology: Animal Behavior Processes, 17,* 273–286.

Santiago, H.C., & Wright, A.A. (1984). Pigeon memory: Same/different concept learning, serial probe recognition acquisition, and probe delay effects on the serial-position function. *Journal of Experimental Psychology: Animal Behavior Processes, 10,* 498–512.

Schneider, W., & Shiffrin, R.M. (1977). Controlled and automatic human information processing: I. Detection, search, and attention. *Psychological Review, 84,* 1–66.

Shaw, M.L., & Shaw, P. (1977). Optimal allocation of cognitive resources to a spatial location. *Journal of Experimental Psychology: Human Perception & Performance, 3,* 201–211.

Shiffrin, R.M., & Schneider, W. (1977). Controlled and automatic human information processing: II. Perceptual learning, automatic attending, and a general theory. *Psychological Review, 84,* 127–190.

Sidman, M.S., Rauzin, R., Lazar, R., Cunninghamn, S., Tailby, W., & Carrigan, P. (1982). A search for symmetry in the conditional discriminations of rhesus monkeys, baboons, and children. *Journal of the Experimental Analysis of Behavior, 37,* 23–44

Sidman, M.S., & Tailby, W. (1982). Conditional discrimination vs. matching to sample: An expansion of the testing paradigm. *Journal of the Experimental Analysis of Behavior, 37,* 5–22.

Spence, K.W. (1936). The nature of discrimination learning in animals. *Psychological Review, 43,* 427–499.

Thorndike, E.L. (1898). Animal intelligence: An experimental study of the associative process in animals. *Psychology Review Monographs, 2,* 551–553.

Vaughan, W. Jr., & Greene, S.L. (1984). Pigeon visual memory capacity. *Journal of Experimental Psychology: Animal Behavior Processes, 10,* 256–271.

Vreven, D., & Blough, P.M. (1998). Searching for one or many targets: Effects of extended experience on the runs advantage. *Journal of Experimental Psychology: Animal Behavior Processes, 24,* 98–105.

Wasserman, E.A., Hugart, J.A., & Kirkpatrick-Steger, K. (1995). Pigeons show same–different conceptualization after training with complex visual stimuli. *Journal of Experimental Psychology: Animal Behavior Processes, 21,* 248–252.

Wright, A.A. (1997). Concept learning and learning strategies. *Psychological Science, 8,* 119–123.

Wright, A.A., Cook, R.G., Rivera, J.J., Sands, S.F., & Delius, J.D. (1988). Concept learning by pigeons: Matching-to-sample with trial-unique video picture stimuli. *Animal Learning & Behavior, 16,* 436–444.

Chapter 9

# The Search for Relational Learning Capacity in *Cebus Apella:* A Programmed "Educational" Approach

*Romariz da Silva Barros, Olavo de Faria Galvão, and William J. McIlvane*

This chapter describes a novel approach to assessing the cognitive capacities of nonhuman primates. Using capuchin monkeys (*Cebus apella*) doing visual information processing tasks as an illustrative example, we present arguments and data supporting the idea that efficacious assessment of such capacities can be done by implementing methods and principles of programmed instruction. We begin by reviewing selectively available data on assessment of stimulus equivalence and related phenomena in this species and selected other species, including recent studies from our laboratory. In this context, we will then briefly discuss studies that have employed programmed instructional methods to assess the cognitive capabilities of children with developmental disabilities. Those methods have been used to solve conceptual and methodological problems that are virtually identical to those encountered in the area of nonhuman cognition. To conclude, we describe a new program of research that more fully implements programmed instructional procedures to better reveal the relational learning capacity of capuchins, and perhaps to serve as a helpful model for work with other species.

## STUDIES OF THE RELATIONAL LEARNING CAPACITY OF NONHUMAN PRIMATES

Much of the research in animal cognition has been conducted to assess the abilities of nonhuman species to learn generalized relations between stimuli or relations between relations (e.g., Rumbaugh, 1990; Thompson, Oden, & Boysen, 1997), typically presented in some sort of visual–visual relational discrimination task. A defining characteristic of this work is a search for "emergent" and/or "generalized" performances, that is, performances that cannot be accounted for by direct conditioning. The ability to respond to generalized stimulus–stimulus relations is one of the hallmarks of intelligent behavior.

## STIMULUS EQUIVALENCE

A particular interest of animal cognition researchers is to assess the degree to which various species have the potential to function symbolically. Deacon (1997) posed the basic question as follows:

> [I]s there some fundamental difference between the way my dog and I understand the same spoken sounds? Common sense psychology has provided terms for this difference. We say that the dog learned the command "by rote," whereas we "understand" it. But this is a notoriously difficult difference to specify (p. 52).

Research in behavior analysis offers a way to differentiate symbolic relations from rote paired-associates, the task that Deacon (1997) termed so difficult. Sidman and Tailby (1982) distinguished relations learned as discrete pairs (*conditional* relations) from "emergent" (*equivalence*) relations, a distinction directly analogous to the distinction between simple paired-associates and true symbolic relations. How might Sidman's analysis contribute to a better operational definition of symbolic behavior? His analysis is unique in the specification of precise criteria for inferring symbolic relations, based on mathematical definitions of equivalence. The properties of an equivalence relation are reflexivity, symmetry, and transitivity.

In stimulus equivalence research, the procedure used most often is matching to sample. In this procedure, the participant chooses among various comparison stimuli (often three per trial) on the basis of another stimulus, the sample; the sample varies across trials, as does the comparison stimulus that is to be chosen. In a typical stimulus equivalence experiment, the participant is taught arbitrary matching relations among three sets of physically dissimilar stimuli, *A, B,* and *C;* relations *AB* and *BC* are taught directly, and then various tests are conducted to assess the nature of the matching relations that were established during the training. Reflexivity tests assess whether each stimulus relates to itself (*AA, BB,* and *CC*), that is, whether the participant will match each member of the various sets to itself on identity matching-to-sample tests. Symmetry tests assess functional sample-comparison reversibility (*BA* and *CB*), that is, whether the participant taught to match *A* to *B* and *B* to *C* will subsequently match *B* to *A* and *C* to *B*. Transitivity tests assess whether a participant taught to match *A* with *B* and also to match *B* with *C* will subsequently match *A* with *C* without further training. The *CA* test is a combined test for several properties. If the participant passes all of the tests just described, then one is justified in describing the behavioral relations as equivalence (or "true" symbolic) relations (Sidman & Tailby, 1982).

## Studies of symmetry and transitivity

In an influential, widely cited study, Sidman, Rauzin, Lazar, Cunningham, Tailby, and Carrigan (1982) failed to demonstrate symmetry in a multi-experiment study conducted with macaques and baboons. The failure was noteworthy because high-accuracy, arbitrary matching-to-sample (MTS) baselines were developed prior to critical symmetry tests and procedures were varied in an effort to encourage stimulus control by task-relevant stimuli. Subsequent research has further explored methodological variations that might foster positive demonstrations of defining properties of stimulus equivalence in a variety of nonhuman primates.

D'Amato, Salmon, Loukas, and Tomie (1985) conducted experiments with capuchins using a zero-delay matching-to-sample procedure that was intended to discourage control by sample-comparison stimulus compounds.[1] Despite the zero-delay procedure, symmetry test results were universally negative. Transitivity tests yielded positive results. However, these positive results were difficult to interpret, raising a variety of questions about the nature of the stimulus–stimulus relations that had been generated by the training contingencies. Was it possible, for example, that unmeasured sample-specific behaviors had been established to the *A* and *B* samples such that the animals' behavior became the effective samples, and that shared "coding" responses rather than "true" transitive stimulus–stimulus relations determined the positive test results (cf. Cohen, Brady, & Lowry, 1981)? Were the transitivity outcomes positive because those tests did not involve presenting stimuli in new locations as occurred on the symmetry tests (see the following "Minimizing stimulus control by location")?

Findings suggesting a possible role for sample-specific behaviors in producing positive equivalence test outcomes were reported by McIntire, Cleary, and Thompson (1987). Their macaques were trained to emit one response pattern to members of one potential equivalence class and another pattern to the members of another potential class. The animals subsequently achieved high scores on tests that were formally identical to symmetry and transitivity tests. As Saunders (1989) pointed out, however, the class-specific differential response training compromised the tests: Behavioral stimuli associated with the class-specific differential response patterns may have been the effective sample stimuli on all trials; thus all performances exhibited on the test trial may actually have been directly trained.

With chimpanzees who were studied in the Rumbaughs' well-known language training studies, results have been similarly difficult to interpret. Cerutti and Rumbaugh (1993) reported that Sherman and Austin, the subjects in the study, had shown equivalence relations with lexigrams related to foods and tools. It seems plausible that use of familiar stimuli during training helped direct the participants' attending to the stimulus aspects that were relevant to the task. Complicating the interpretation, however, was a recently reported study by Dugdale and Lowe (2000). They gave Sherman a symmetry test using procedures like those of Sidman and colleagues (1982), and he failed it, despite his extensive extraexperimental history. In another equivocal finding, Tomonaga, Matsuzawa, Fugita and Yamamoto (1991) reported evidence of symmetry in one of three chimpanzees (*Pan troglodytes*). In summary, none of the reported studies with nonhuman primates has provided data that strongly support claims that these species can pass symmetry or transitivity tests as defined by Sidman and Tailby (1982).

## Studies of generalized identity matching (reflexivity)

A number of identity-difference relational learning studies have been conducted with nonhuman primates. Among the best known was work by Premack (e.g., 1983) and his colleagues in studies of the ability of chimpanzees to acquire language-like performances. Results from that program suggest that chimpanzees may be capable of generalized same–difference judgments. For example, Oden, Thompson, and Premack (1990) obtained positive results using a preferential-looking technique similar to that used extensively in testing human infants. Other supporting data came from a 1988 study by this same team of investigators, also studying infant chimpanzees. During an initial familiarization phase, the animals were taught to place objects in baking tins on command. During a subsequent training phase, the task was systematically changed to identity matching to sample; animals learned to place in the tins objects that matched sample objects. On subsequent transfer tests with novel objects, scores were typically high, thus suggesting generalized identity matching.

Published results with capuchins have been equivocal. For example, D'Amato, Salmon, and Columbo (1986) have reported savings on acquisition of new identity-matching performances after training on an original identity matching problem. No convincing demonstration of generalized identity matching with this species has yet been reported. However, new data from our laboratory may change that picture. We will describe those data after discussion of methodological issues that could be important in understanding why it may be difficult to demonstrate generalized stimulus–stimulus relations in nonhumans.

### *Minimizing stimulus control by location*

Recent studies in the primate laboratory of the Universidade Federal do Pará (UFPA) have investigated the possibility that improvements in teaching methodology may allow capuchins to demonstrate generalized identity matching and perhaps, ultimately, stimulus equivalence. One source of inspiration for this line of work came from earlier work with other species. With macaques, Iversen, Sidman, and Carrigan (1986) demonstrated that invariant sample location (the typical procedure in equivalence research) may encourage undesired control by sample stimulus–sample location compounds rather than control by the experimenter-specified sample. Briefly, macaques were trained on an identity matching-to-sample task with both colors and line orientations. During training, the location of the sample stimulus was invariant (a key in the center of the display). On subsequent test trials, the sample was presented on side keys. That procedure disrupted well-learned identity matching with the lines (but not the colors). Thus, these investigators demonstrated that the effective sample in the line-matching task was line orientation plus line location (e.g., a vertical line on the center key rather than merely a vertical line). Iversen (1997) and Lionello and Urcuioli (1998) reported similar findings with rats and pigeons. In conceptually related work, Washburn, Hopkins, and Rumbaugh (1989) reported that performance on an identity matching-to-sample task was enhanced when sample stimuli changed location both within and across trials. Varying stimulus location, in fact, enhanced performance on a number of visual information processing tasks.

In every equivalence study published so far, experimenters took for granted that sample location was not part of the participants' stimulus definition. As the work by Iversen and colleagues (1986) shows, however, one cannot safely make the same assumption in all cases. Given that, what evidence exists that undesired control by stimulus location can be eliminated by presenting discriminative stimuli in continuously varying locations? Iversen and colleagues (1986) showed that it was feasible to present sample stimuli in varying locations in a matching-to-sample paradigm. However, these investigators did not incorporate any explicit training procedures to prepare their subjects for varying locations. The feasibility of such training was demonstrated in a subsequent study by Lionello-DeNolf and Urcuioli (2000), who showed that pigeons could be taught arbitrary matching performances with continuously varying sample locations. In unpublished work conducted at UFPA, we have shown that capuchin monkeys can master identity- and arbitrary-matching performances when sample and comparison stimuli vary continuously over nine locations presented on a touchscreen-equipped computer monitor (Barros, 1998). It appears feasible to conduct tests for relational learning capacity in which unwanted stimulus control by location is systematically managed.

### *Minimizing other irrelevant sources of stimulus control*

Minimizing unwanted control by location is not sufficient by itself to verify that the animal detects those aspects of the stimuli that the experimenter defines as relevant. Animal cognition researchers have long recognized that it is necessary also to assure that animals do not respond to other irrelevant stimulus features. We already mentioned the zero-delay procedure reported by D'Amato and colleagues (1986) to control for stimulus compounding. Procedures are also needed to prevent neophobic responding (i.e., avoiding novel stimuli) (e.g., Zentall, Edwards, Moore, & Hogan, 1981). Ideally, the animal should have a history with the test stimuli before the critical tests. That history should not only adapt the animal to stimulus novelty but also provide a history of reinforcement with test stimuli. Without such a history, one cannot separate test failures due to lack of relational learning capacity from failures due to avoiding stimuli that do not have the same reinforcement valence (or potential valence) as the baseline relations on which the emergent relations depended.

The study by Iversen and colleagues (1986) also made an important point about the importance of stimulus variables. Recall that varying sample location disrupted well-established identity-matching with line orientations but not similar performances with colors. The colors were presumably easier to discriminate than the lines; the latter have several physical features in common (e.g., similar shape, color, length, etc.; cf. Carter & Eckerman, 1975). Such data help make the point that appropriate tests for generalized identity-matching should present sample and comparison stimuli that are readily discriminable by the subject. The more physically similar the elements of the test set, the more likely is it that generalization will artifactually suppress test scores. Unfortunately, stimulus similarity is not always easily detected by the experimenter when choosing test stimuli. The preceding does not exhaust the list of those irrelevant sources of control that can compromise a test for relational learning capacity. Providing adequate motivational supports during training and testing is also important, assuring that training procedures generate sample-*S*+ rather than sample-*S*– control (cf. Sidman, 1987), and providing overtraining of baseline relations prior to testing.

### *Comparisons with developmentally limited humans*

When human children are exposed to tests for generalized identity matching, the experimenter and participant have the advantage of a vast extraexperimental learning history. For example, children learn to attend to relevant features of the environment via language training and many other formal and informal teaching opportunities. For example, children are often explicitly taught to attend to relevant stimuli and stimulus classes (as in the implicit shaping entailed in "motherese"). Typical preschool children have virtually daily exposure to the symbolic nature of letters, numbers, and other forms that serve as symbols, *if only through watching television.* In addition, fortunate human children are immersed in an environment that provides a wealth of social and other consequences for learning, and the absence of such an environment may lead to lifelong learning problems (Ramey & Ramey, 1998). When evaluating the relational learning capacity of nonhuman primates, is it reasonable to expect phylogenically lower species to meet the human standard in the absence of these potentially relevant experiences? We think not.

## A PROGRAMMED EDUCATIONAL APPROACH

Is it possible to provide nonhumans with learning experiences that help to compensate for their relative lack of extraexperimental experience as compared to human children? The UFPA laboratory has initiated a research program that is directed at answering this question. Work thus far is of a preliminary nature, but it has progressed enough that we may provide a provisional "Yes" to the question. In the current work, we are approaching capuchins not merely as animals to be tested but also as students to be instructed. In a programmed educational approach, one begins with a careful analysis of the behavioral prerequisites for success on a given task. Thereafter one designs training procedures to supply the critical behavioral prerequisites. The adequacy of those procedures is evaluated by learning outcomes. If those outcomes are positive, then the initial analysis is confirmed. However, if learning outcomes are not positive, then it is the experimenter's responsibility to rethink the analysis and procedures to better instantiate the necessary behavioral prerequisites—just as a regular or special education teacher might do.

This approach is unusual in experimental animal cognition research. Many prior experimental studies have taken a more incrementalist approach. When animals fail to exhibit performances of interest in one experiment, another is performed with technical improvements that the experimenter hopes will succeed. The Sidman and colleagues (1982) study provides an excellent example of the incremental approach. This approach is eminently reasonable and understandable from the perspective of the curious scientist who wants to know precisely what procedural variation(s) "did the trick." However, the incrementalist approach may have certain disadvantages when the goal is trying to establish the behavioral capacities of a given species. Because many experiments may have to be done, it may be a slow and perhaps unnecessarily costly enterprise.

In a programmed educational approach, by contrast, the goal from the beginning is to provide the animal with every potentially helpful procedural feature that one can devise based on what is known about effective teaching (cf. Barros, et al., in press). The approach is similar to the "treatment package" approach that is often used to address clinical problems in behavioral intervention research. No attempt is made initially to identify specific procedural features that may be critical. If such analyses are of interest, they are done via subsequent component analyses—after the performance of interest has been generated by the comprehensive treatment package.

One advantage of a programmed approach is that it may reduce the need to conduct numerous unsuccessful experiments before achieving success (or not). Indeed, the incremental approach may require an unusually dedicated investigator who has the time and support to proceed with caution, and who is not unduly discouraged by negative findings. One risk of a programmed approach is that the experimenter/teacher may devise procedures that generate performances phenotypically similar to the performance of interest, but controlled by diferent variables than those inferred by the experimenter, as in the study by McIntire and colleagues (McIlvane & Dube, 1992). In both the incrementralist and the programmed approach, careful analysis of the behavioral prerequisites for a given task can minimize such problems. Moreover, if the prerequisite analysis is reasonably comprehensive and if the best available teaching methodology fails to generate encouraging findings, then one can be more confident that the capacity of interest may not be present or at least not demonstrable with techniques now available.

### *Precursors in the current literature*

A programmed educational approach to studying nonhuman primate cognition has many precursors. For example, there has been a number of impressive training studies that sought to provide animals with experiences that might allow them to display language-like and other advanced cognitive performances (e.g., Premack, 1983; Pepperberg, 2000). One can readily detect in these studies concern for effective teaching methods, although analysis of behavioral prerequisites was often implicit rather than explicit. Notably, most of these studies have produced encouraging results. However, there is an inherent problem in interpreting data from quasi-naturalistic training settings. Has the methodological rigor been adequate? It would be helpful if seemingly advanced cognitive abilities were also demonstrated under unquestionably rigorous laboratory conditions.

With respect to stimulus equivalence and related phenomenon, we know of only one other laboratory that has combined laboratory methodology with an instructional approach similar to what we are advocating. Schusterman and his colleagues (e.g., Schusterman & Kastak, 1993; Schusterman, Gisiner, Grimm, & Hanggi, 1993) have used the basic approach to provide impressive data on the relational learning potential of the California sea lion. Those studies have demonstrated stimulus equivalence, generalized identity matching, and exclusion (emergent matching-to-sample performances that are procedurally identical to the "fast mapping" phenomenon reported in child language research; cf. Wilkinson, Dube, & McIlvane, 1997). The approach is exemplified in the Schusterman and Kastak 1993 equivalence study. The investigators gave the animal "a programmed instructional sequence which gave the animal a greater breadth of MTS experience with examples of sample and comparison switching roles *prior* (italics original) to tests of novel symmetrical relations" (p. 825). Also, they reinforced test trial performances. Thus, they did not require the sea lion to demonstrate equivalence relations in the face of extinction conditions.

## A PRELIMINARY IMPLEMENTATION WITH THE CAPUCHIN

### Generalized identity matching

Recent data from the UFPA laboratory suggests that capuchins may in fact be capable of generalized identity matching. Barros, Galvão, and McIlvane used a variety of procedures designed to encourage the development of stimulus control by relevant stimulus differences. A zero-delay identity-matching procedure was used to discourage stimulus compounding. Stimuli to be discriminated on matching trials were given a history on simple discrimination trials to reduce neophobic reactions and to provide a history of reinforcement for responding to them. Correct matching responses on all trials were followed by reinforcers. Sample and comparison locations were varied continuously across trials to minimize development of location-based controlling relations. Three-choice MTS was employed to reduce the likelihood of accurate performance based on sample-S- relations. Using these procedures, many near-perfect scores were obtained on tests of generalized matching of colors and forms. These data suggested that test-score performance might be enhanced if the capuchins were first exposed to a procedure in which stimuli to be matched on critical tests were first presented on simple simultaneous discrimination trials in which the stimuli defined as *S+* and *S–* were repeatedly reversed (cf. Schusterman & Kastak, 1993). Such reversal training provided potentially relevant behavioral prerequisites for conditional identity matching-to-sample tests; the reversal trials not only assured that the monkeys could discriminate each of the stimuli from the other but also gave each of the stimuli a history as both *S+* and *S–*.

Follow-up data from the original study are presented in Figure 9.1. Each of the bars shows scores on initial identity-matching tests with two sets of abstract forms (e.g., Greek letters). Each of three capuchins exhibited test scores that were substantially above chance on all four tests. Moreover, scores were substantially higher when the stimulus sets had reversal histories. Given reversal histories, one animal achieved perfect scores on both identity-matching tests. Thus, the data from Barros, Galvão, and McIlvane, and those reported here, encourage further studies of the relational learning potential of this New World monkey.

## Arbitrary matching to sample

Other research at UFPA asked whether procedures like those just reported can be used to develop arbitrary matching-to-sample baselines. Two animals were exposed to a zero-delay arbitrary matching-to-sample procedure in which three forms (Set *A*) and three colors (Set *B*) served as samples and comparison stimuli, respectively (*AB* matching). Those same stimuli also appeared on zero-delay identity-matching trials. Correct matching responses on all trials were followed by reinforcers. After protracted training, both animals ultimately acquired the *AB* matching baseline to an accuracy criterion of 18 successive correct responses. *BA* symmetry tests followed. Although one animal displayed "chance"-level performance, the symmetry-test performance of the other animal was unusual and possibly encouraging. The latter animal's performance on four successive *BA* symmetry tests is shown in Figure 9.2.

**Figure 9.1**
Results of Follow-up Tests Conducted after Barros, Galvão, and McIlvane

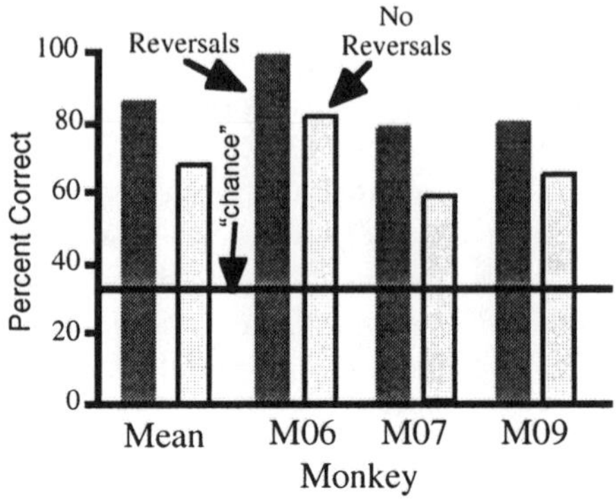


The matching-to-sample task presented three comparisons, hence "chance" correct performance line is placed at 33% correct.

The animal displayed a surprising avoidance of the comparison stimulus that was to be matched to the sample. Although the high "chance" level of the procedures made the findings difficult to interpret, this animal may have been exhibiting some form of conditional stimulus control, which was maintained over four tests, despite the absence of reinforcers for nearly all test-trial selections. A trial-by-trial analysis of the data did not reveal any consistent response patterns other than avoiding the "matching" comparison. Moreover, the *A* and *B* stimulus sets were physically dissimilar (colors and forms), and one could not attribute the seeming avoidance to undetected matching of similar stimulus features. Although the procedures and data indicate extreme caution in drawing firm conclusions, they support an interesting speculation: Perhaps the training did establish some potential for emergent behavior, but based on stimulus control topographies that differed from those that we hoped to establish (see Sidman [1980] and McIlvane, Serna, Dube, & Stromer [2000] for discussions of relevant issues in interpreting conditional discrimination accuracy scores).

### Critique of the work thus far

These UFPA capuchin studies represent only a limited implementation of a programmed educational approach, due in part to limitations of current apparatus and computer programs. The most comprehensive implementation has been in the work on generalized identity matching, and those studies have produced the most encouraging results. Nevertheless, the laboratory is not yet fully practicing what we have been preaching here. Recognizing that, we now

**Figure 9.2**
Symmetry Test Results from One Monkey

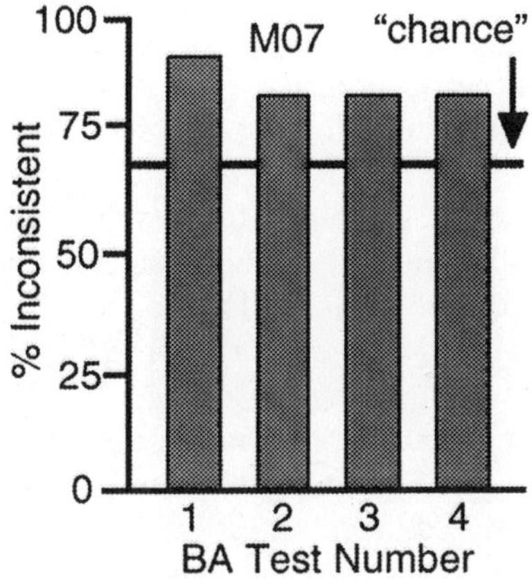


Bars show the percent of selections that were inconsistent with symmetry. The matching-to-sample task presented three comparisons, hence "chance" inconsistent performance line is placed at 67% correct.

prefer to think of the UFPA laboratory as an experimental "school" for capuchins (Barros & Galvão, submitted). The goal of the "schooling" is to help the capuchins maximize their potential to inform us about what they know and can learn to do. With this change of approach, the primate laboratory at the Universidade Federal do Pará is collaborating with the Behavioral Technology Group at the Shriver Center. This group has pursued a long-standing program of research on teaching technology for individuals with severe mental retardation. At Shriver, a programmed educational approach has been evident since pioneering work by Sidman and Stoddard (1966) first revealed the potential of effective instructional programming for this population. More recent work there has led to a number of techniques that may be potentially helpful to researchers interested in assessing the relational learning potential of nonhuman primates and possibly other species.

## RELEVANT STUDIES WITH MINIMALLY VERBAL HUMANS

In this part, we first give a highly selective history of research on relational learning in humans with severe developmental limitations. Our goals are to (1) establish context for presenting subsequent material on teaching approaches and (2) illustrate that the scientific interest of investigators working in this area directly parallels that of certain researchers in the field of animal cognition.

### Stimulus equivalence in minimally verbal individuals

This topic has been of interest throughout the history of this research area (e.g., Sidman, Cresson, & Willson-Morris, 1974; Sidman, Willson-Morris, & Kirk, 1986). The interest has been partly to determine the relational learning capabilities of such individuals, and further, to assess the role that language acquisition might play in permitting and/or enhancing the ability of individuals to display equivalence classes. For example, a widely cited study by Devany, Hayes, and Nelson (1986) led initially to the conclusion that equivalence classes might be demonstrable only in individuals who exhibited some verbal skills. Data supporting this possibility also came from the Bangor group who studied typically developing children (e.g., Horne & Lowe, 1996). However, the picture has begun to change. Carr, Wilkinson, Blackman, and McIlvane (2000) reported positive outcomes in four participants with severe mental retardation, no oral naming skills, and very limited listening skills. Unpublished follow-up studies have replicated these results with two additional participants with similar behavioral repertoires, showing equivalence with individuals who have only the barest rudiments of language.

### Related methodological investigations

The 1980s were a time of unprecedented growth of research on stimulus equivalence and other relational learning phenomena. As part of this growth, equivalence procedures were being replicated with participants who functioned at lower behavioral levels than those typically studied in the seminal work by Sidman and his colleagues (Sidman, 1971; Sidman, Kirk, & Willson-Morris, 1986). Failures to establish the necessary baselines were frequent with this population. What was needed for relational learning research was a reliable methodology for routinely establishing matching-to-sample baselines within practical time limits. Methodological research with that goal has been conducted by several laboratories (e.g., Dube, Ienacco, & McIlvane, 1993; Saunders & Spradlin, 1990; Zygmont, Lazar, Dube, & McIlvane, 1992).

## PROGRAMMING TECHNIQUES FOR NONVERBAL POPULATIONS

For illustrative purposes, we will concentrate on procedures for teaching identity matching to sample, in which a programmed educational approach has been the most extensively implemented. Specifically, we offer a detailed summary of methodology that has reliably established generalized identity matching in an ever-growing population of participants functioning at low behavioral levels. We refer the reader to other sources (e.g., Carr et al., 2000) for information on procedures for teaching arbitrary matching, which are somewhat less extensively researched.

As we suggested earlier, the beginning point of a programmed educational approach is to develop a taxonomy of discrimination skills necessary to build identity-matching repertoires and an analysis of the stimulus control requirements of those skills. For example, one can draw a distinction between simple and relational discrimination. In a simple simultaneous discrimination problem, for example, the participant may be given trials presenting a choice between stimuli *X* and *Y*, with selections of *X* the reinforced behavior. To receive a reinforcer on every trial, the participant must merely locate *X* and touch it. By contrast, matching to sample requires a relational discrimination; the participant must locate at least two discrete stimuli before responding (e.g., the sample and the positive comparison).

In analyzing matching-to-sample requirements, we have also found it useful to distinguish between nonconditional and conditional matching problems (Dube, McIlvane, & Green, 1992). In nonconditional matching procedures, the *S*+ and *S*– functions of comparison stimuli do not change from trial to trial. In other words, the participant is never confronted with a situation in which a recently correct comparison stimulus is now incorrect or vice versa. In conditional matching, by contrast, the *S*+ and *S*– functions of comparison stimuli do change from trial to trial, conditionally upon the sample. Distinctions drawn between simple versus relational discrimination, and conditional versus nonconditional matching are important; there is good evidence that such procedures may have very different behavioral outcomes (see Dube et al., 1993). In a programmed teaching approach, going from simple to relational discrimination and from nonconditional to conditional discrimination are major program goals.

A decade-long program of methodological research in the Shriver laboratories sought to develop the capability for routinely producing generalized identity matching of abstract two-dimensional forms in individuals with low mental age (MA) scores. The research has focused on individuals with low MAs because past studies have reported that such individuals may present difficult challenges in securing stimulus control by tasks that assess same–difference judgments (e.g., House, Brown, & Scott, 1974; cf. Soraci & Carlin, 1992). In particular, the Shriver labs have been interested in building the capability for effective teaching of individuals who do not respond well to syntactically complex verbal instructions. Nonverbal instructional technology has been emphasized to reach even individuals with minimal verbal behavior.

There are several reasons for a focus on abstract, nonrepresentative stimuli:

(1) A major goal of the Shriver program has been to develop broadly useful teaching and assessment procedures that do not depend on the type of stimuli used or their preexperimental history (cf. Sidman et al., 1986). Discrimination of abstract forms has historically presented difficult challenges for people with intellectual disabilities (e.g., Zeaman & House, 1979). A training approach that could overcome these challenges would be likely to prove broadly useful.

(2) There has been a long-standing interest in developing a reliable methodology for establishing the behavioral prerequisites for rudimentary reading in individuals

who do not ordinarily learn these skills. For example, Sidman's widely publicized research on stimulus equivalence and related phenomena began with efforts to teach relations among dictated English words, representational pictures, and printed English words (e.g., rudimentary reading). Discrimination of printed letters and letter combinations obviously requires discrimination of abstract, two-dimensional forms.

(3) Abstract forms are routinely used in experimental work that seeks to isolate the effects of experimental training contingencies from extraexperimental influence. For example, as Sidman's equivalence research program progressed, it was deemed necessary to employ abstract stimuli for purposes of experimental control in certain experiments (a practice that has been emulated by many other laboratories).

## USE OF COMPUTERS IN TEACHING

Consistent with the goal of developing broadly useful procedures, assessment and teaching programs are implemented on a microcomputer equipped with a touch-sensitive screen. Microcomputer presentation allows a high degree of rigor and precision in both stimulus presentation and response recording. These attributes, as well as the ease with which complex programming can be presented, are likely to contribute to a more broadly applicable and replicable teaching and testing technology.

## OVERVIEW OF TRAINING PROCEDURES

Figure 9.3 illustrates the Shriver approach for assessing and teaching generalized identity matching (see Serna, Dube, & McIlvane, 1997 and Dube & Serna, 1998 for further details). Participants are first familiarized with the teaching materials and/or apparatus; then they are assessed for readily demonstrable identity matching. Those who show such performance (initial assessment, Figure 9.3) are then tested for generalized matching with a variety of other stimuli (Figure 9.3, rightmost path). Individuals who fail initial assessments are then exposed to one or more of the five teaching components shown in the left and lower portions of Figure 9.3. Success within any one component leads to a reassessment of identity matching; failure results in training with the subsequent component.

The five training components are arranged from the least to most elaborate, in terms of time and effort required for programming. In this approach, the first three components are designed to screen out individuals who can be successful with quasi-programmed teaching of the type that characterizes instructional programming in good education programs; the remaining two components represent more elaborate interventions designed for more challenging cases. A brief summary follows of the essential features of the components and the research that supports them.

**Figure 9.3**
Diagram of the Procedures Reported by Serna, Dube, and McIlvane (1997) for Assessing and Teaching Generalized Identity Matching-to-Sample

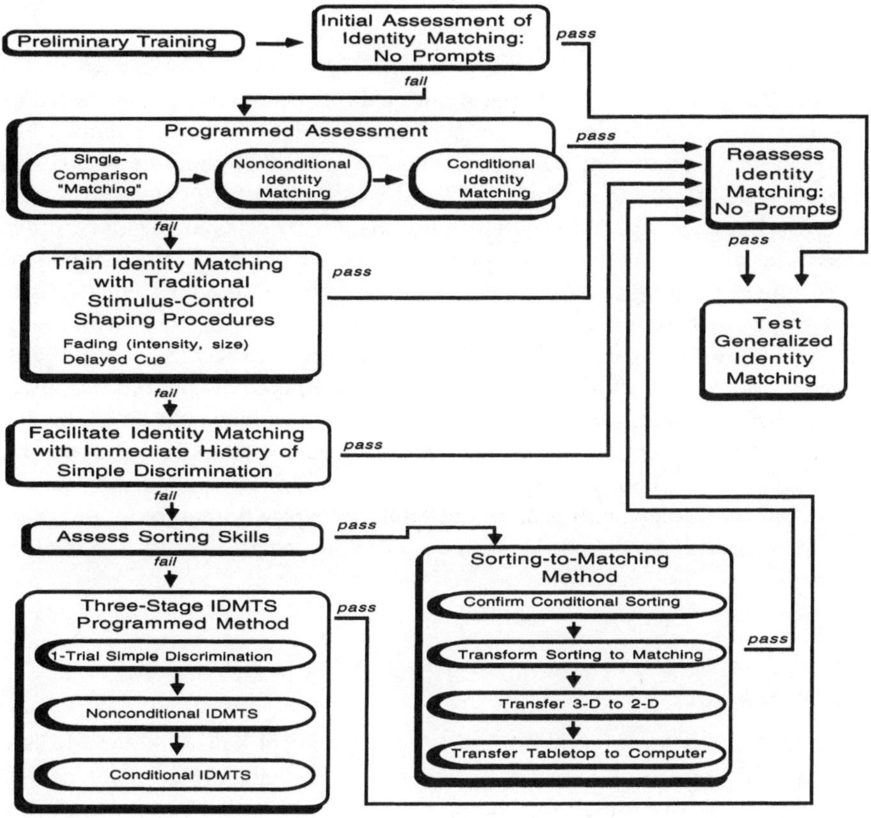


## Programmed assessment

Participants progress through three steps presenting increasingly complex stimulus-control requirements. For example, in the first step, only the sample and a comparison that is identical to the sample are presented. In the second step, responses to the sample result in the presentation of two or more comparison stimuli, one of which is identical to the sample. In this step, different stimuli appear on each trial. In the final step, the same comparisons are presented on every trial, and the stimuli from the set alternate irregularly over successive trials as the sample. Participants progress through the steps according to programmed criteria: High accuracy advances the participant to the next step, intermediate accuracy results in re-presentation of the step, and low accuracy results in a return to the previous step. Dube and colleagues (1993) found that this quasi-programmed technique led to generalized identity matching with two-dimensional abstract forms in 18 of 31 individuals with severe mental retardation. Another laboratory reported similar findings (Saunders, Johnston, Tompkins, Dutcher, & Williams, 1997).

### Stimulus control shaping

Students who do not pass the programmed assessment are then exposed to one or more stimulus-control-shaping procedures. This generic term describes techniques that teach new discriminations via graduated stimulus changes in existing ones (McIlvane & Dube, 1992). Also termed "errorless learning" procedures, they include "fading," "stimulus shaping," "delayed cue," and other procedures. Beginning with the pioneering work of Sidman and Stoddard (1966), these techniques have been profitably applied in teaching and evaluating children with severe retardation and typically developing children as young as eighteen months (e.g., McIlvane, 1992; Stoddard, McIlvane, & Serna, 1994). Recent advances in computer technology that have greatly expanded the possibilities for manipulating stimuli (e.g., via helpful techniques like "morphing") have supported this effort. Straightforward implementation of simple shaping techniques can be highly effective with some children (e.g., Saunders, et al., 1997), but individuals with profound disabilities typically require more elaborate shaping procedures (e.g., Dube et al., 1993), which are described in the following part.

### Facilitating IDMTS with simple discrimination performance

Some research suggests that training with less complex discriminations may facilitate the acquisition of matching. Dube and colleagues (1993) reported that extended simple discrimination training seemed to facilitate identity matching acquisition in some participants. As discussed above, the demonstration of generalized IDMTS in the UFPA laboratory was apparently facilitated by exposing the animals to repeated shifts of simple discrimination.

### "Synthesizing" IDMTS from simple discrimination performance

As previously mentioned, procedures described so far may be seen as means for identifying individuals who do not need protracted and/or procedurally complex training to acquire IDMTS. The quasi-programmed methods and elementary stimulus-control-shaping procedures have proven effective with a surprising number of individuals. Nevertheless, a significant subset of individuals clearly do require more intensive training. For them, a training program has been developed that synthesizes generalized identity matching from simple discrimination performances. This method (Three-Stage IDMTS Programmed Method, Figure 9.3) uses programmed instructional procedures such as fading to establish the first instances of simple discrimination of visual stimuli, and then builds upon and systematically elaborates performance to the point of conditional identity matching.

The teaching program consists of three major stages:

*Stage I* is designed to generate rapid, generalized, simple discrimination learning, with arbitrary visual forms. The procedures combine the methods of learning-set research (Harlow, 1949) with those of programmed instruction. Initially, participants learn a series of simple simultaneous discriminations with a standard prompting procedure such as fading. Over a large number of different discrimination problems with different stimuli, the amount of prompting per discrimination is gradually decreased ("fading out the fading"). Successful completion of the program establishes a two-trial performance: On the first trial, a single stimulus ($S+$) is displayed and its selection is reinforced. On the second trial, two stimuli are displayed, $S+$ from the previous trial and a different stimulus ($S-$), and selection of the previously displayed $S+$ is again reinforced. Because different stimuli appear on each two-trial problem, consistently accurate performance on the second trial of each problem demonstrates acquisition of new discriminations after exposure to a single training trial (i.e., the first trial of each problem, when $S+$ was presented alone).

*Stage II* changes the two-trial procedure. When a new stimulus is first presented alone as $S+$, selecting it no longer produces a reinforcer. Instead, the stimulus remains displayed as a sample stimulus, and $S+$ and $S-$ comparison stimuli appear; the $S+$ comparison is identical to the sample. This change is accomplished over several steps. The result is nonconditional identity matching to sample.

*Stage III* introduces reversals of $S+$ and $S-$ stimulus functions into the baseline of nonconditional identity matching. The number and frequency of reversals is increased gradually over sessions, and the number of different stimuli displayed in each session is reduced. Ultimately, only two stimuli appear as comparisons in each session and the participant has to choose between them on every trial; the $S+$ and $S-$ functions of these stimuli alternate irregularly over trials conditionally upon sample stimuli (conditional identity matching-to-sample).

This careful, exhaustive programming is not necessary for most individuals. Indeed, other methods are preferable in educational programs (e.g., the "Sorting-to-Matching" program branch in Figure 9.3 that endeavors to exploit extraexperimentally acquired sorting skills). However, the exhaustive programming has succeeded with virtually everyone who has required it. The findings make an important point. Prior to the Shriver methodological studies, the literature on human cognitive development would have predicted that very low-functioning humans were incapable of generalized identity matching of abstract two-dimensional forms. Long-standing assumptions proved to be false. What is probably true, however, is that without a sustained methodological effort of the type undertaken, these capabilities would never have been revealed.

## CONCLUSION: EXTENSIONS TO NONHUMAN PRIMATE POPULATIONS

From the material presented in this chapter, it is clear that laboratories interested in relational learning potential of nonhuman primates and nonverbal humans have many interests in common. We suggest that it will be of considerable benefit to adapt the effective teaching methods that have been developed for the latter population to the former. In particular, we suggest that stimulus-control-shaping techniques, rarely used with nonhuman primates, be systematically explored. Not only might these techniques speed acquisition, but perhaps would also assist in minimizing unwanted stimulus control that tends to develop with simple differential reinforcement techniques.

In the UFPA lab, we are endeavoring to implement a fuller range of the techniques that have been developed at the Shriver Center and elsewhere. Particularly helpful, we believe, will be shaping techniques for teaching one-trial discrimination learning and arbitrary matching baselines. Perhaps more important than implementing specific techniques, however, is approaching our animals in a new way. As a "school for primates," we are now undertaking studies that fully implement a programmed educational approach. The goal of the work is to assess whether this approach will indeed reveal formerly unsuspected behavioral capacities in our capuchins. All of the techniques described in this chapter will be implemented. Also to be implemented are techniques that we have not described here for reasons of space (e.g., beginning discrimination training with highly salient stimuli such as food items [Stoddard, 1982], using outcome-specific differential reinforcement techniques [Litt & Schreibman, 1981], implementing procedures to build behavioral momentum during baseline development [Nevin, 1992; Dube & McIlvane, in press], etc.).

We recognize that the programmed educational approach has few direct precedents in the animal cognition literature. Nevertheless, we are encouraged by the many successes that the approach has brought in studies of humans with severe developmental limitations. We are similarly encouraged by the results of the limited implementation thus far in the UFPA laboratory. Of course, the value of our approach will be judged by the degree of success that we achieve. If our exhaustive programming efforts do not lead to demonstrations of stimulus equivalence and related phenomena, then our colleagues will no longer have to wonder whether an effort like this could have succeeded. If we are successful, however, then we will contribute a set of techniques that can be replicated in other laboratories and extended to other species. Replication will be possible because all of the work will be implemented under precisely controlled conditions and with replicable apparatus and computer programs.

The approach articulated here may set the stage for a novel interdisciplinary approach to comparative cognition. Just as work with nonhuman primates has influenced techniques for neuropsychological evaluation of humans with limited behavioral development (e.g., Diamond, 1990), so too may the wealth of information now available in teaching developmentally limited humans influence practices in the animal cognition laboratory. Such interdisciplinary activity may lead to better animal models of human behavior for neuroscience research (cf. Crnic & Nitkin, 1996; McIlvane & Cataldo, 1996), because animals can be given experiences that are arguably more comparable to those available to humans. Also, consistent with our general educational interests, comparable animal models may also provide useful preparations for developing new approaches for teaching humans with developmental limitations.

## NOTE

Author order was determined alphabetically. Preparation of this chapter was supported by CNPq (Brasil) via a PRONEX grant and by NICHD (U.S.) grant number HD 25995 and HD 39816. Statements made in this chapter reflect the authors' own views, and do not represent policy statements from their funding institutes. Correspondence may be directed to Olavo Galvão (ofg@cpgp.ufpa.br) or William McIlvane (wmcilvane@shriver.org).

1. Animal cognition researchers have long understood that matching-to-sample problems can be solved by learning specific responses to specific stimulus configurations (e.g., Lashley, 1938; Cumming & Berryman, 1965). For example, in a two-stimulus, two-comparison matching-to-sample baseline involving red (R) and green (G) stimuli, the subject could learn the following four performances: Given RRG, go left; given RGG, go right; given GGR, go left; and given GRR, go right. Accordingly, neither red nor green is "separable" from the larger stimulus compound (see Stromer, McIlvane, & Serna [1993] for a discussion of stimulus separability and Sidman [1992] for an empirical demonstration of configural control). When procedures establish compound stimulus control, there are no stimulus-stimulus relations to assess and there cannot be a valid test for generalized identity matching.

## REFERENCES

Barros, R.S. (1998). Controle do comportamento por relacões entre estímulos em *Cebus Apella*. Unpublished doctoral dissertation. Universidade Federal do Pará, Brasil.

Barros, R.S., Galvao, O.F., & McIlvane, W.J. (in press). Generalized identity matching-to-sample in *Cebus Apella. The Psychological Record.*

Barros, R. S. & Galvão, O. F. (submitted). Escola experimental de primatas. Estudos em Psicologia. (Experimental school for primates)

Carr, D., Wilkinson, K.M., Blackman, D.E., & McIlvane, W.J. (2000). Equivalence classes in individuals with minimal verbal repertoires. *Journal of the Experimental Analysis of Behavior, 74,* 101–114.

Carter, D.E., & Eckerman, D.A. (1975). Symbolic matching by pigeons: Rate of learning complex discriminations predicted from simple discriminations. *Science, 187,* 662–664.

Cerutti, D.C., & Rumbaugh, D.M. (1993). Stimulus relations in comparative primate perspective. *The Psychological Record, 43,* 811–821.

Cohen, L.R., Brady, J., & Lowry, M. (1981). The role of differential responding in matching-to-sample and delayed matching performance. In M.L. Commons and J.A. Nevin (Eds.), *Quantitative analysis of behavior: Vol. 1. Discriminative properties of reinforcement schedules* (pp. 345–364). Cambridge, MA: Ballinger.

Crnic, L.S., & Nitkin, R.M. (1996). Animal models of mental retardation: An overview. *Mental Retardation Developmental Disabilities Research Review, 3,* 185–187.

Cumming, W.W., & Berryman, R. (1965). The complex discriminated operant: Studies of matching-to-sample and related problems. In D.I. Mostofsky (Ed.), *Stimulus generalization* (pp. 284–330). Stanford, CA: Stanford University Press.

D'Amato, M.R., Salmon, D., & Columbo, M. (1986). Processing of identity and conditional relations in monkeys (*Cebus apella*) and pigeons (*Columba livia*). *Animal Learning and Behavior, 14,* 365–373.

D'Amato, M.R., Salmon, D., Loukas, E., & Tomie, A. (1985). Symmetry and transitivity of conditional relations in monkeys (*Cebus apella*) and pigeons (*Columba livia*). *Journal of the Experimental Analysis of Behavior, 44,* 35–47.

Deacon, T.W. (1997). *The symbolic species: The co-evolution of language and the brain.* New York: W.W. Norton.

Devany, J.M., Hayes, S.C., & Nelson, R.O. (1986). Equivalence class formation in language-able and language-disabled children. *Journal of the Experimental Analysis of Behavior, 46,* 243–257.

Diamond, A. (1990). The development and neural bases of memory function, as indexed by the AB and delayed response tasks in human infants and infant monkeys. *Annals of the New York Academy of Sciences, 608,* 637–676.

Dube, W.V., Iennaco, F.M., & McIlvane, W.J. (1993). Generalized identity matching to sample of two-dimensional forms in individuals with intellectual disabilities. *Research in Developmental Disabilities, 14,* 457–477.

Dube, W.V. & McIlvane, W.J. (in press). Behavioral momentum in computer-presented discriminations of individuals with severe mental retardation. *Journal of the Experimental Analyis of Behavior.*

Dube, W.V., McIlvane, W.J., & Green, G. (1992). An analysis of generalized identity matching-to-sample test procedures. *The Psychological Record, 42,* 17–28.

Dube, W.V. & Serna, R.W. (1998). Reevaluation of a programmed method to teach generalized identity matching to sample. *Research in Developmental Disabilities, 19,* 347–379.

Dugdale, N., & Lowe, C.F. (2000). Testing for symmetry in the conditional discriminations of language-trained chimpanzees. *Journal of the Experimental Analysis of Behavior, 73,* 5–22.

Harlow, H.F. (1949). The formation of learning sets. *Psychological Review, 56,* 51–65.

Hayes, S.C. (1991). A relational control theory of stimulus equivalence. In L.J. Hayes & P.N. Chase (Eds.), *Dialogues on verbal behavior* (pp. 19–40). Reno: Context Press.

Horne, P.J., & Lowe, C.F. (1996). On the origins of naming and other symbolic behavior. *Journal of the Experimental Analysis of Behavior, 65,* 185–241.

House, B.J., Brown, A.L., & Scott, M.S. (1974). Children's discrimination learning based on identity or difference. In H.W. Reese (Ed.), *Advances in child development and behavior: Vol. 9.* (pp. 1–45). New York: Academic Press.

Iversen, I.H. (1997). Matching-to-sample performance in rats: A case of mistaken identity? *Journal of the Experimental Analysis of Behavior, 68,* 27–46.

Iversen, I.H., Sidman, M., & Carrigan, P. (1986). Stimulus definition in conditional discrimination. *Journal of the Experimental Analysis of Behavior, 45,* 297–304.

Lionello, K.M., & Urcuioli, P.J. (1998). Control by sample location in pigeons' matching to sample. *Journal of the Experimental Analysis of Behavior, 70,* 235–251.

Lionello-DeNolf, K.M., & Urcuioli, P.J. (2000). Transfer of pigeons' matching to sample to novel sample locations. *Journal of the Experimental Analysis of Behavior, 73,* 141–161.

Litt, M.D., & Schreibman, L. (1981). Stimulus-specific reinforcement in the acquisition of receptive labels by autistic children. *Analysis and Intervention in Developmental Disabilities, 1,* 171–186.

McIlvane, W.J. (1992). Stimulus control analysis and nonverbal instructional technology for people with mental handicaps. In N.R. Bray (Ed.), *International review of research in mental retardation: Vol. 18* (pp. 55–109). New York: Academic Press.

McIlvane, W.J., & Cataldo, M. (1996). On the clinical relevance of animal models for the study of mental retardation. *Mental Retardation Developmental Disabilities Research Review, 3,* 188–196.

McIlvane, W.J., & Dube, W.V. (1992). Stimulus control shaping and stimulus control topographies. *The Behavior Analyst, 15,* 89–94.

McIlvane, W.J., Serna, R.W., Dube, W.V., & Stromer, R. (2000). Stimulus control topography coherence and stimulus equivalence: Reconciling test outcomes with theory. In J. Leslie & D.E. Blackman (Ed.), *Issues in experimental and applied analyses of human behavior.* Reno: Context Press.

McIntire, K.D., Cleary, J., & Thompson, T. (1987). Conditional relations by monkeys: Reflexivity, symmetry, and transitivity. *Journal of the Experimental Analysis of Behavior, 47,* 279–285.

Nevin, J.A. (1992). An integrative model for the study of behavioral momentum. *Journal of the Experimental Analysis of Behavior, 57,* 301–316.

Oden, D.L., Thompson, R.K.R., & Premack, D. (1988). Spontaneous transfer of matching by infant chimpanzees. *Journal of Experimental Psychology: Animal Behavior Processes, 14,* 140–145.

Oden, D.L., Thompson, R.K.R., & Premack, D. (1990). Infant chimpanzees spontaneously perceive both concrete and abstract same/different relations. *Child Development, 61,* 621–631.

Pepperberg, I. (2000). *The Alex Studies: Cognitive and communication abilities of grey parrots.* Cambridge, MA: Harvard University Press.

Premack, D. (1983). Animal cogntion. In M.R. Rosenweig & L.W. Porter (Eds.), *Annual Review of Psychology, 34,* 351–361.

Ramey, C.T., & Ramey, S.L. (1998). Early intervention and early experience. *American Psychologist, 53,* 109–120.

Rumbaugh, D.M. (1990). Comparative psychology and the great apes: Their competence in learning, language, and numbers. *The Psychological Record, 40,* 15–39.

Saunders, K.J. (1989). Naming in conditional discrimination and stimulus equivalence. *Journal of the Experimental Analysis of Behavior, 51,* 379–384.

Saunders, K.J., Johnston, M.D., Tompkins, B.F., Dutcher, D.L., & Williams, D.C. (1997). Generalized identity matching of two-dimensional forms in individuals with moderate to profound retardation. *American Journal on Mental Retardation, 102,* 285–291.

Saunders, K.J., & Spradlin, J.E. (1990). Conditional discrimination in mentally retarded adults: The development of generalized skills. *Journal of the Experimental Analysis of Behavior, 54,* 239–250.

Schusterman, R.J., Gisiner, R., Grimm, B.K., & Hanggi, E.B. (1993). Behavior control by exclusion and attempts at establishing semanticity in marine mammals using match-to-sample paradigms. In H. Roitblat, L. Herman, & P. Nachtigall (Eds.), *Language and communication: Comparative perspectives.* Hillsdale: Erlbaum.

Schusterman, R.J., & Kastak, D. (1993). A California Sea Lion (*Zalophus californianus*) is capable of forming equivalence relations. *The Psychological Record, 43,* 823–844.

Schusterman, R.J., Kastak, D., & Reichmuth, C.J. (1997). What's in a name? Equivalence by any other name would smell as sweet. *Journal of the Experimental Analysis of Behavior, 68,* 252–257.

Serna, R.W., Dube, W.V., & McIlvane, W.J. (1997). Assessing same/different judgments in individuals with severe intellectual disabilities: A status report. *Research in Developmental Disabilities, 18,* 343–368.

Sidman, M. (1971). Reading and auditory-visual equivalences. *Journal of Speech and Hearing Research, 14,* 5–13.

Sidman, M. (1980). A note on the measurement of conditional discrimination. *Journal of the Experimental Analysis of Behavior, 33,* 285–289.

Sidman, M. (1987). Two choices are not enough. *Behavior Analysis, 22,* 11–18.

Sidman, M. (1992). Adventitious control by the location of comparison stimuli in conditional discriminations. *Journal of the Experimental Analysis of Behavior, 58,* 173–182.

Sidman, M., Cresson, O., Jr., & Willson-Morris, M. (1974). Acquisition of matching to sample via mediated transfer. *Journal of the Experimental Analysis of Behavior, 22,* 261–273.

Sidman, M., Rauzin, R., Lazar, R., Cunningham, S., Tailby, W., & Carrigan, P. (1982). A search for symmetry in the conditional discrimination of rhesus monkeys, baboons, and children. *Journal of the Experimental Analysis of Behavior, 37,* 23–44.

Sidman, M., & Stoddard, L.T. (1966). Programming perception and learning for retarded children. In N.R. Ellis (Ed.), *International Review of Research in Mental Retardation: Vol. 2.* (pp. 151–208). New York: Academic Press.

Sidman, M., & Tailby, W. (1982). Conditional discriminations vs. matching-to-sample: An expansion of the testing paradigm. *Journal of the Experimental Analysis of Behavior, 37,* 5–22.

Sidman, M., Willson-Morris, M., & Kirk, B. (1986). Matching-to-sample procedures and the development of equivalence relations: The role of naming. *Analysis and Intervention of Developmental Disabilities, 6,* 1–20.

Soraci, S.A., & Carlin, M.T. (1992). Stimulus organization and relational learning. In N.W. Bray (Ed.), *International Review of Research in Mental Retardation: Vol. 18* (pp. 29–53). San Diego: Academic Press.

Stoddard, L.T. (1982). An investigation of automated methods for teaching severely retarded individuals. In N.R. Ellis (Ed.), *International review of research in mental retardation* (pp. 163–207). New York: Academic Press.

Stoddard, L.T., McIlvane, W.J., & Serna, R.W. (1994). A note on stimulus control shaping and one-trial learning in two- and three-year-old children. *The Psychological Record, 44,* 289–299.

Stromer, R.L., McIlvane, W.J., & Serna, R.W. (1993). Complex stimulus control and equivalence. *The Psychological Record, 43,* 585–598.

Thompson, R.K.R., Oden, D.L., & Boysen, S.T. (1997). Language-naive chimpanzees (*Pan troglodytes*) judge relations between relations in a conceptual matching-to-sample task. *Journal of Experimental Psychology: Animal Behavior Processses, 23,* 31–43.

Tomonaga, M., Matsuzawa, T., Fugita, K., & Yamamoto, J. (1991). Emergence of symmetry in a visual conditional discrimination by chimpanzees (*Pan troglodytes*). *Psychological Reports, 68,* 51–60.

Washburn, D.A., Hopkins, W., & Rumbaugh, D. (1989). Video-task assessment of learning and memory in macaques (*Macaca mulatta*): Effects of stimulus movement on performance. *Journal of Experimental Psychology, 15,* 393–400.

Wilkinson, K.M., Dube, W.V., & McIlvane, W.J. (1997). Fast mapping and exclusion in developmental language, behavior analysis and animal language research. *The Psychological Record, 48,* 407–422.

Zentall, T.R., Edwards, C.A., Moore, B.S., & Hogan, D.E. (1981). Identity: The basis for both matching and oddity learning in pigeons. *Journal of Experimental Psychology, 7,* 70–86.

Zygmont, D.M., Lazar, R.M., Dube, W.V., & McIlvane, W.J. (1992). Teaching arbitrary matching via sample stimulus-control shaping to young children and mentally retarded individuals: A methodological note. *Journal of the Experimental Analysis of Behavior, 57,* 109–117.

# Part III

# Visualism: Philosophical Approaches

Chapter 10

# Visualism in Science

*Don Ihde*

It will be the contention of this chapter that one of the cultural habits of the sciences is to produce, display and reiterate what counts for evidence in visual form. I call this "science's visualism." This cultural habit has been accelerated in late modernity through the sophisticated development of imaging technologies which now transform ranges of phenomena which include, but also exceed, all human perceptual capacities and translate these phenomena into visual forms. These techniques, I contend, constitute a *visualist hermeneutic* which has become the favored mode of evidence display within most sciences.[1]

By characterizing science's preference for visualism as a cultural habit, there is an implicit claim here that this cultural "choice" is not a necessary one, but a perceptually and culturally contingent one. I shall try to show how this is the case through three strategies. One strategy is to take some account of what I term "whole body perception" by means of both phenomenological and non-phenomenological accounts of perception. The other strategy is to make selective reference to incidents in the history of science as it increasingly condenses its evidence presentations into visual forms. The third strategy is to locate much of the production of visual displays in science's technologies or instruments and to show how the various information-gathering devices are increasingly developed to make just such visual displays.

## WHOLE BODY PERCEPTION

I begin here by drawing from phenomenological traditions which explicate perceptual phenomena through bodily actions. Phenomenologists from Maurice Merleau-Ponty through Hubert Dreyfus claim that human bodily action is the focal and necessary basis for human embodied intelligence.[2] I shall here follow this style of descriptive analysis:

- In actual experience there is a constancy of what I shall call *whole body perception,* in the sense that our perceptions occur as a plenary Gestalt in relation to an experienced environment. We interact with the "world" around us. It is also the claim here that our whole body perceptions are sensorily synthesized in our interactions with a "world." Unlike the older traditions of discrete and separable "senses," phenomenology holds that I never have a simple or isolated visual experience. My experience of some seen object is simultaneously and constantly also an experience which is structured by all the "senses." It takes some deliberate constructive manipulation or device to produce the "illusory" abstraction which could be called vision by itself—in empirical psychological experiments, when various color illusions or Gestalt deformations are inspected, the experiment deliberately designs the situation to dampen and displace the ordinary whole body engagement with surroundings. To look through a tube, or to be seated in a darkened and silenced environment, in effect, produces a quasi-illusion of seeing itself or an abstract seeing.
- There is here, too, a doubled phenomenon. On the one hand, the experimental "abstraction" that focuses upon vision alone is also a situation shaping into special form, attention or focal action itself. Whole body, ordinary, activities occur within complex environments in which much is simultaneously occurring. This is the perceptual field which constitutes our immediate environment and it is never simple. We therefore must focus our attentions on those aspects of this complex field in such a way that our "intended" actions may be carried out: If I am a bird watcher, I will focus upon the goshawk chasing the kingfishers over my pond, rather than attending to the music coming from the floor below (although the focus cannot eliminate the sound of the music, it merely pushes it into the background). Similarly, although in a more controlled form, the instructions that determine what I am to look at or see in the constructed experiment relies upon this ability to perceptually focus. In this case the very ability to focus helps to enhance the quasi-illusion of a "pure" visual phenomenon by subduing the other sensory dimensions.
- Overall, this analysis aims to show that regardless of deformations or manipulations, whole body or multi-sensory dimensional perceptions remain constant in environmentally interactive situations. The focal attentions that concentrate upon visual aspects of phenomena are nevertheless situated within the more primary, synthesized, whole body perceivings.

If, now, I turn to some examples of scientific practice which entail whole body perceptual actions related to concrete scientific work, one can see how this plenary or multidimensional perceptual action can and often does come into play. Imagine a physical anthropologist searching the ridge of some African plain: there is simultaneously the bodily motility of the anthropologist as he or she walks along the ridge engaging his or her active visual scanning, focally searching the field for some small indication of a human fossil. There! It looks like a tooth or some teeth amid the stones and even possible nonhuman animal remains littering the surface. The anthropologist kneels, both for a closer look and then to feel and probe into the dirt—the tooth protruding is part of a jawbone, so with care, perhaps now with a tool, the probing begins with fully engaged bodily-plus-tool actions at the site. In this illustration, it is ordinary, full-body engagement which characterizes the discovery situation. Informed vision is employed along with tactility, bodily motility, and the rest of the interaction with the immediate environmental "world" of this situation. The initial point here is that the scientific activity is occurring within precisely this full body perceivability of the action of the anthropologist and is not exclusively or even necessarily primarily visual in character. The tooth–jawbone is seen, felt, handled by the anthropologist, and, except for the highly trained and critical perception involved, no different from cases of ordinary objects.

A more extreme example can be found in those rare cases of scientists who have risen to the top of their fields although blind. In the contemporary world, Geerat J. Vermeij is a good example. Vermeij is a leading molluskologist whose work on the evolution of shells was conducted tactilely through his highly trained sense of feeling the surfaces of mollusks. (1995.)

We turn to a different science practice—contemporary astronomy—for a somewhat different pattern. Here the scientific object must be experienced very differently because it does not have any direct bodily presence at all. For instance, a recent *Science* report indicates that earlier Hubble Deep Field exposures (yielding *images)* seemed to suggest that as the telescopic probe got deeper into space–time, indicators of star formation seemed to slow down. A new report, however, showed intense gamma ray blasts, suggesting star formation activity as far back as detectable (13 billion light years), thus contradicting the Hubble observations.[3] While both instrument sets yield visualizable images as part of the evidence, the visual displays are results of instrumentally constructed or mediated processes and are, moreover, translations into visual forms of phenomena themselves not directly available to human perception (gamma rays, or in the case of HDF, indications of an "empty" field). In short, the whole body perceivability situations are quite different in these two sciences. Yet, in both instances visual displays continue to become part of the eventually developed evidence for claims being made.

## PERCEPTUAL VARIANTS

I have so far emphasized the way visualization lies embedded in scientific practices and are enhancements of certain dimensions of whole body perception, I wish to shift ground a bit, now, to deal with what I see as a traditional prejudice concerning a presumed intrinsic perceptual superiority of vision over any other human "sense," thus implicitly justifying science's cultural choice as in some way "natural." As a long-time researcher into auditory phenomena, I acknowledge that this sensory dimension has some counterpart claim to discriminatory equivalence. At the physiological level, there is no distinctive difference between seeing and hearing with respect to detecting duration (both at 0.01 sec.), but the sense of hearing actually carries longer before nerve dampening occurs with respect to a single stimulus. The discriminations trained musicians can make are as fine or finer than those for trained sight, and there is no counterpart for "perfect pitch" in sight such as there is for hearing.[4] The range of auditory perception with respect to the continuum of wave phenomena (for humans, from 20–20,000 hertz) is vaster than for seeing within the optical range of light waves.[5] Yet another indicator of auditory equivalence may be found in recent experiments that have created auditory equivalents to inverting glasses [MP]. (Dutch researchers have recently fashioned distorting "ears" that deflected sounds from a speaker on a robotic arm inverting locations (above/below; right/left) yet; in an exact parallel with inverting glasses, the "ear wearers had compensated and were able to make correct judgments. The surprise came when they finally removed the ears: they were immediately able to adjust to their old ears. 'That's quite astounding,' says Fred Wightman, psychologist at the University of Wisconsin. . . . 'it may mean that locating objects by sound is a more complicated cognitive function than with vision.'"[6]

Culturally, many groups are often focused more on auditory than visual metaphors and descriptions in both perception and language. Steven Feld, for example, produced an in-depth study of the "acoustemology" of a Papua New Guinean people, showing how in dense rain forests, auditory experience is often the primary perceptual variable for place, location, and space.[7] If I were able to trace these implications further, I would contend that it is nothing more than a sometimes self-serving cultural bias to claim the intrinsic superiority of sight over hearing. This is not to claim that visualization lacks certain advantages that make it fit neatly into science practice.

Turning from perceptual variations to instrumentally mediated perceptions, it may be noted that scientific instrumentation itself has not always been a matter of visualization either. Bettyann Kevles' study of medical imaging, *Naked to the Bone* (1997), indicates that earlier in the century, after the introduction of the phonograph that captured transient sensual experience, Edward Bellamy suggested that, "hearing is the sense of the future, and coming none too soon to rescue eyesight, which 'was indeed terribly overburdened previous to the introduction of the phonography, and now that the sense of hearing is beginning to assume its proper share of the work, it would be strange if an improvement in the condition of people's eyes were not noticable.' "[8] She goes on to point out that medical technologies were also often more auditory than visual—the stethoscope, percussion of the chest, and even recorded phonographic sounds.[9] Similarly, sonic transponders and other early submarine instruments often were "read" auditorily prior to visual graphing translations which were developed later. Of course, this brief hope for auditory modeling preceded the invention of motion pictures and later, motion pictures with sound and listening gear was replaced by graphic radar-like screens for transponders.

## SCIENCE'S VISUALIST HISTORY

I have tried to show that visual displays, at least as contrasted to either whole body perceptions or to possible auditory ones, had to have been something of a "choice." But this is not to say that such a choice is individual—it is more a historical–cultural event. It would be impossible within the confines of a chapter to do justice to what is needed here, but impressionistically it is possible to point out several of the outstanding moments of this cultural history. In what follows, I will try to interrelate the trajectory toward visualism with its embodiment in the technologies or instrumentarium of early modernity:

- If we take Leonardo da Vinci (ca. 1500) as our example of this early modern shift to visualism, we see a double transformation of how visualization occurs. First, there is what I call a shift to the visual, that is an enhancement of the visual over and above, and often to the detriment, of either full perceptual or nonvisual perceptions, indicated in da Vinci's early "anatomy," which visually depicts tendons, muscles and veins. There are two particularly important points here: first, descriptive anatomy at the time was often made in terms of tactile and olfactory terms, referring to how an organ felt (hard, soft, pliant, etc.) or smelled (putrid, metallic, etc.). Da Vinci "reduces" this anatomy to a structural and analytical set of drawings (he was later followed by Vesalius' famous anatomical studies, ca. 1540). But, equally important although more often overlooked, the new "realism" of da Vincian drawing with its presumed isomorphism was itself part of the Renaissance "technologization" of vision.
- It is well known that one of the favorite visual "toys" of the Renaissance was the *camera obscura*. What often goes unremarked was the highly important role this optical instrument later played in the very development of so-called "Renaissance perspective." Alberti (ca. 1437) apparently used the camera obscura quite regularly. He may have been among the first to "draw by the lines."[10] Note that the camera obscura "automatically" reduces three-dimensional objects to a two-dimensional image. That is, this isomorphic "reduction" is an artifact of an early imaging technology. Da Vinci, in turn, was the first Renaissance describer of a detailed camera obscura (ca.1531).[11] He went on to extend this possibility of isomorphic, analytic visual display in drawing practice to a whole range of possible objects, from machines to human anatomy. We now have what I am claiming is a doubled transformation, first the transformation to vision or the visualizable alone, but also a transformation of vision through the camera obscura-like reduction of vision to its produced image.
- Next in this impressionistic history is Galileo Galilei (ca. 1610 for telescope). But the Galileo of this history is the visualizing Galileo, the Galileo of the telescope. Himself a maker of many optical instruments—before his discovery of the telescope through Paolo Sarpi (ca.1609)—he had already been the maker of surveying instruments, and later also used the microscope.[12] It was, however, through the results of his telescopes, of which he made approximately a hundred, that his form of science's visualization entered this trajectory. Although it apparently took him some time from his first telescopic results for him to turn the new instrument to the skies, once he did he was, within three months, publishing his results in *Sidereal Nuncius*. He was quite immediately able to formulate a simple "instrumental realism" by recognizing that the new

visual phenomena he was seeing were real. His lasting observations that have stood the test of time include the surface features of the moon, sunspots, the four main satellites of Jupiter, and the phases of Venus. Not only were these heretofore unseen phenomena, but the new instrumental mediation made possible the transformation of perceived space–time (for example, reduction of apparent distance; magnification of apparent motion), not possible before the telescope.

- Next in this simplified cultural history is the rapid scientific acceptance of photography. This imaging technology was, in fact, one of the most rapidly accepted and adapted imaging processes in the history of science. Daguerre's improvement on Niepce's process was publicized in 1839 and by 1840, first in a photo of the gibbous moon, then, by 1842, photographs being made through spectrographs, micrographs and other compound imaging technologies. With faster exposure and shutter speeds, time-stop photography began to depict motion studies of interest to nineteenth-century science. But why was photography such a breakthrough? There are at least two aspects to an answer of interest here: first, while the optical technologies that magnified—even, early, beyond the limits of ordinary vision—phenomena yielded new and unexpected entities, to show or demonstrate these, one either had to have each observer take his or her own sighting, or, more commonly, make a drawing of the object. This meant that two matrices for "subjectivity" had to be surpassed; the first was that of skilled vision through the instrument, the second, of skilled reproduction by drawn representation. Thus, although Galileo did observe the "rings" of Saturn, he did not correctly identify these phenomena as rings. That remained for Huygens some forty-nine years later.[13] Similarly, whereas Leeuwenhoek was the first to report the microscopic sighting of spermatozoa, again the problem was one of a drawn representation. Photography, not unlike the camera obscura, automatically reduced the object to an isomorphic and realistic fixed image on the photographic plate. And, it accomplished this "without subjectivity," by means of a mechanical process. (I deliberately add the nineteenth-century concern for subjectivity, which was not a topic of discussion earlier.) Technologies, traditionally taken as nonhuman, thus serve to standardize perceptual results in imaging processes.
- Before leaving photography, there is one more facet to this process which makes it scientifically interesting: Although the early processes called for long exposure times, necessarily limiting imaging to stable objects, more and more rapid exposure times allowed later photographs to manipulate or transform time. Muybridge's 1878 photos of animal and human locomotion were used as evidence for previously unknown dynamic phenomena; the Mach brothers showed shock waves for the first time; and much later strobe techniques "stop time" to reveal even more microfeatures of dynamic phenomena. This accumulated success of photographic imaging is noted by Kevles: "By the 1890's photographs had become the standard recorders of objective scientific truth."[14]

I wish now to temporarily break off from this narrative history of scientific visualism. I note in doing so that I have quite deliberately emphasized not only the turn to emphasizing visual phenomena (including their development in imaging technologies), but also what I call their largely "modern" features, including *realistic isomorphism, representability in images,* and as *analogues to ordinary visual phenomena.* I am not contending here that this is all that happens in the history of science prior to the twentieth century, but is perhaps the dominant set tendencies, at least regarding the incorporation of perceptibility within science.

Special reference should be made here, too, to the ways in which imaging technologies from the beginning transform perceptibility. Even the simple visual transformations of the camera obscura are significant (reduction of three dimensions to two dimensions, inversion of the image in relation to the referent object, etc.), those of the earliest magnification technologies (microscope and telescope) to human vision beyond its ordinary bodily limits—while retaining the obvious analogue qualities of ordinary vision (mountains on the moon or "animacules" within rainwater), making the sometimes strange phenomena nevertheless recognizable—are even more so. This "mechanical," reproduced isomorphism initially gave imaging its scientific advantages. As early as 1888, P.J.C. Janssen claimed,

> The sensitive photographic film is the true retina of the scientist… for it possesses all the properties which Science could want: it faithfully preserves the images which depict themselves upon it, and reproduces and multiplies them indefinitely upon request: in the radiative spectrum it covers a range more than double that which the eye can perceive and soon perhaps will cover it all; finally, it takes advantage of that admirable property which allows the accumulation of events, and whereas our retina erases all impressions more than a tenth of a second old, the photographic retina preserves them and accumulates them over a practically limitless time."[15]

Here, however, I shall make a break in the narrative to consolidate some of the features and analogues to phenomenological vision that are implicit in scientific visualism and that play a major, if sometimes not explicitly noticed role, in this visualism.

## STRUCTURES OF VISION

The primary model of visual perception I have been following here is one derived from phenomenology, which holds to the primacy of an actional body–environment relativity. Many of the insights of Gestalt psychology are also related to the phenomenology of perception (Koffka and Kohler were students of Husserl). There are some salient features to help underline the rationale for science's visualism:

- Both gestalt and phenomenological psychologies emphasize the importance of the variable figure–ground relationship. Figures or focal objects only appear within and against a ground, but figures are variable within any given ground: the ability to pick out a comet or a supernova, either from observation or from alternating photographic images, utilizes this visual skill.

- Full, instananeous gestalt pattern recognition is another visual structural skill that comes into play. A personal example here shows the power of this form of visualization. My oldest son and his wife recently gave birth to their first child; they quickly, in the fashion of high-tech households, e-mailed a digital set of photos to me. My computer, however, had text-only capacity, so that some twenty pages of "gobbledegook" came forth. One simply could not make out what the referent object was from the data itself. When sent to my wife's better machine, the picture on a single page emerged with the aforesaid instant recognition. A picture is worth far more than a thousand words these days.
- Assuming the full panoply of contemporary imaging previously noted, reiterability is yet another valuable feature of scientific visualism. One can return, again and again, to the image to detect features overlooked or previously unnoted. It is the reiterable and instant feature of the visualization that makes it valuable for science.
- Pattern shifts or repetitions themselves are part of the recognition process involved. In a parallel to the data–photo example above, the scaled repetition-patterns of such imaging as fractal sets immediately display patterns that either would be quite indiscernible in the data or misunderstood.

To this point I have held to a parallelism between full, actional, ordinary perceptions and their scientifically guided, but merely enhanced perceptions. For that reason, I have restricted the discussion to analog visual phenomena. A scientific perception within this limitation may be a more informed perception, may be more skilled in various focal concentrations, and may "see more" hermeneutically within any given object seen, but it is a perception that still retains its gestalt and patterned qualities.

Galileo, upon seeing the four satellites of Jupiter, virtually instantaneously (the "aha" phenomenon) recognized that here was another body in the heavens that had "planets," as he called them, circling it. Thus at the very least, the Earth was not the only such body that had satellites. Yet, there was simultaneously a much bigger transformation of what would count for scientific vision implicit in these early observations.

## "SECOND SIGHT"

The transformation of vision is what happens to vision through the technologies of vision or instruments. I have suggested that at least since early modernity, the role of instruments in science has been crucial—but this is also a perceptually crucial factor in the constitution of science's visualism. If the camera obscura geometrically rearranged visual objects—and it belongs to the earliest movements of the geometrical method, explicitly playing a role in both Descartes and Locke[16]—it was lensing that opened the way to the much more radical transformations to follow. Lensing, in the telescope and microscope, transformed the phenomenological sense of space–time. What is called apparent distance in technical language today is a description of the changed relative distance between the observer and the observed. It is relativistically equivalent to say that the telescope brought the mountains of the moon "closer" to Galileo (magnification) or Galileo "closer" to the mountains. It was in the difference of direct and telescopically mediated distances that the sense of space was transformed. Similarly, the microscope repeats the same feat for the miniscule of transformed sense of distance, and phenomenologically speaking, it may be said to be equivalent regarding telescope and microscope. (Moon mountains and paramecia both take up focal places within the now instrument-mediated apparent space, and are the "same sized" objects within central vision.) These transformations, partially noted by Galileo and his detractors, were either accepted or rejected precisely because of the transformational changes to vision. Eventually, of course, the transformations won out and science became, in modernity, a process of instrumental realism. What could be seen through the lensing systems was taken as real—but in part taken as real because it retained its analog qualities to unmediated vision.

Magnification, whether micro or macro, already holds the secret of getting "beyond vision" insofar as it is limited to direct and ordinary embodied vision. But, I would contend, it is only by way of introducing a much more drastic set of variables; what I call "second sight" emerges. Galileo accidentally stumbled on one such phenomenon in his invention of a heliograph through which he imaged the first observed sunspots. (A heliograph in this instance was a screen on which the telescope could cast an image, a camera obscura-like device attached to the telescope.) Sunspots, unlike mountains or satellites, entail more than relative spatial distancing. To observe a sunspot with the naked eye is to incur blindness—a different form of embodiment is needed to see a sunspot and that is what the heliograph performs for vision. It not only enhances analog vision, it displaces it with a type of "second sight." This "second sight" becomes a unique, instrumentally constituted "scientific object" for sight.

## "SECOND SIGHT" AND TECHNOCONSTRUCTION

Now I will follow—in a limited way—the trajectory I have opened up with "second sight," or more radically technologically constituted scientific vision. I have previously noted that in the very earliest days of scientific photography, compound instruments were being used (the spectrograph in 1842). A spectrograph does not image an ordinary analog visual object (a star, for example); rather, it depicts the spectrum of light emitted and displays it as a spectrum of colors (that can be analyzed as signs of chemical composition). It is this deliberate set of manipulations that becomes, in fact, more typical for late modern science:

- Imaging in the twentieth century becomes "second sight" imaging that employs imaging from the infrared and ultraviolet ranges of the optical spectrum beyond ordinary visual capacities but instrumentally translated into visible patterns.

- Once beyond the optical ranges, the same translation capacities are employed for wave phenomena ranging from gamma to radio waves; again, these are made visible through the translation into visibility.
- False color manipulations are used under different conventions to visually depict degrees of intensity or wave phenomena well beyond ordinary human bodily capacities to envision.
- Deliberate enhancement and contrast techniques are employed, particularly in computer tomographic processes to better display easy-to-miss features of the target object (techniques used in scientific fields from astronomy to medical imaging).

I am terming all of these visualizations a form of "second sight" that lies beyond ordinary or whole body engagements, yet remains a visualization. "Second sight" is a hermeneutic style of envisioning phenomena. It retains all the advantages of gestalt and phenomenological visions, yet is a translation into the visible from phenomena which lie beyond literal vision.

## THE ANTHROPOMORPHIC INVARIANT AND THE LATOUREAN LABORATORY

I will draw two implications from this narrative of scientific visualism. The first regards what I call the *anthropomorphic invariant,* relating to the necessity of the bodily perceiving person who performs the pertinent science. I began with a claim that the culture of science is one that "chooses" to carry out its operations within a certain style of perceiving that I call visualism. Yet what is invariant within this cultural style is the necessity for there to be a bodily perceiver. At first, and at bottom, this perceiver is an ordinary and direct perceiver; this is to say that there could be a sort of lifeworld science in the old-fashioned Husserlean sense. But, historically, science since its earliest modern forms became more than directly bodily perceiving in its observations; it became technologically embodied through its instruments, although these have frequently been imaging instruments. Yet, with, through, and among these instruments, the scientist also remains a bodily perceiver—that is the reflexive retroreferent of scientific activity. And, this perceiver remains the anthropomorphic invariant throughout the entire spectrum of observation, from direct to translated and technoconstituted imaging, from "first sight" through "second sight."

The second conclusion I make has to do with the social–cultural construction of science in its current forms. Insofar as it retains the visualist trajectory it took upon itself in early modernity, it now functions very much like what I have called a "Latourean Laboratory," meaning that the hermeneutic insight employed by Bruno Latour (1987), remains operational here as well. Laboratories are, for Latour, the places where scientists work and where instruments are employed. But instruments, for Latour, are inscription-making devices which produce visual displays: "I will call an instrument (or inscription device) any setup, no matter what its size, nature and cost, that provides a visual display of any sort in a scientific text."[17]

The invariant of the bodily perceiver, the anthropomorphic standard, is linked to the complex hermeneutic devices of instruments or inscription-making devices that produce visual displays, some simple and direct, others complex and indirect (translation devices), but that all produce the visualizations that count as evidence in late modern science. And all retain the structural advantages of gestalt vision enacted through human bodily engagement with now the specialized world of the sciences.

## NOTES

1. The theses elaborated here may be found in more complete form in Don Ihde, *Expanding Hermeneutics: Visualism in Science* (Evanston: Northwestern University Press, 1998).

2. The basic text that outlines a phenomenological theory of bodily active perception is Maurice Merleau-Ponty, *The Phenomenology of Perception* (Routledge/Keegan Paul, 1962). Hubert Dreyfus's famous critique of artificial intelligence is based on this same bodily action theory of perception and fits into a more contemporary context; see his *What Computers Can't Do* (Harper Colophon Books, 1979).

3. *Science,* 282, p. 1608.

4. *Britannica Macropedia,* (Chicago: Enclyclopedia Britannica, 1994), Vol. 27, pp. 568–569.

5. *Macropedia,* p. 569.

6. *Science,* 281, p. 1597.

7. Steven Feld, "Waterfalls of Song: An Acoustemology of Place Resounding in Bosavi, Papua New Guinea," in *Senses of Place* (SAR Press, 1996), 91–136.

8. Bettyann Kevles, *Naked to the Bone: Medical Imaging in the Twentieth Century* (New Brunswick: Rutgers, 1997), 13.

9. Kevles, *Naked to the Bone,* 13.

10. *Encyclopedia Britannica,* (1929),Vol. 4, p. 659.

11. There remains a paucity of detail regarding so much of classical Islamic science in the literature, although historians such as David C. Lindberg, *The Beginnings of Western Science* (Chicago, 1992), has begun to rectify this situation. The recent publication of the *Encyclopedia of the History of Science, Technology, and Medicine in Non-Western Cultures,* ed. H. Selin (Kluwer, 1997) makes a major contribution, particularly with regard to its extensive work on Arabic philosophers, see Alhazen, pp. 405–408.

12. Daniel J. Boorstin, *The Discoverers* (New York: Vintage Books, 1985), 318.

13. Edward R. Tufte, *Visual Explanations* (Cheshire, CT: Graphics Press, 1997), 107.

14. Kevles, *Naked to the Bone,* 15.

15. Jon Darius, *Beyond Vision* (Oxford University Press, 1984), 11.

16. Lee W. Bailey, "Skull's Darkroom: The *Camera Obscura* and Subjectivity," in *Philosophy of Technology* (Kluwers, 1989) pp. 63–79.

17. Bruno Latour, *Science in Action* (Cambridge, MA: Harvard University Press, 1987), p. 68.

Chapter 11

# Technology, Transcendence, and Modernity: Marcel and Jaspers

*Gregory J. Walters*

Gabriel Marcel and Karl Jaspers both interpreted our being-with technology in the situation of modernity. For both philosophers, technology has won the leading role in the theater of modernity, while alterity and transcendence have become its foil. They both lived through the remarkable transformations wrought by the late industrial revolution, the rise of bureaucracy and the modern totalitarian state, two world wars, the rise of mass marketing for the production, distribution, and consumption of goods and services, the early development of computer and information technologies, the invention of the atomic and hydrogen bombs, the success of space travel, the development of the birth control pill, and the discovery of DNA. Their experience during the tumultuous technological transformations of this past century would alone warrant a prima facie respect for their views on technology.

Gabriel Marcel articulates an uneasiness about our being-with technology that is a function, *inter alia,* of his understanding of alterity, that is, the functionalization of persons and loss of the sacred or transcendence because of the domination of technology. Although Marcel and Jaspers share similar concerns with the impact of technology on the metaphysical crisis of modernity and the functionalization of persons, their differing philosophical frameworks lead them to differing overall assessments. Marcel is far more ambivalent toward technology than Jaspers. This essay offers a Jaspersian reading of Marcel's key philosophical themes that address the question concerning technology. The pressing challenge is not so much discussion about Marcel and Jaspers as it is reflection on our situation inspired by their thinking. Technological means may serve creative communication and respectful alterity, but they are no guarantee for securing intersubjective communication in our present information age.

## KEY PHILOSOPHICAL WORKS AND THEMES

Gabriel Marcel's most mature presentation and "approximate synthesis" of his philosophical reflection comes in his 1949–50 Gifford Lectures, *Le Mystère de l'être,*[1] wherein he derides the notion of a philosophical "doctrine" or the generalized thinking of an absolute *Denken überhaupt*. These lectures "quest" after the essence of spiritual reality where "reality" is applied to a body of persons and things bound to one another by relationships. In *Les Hommes contre l'humain,*[2] Marcel invites readers by way of his diffuse *néo-Socratique* style into a conversation on problems concerning technology, fanaticism, the spirit of abstraction, values, war, and pessimism. One of the key themes of this work is "the universal against the masses." The universal is understood as the meeting of intelligence and love over and against the reduction of the masses to a state of alienation and abasement. Marcel's uneasiness with technology and the devaluation of wisdom and common sense are echoed again in 1954 in *Le Décline de la sagesse.*[3] This work rebukes the charge that his thought is "irrationalist" in the face of a humanity dangerously tempted to commit suicide by means of atomic annihilation. In *L'Homme problématique* (1955) Marcel asks "under which conditions has man become, in his entirety, a question for man?"[4] Modernity has been brought about in equal measure by nationalism and technologically based industrialism. These potent forces in Marxist and capitalist countries alike have given rise to the anxious "barracks man" whose ultimate problem is a "genuine necrosis whose principle is metaphysical," that is, the loss of a divine reference for personal existence.[5]

Marcel saw his philosophical work as essentially dramatic or of the nature of a musical meditation. His play, *Le Monde Cassé (The Broken World),* written in 1932, is apropos to all of his dramatic works because they reflect a theatre of the broken world where the breaking is frequently death itself, a death that tears apart the very fabric of human relationality.[6]

### Functionalization versus ontological exigence

Marcel's dramatic and philosophical works express this polarity between a world grown increasingly *functionalized,* a world of the *problematic,* a mechanized world without heart, on the one hand, and a world characterized by *ontological exigence,* or mystery, on the other hand.[7] His 1933 essay "On the Ontological Mystery"—written shortly after reading Jaspers's three-volume systematic philosophy—provides a window into Marcel's earliest reflection on technology. Marcel is uncertain of how much Jaspers's philosophy influenced this essay. He notes that he owes "a real debt to this noble and profound thinker, and I am anxious to acknowledge the inward and almost indefinable influence which he has exercised over my own mind."[8] Marcel sharply contrasts the idea of the functionalization of existence with the mystery of being that is totally resistant to technological determinations and any dominating view of the person as a mere agglomeration of vital functions, whether espoused by historical materialism, psychoanalytic reductionism, or the social functions of consumer, producer and citizen. Marcel personifies "function" in the figure of the subway ticket puncher whose existence, itself, has become an embodied timetable. Sickness breaks his vital schedule, with the hospital functioning merely as a repair shop, while his death becomes the scrapping of an existence that has merely ceased to be functional. Jaspersian boundary or limit situations have no meaning other than to perpetuate sadness and despair in a world where relativism reigns supreme, the tragic dimension of existence and the transcendent are denied, and verification takes precedence over an inward realization of presence through love "which infinitely transcends all possible verification because it exists in an immediacy beyond all conceivable mediation."[9] In a technocratic, functionalized world persons, especially the elderly, are viewed as no longer having any use value, especially when output is measured, as it often is today, with sophisticated software that tracks sick time, vacation time, holiday time, "comp" time, jury time, overtime, log-in time, and punch card time. The cipher of human being becomes mechanized, objectified, a machine, or, as we would say today, a "cyborg" or "person whose physiological functioning is aided by or dependent on, a mechanical or electronic device."[10]

Transformed from a science fiction to a biological to a cultural metaphor, the cyborg has become a North American symbol of cultural modernity which, if Donna Haraway is correct, offers "a way out of the maze of dualism in which we have explained our bodies and our tools to ourselves. This is a dream not of a common language, but of a powerful infidel heteroglossia."[11] Whether contemporary cyberphilosophies ironically mask a poststructuralist, postmodern materialism or betray what Marcel calls the "spirit of abstraction"[12] may be too early to tell. What is clearer is that the dominant, technological cultural cyborg is not yet a cyborg "cipher" that images a grid of instrumental, totalistic technological control on the planet or a *Star Wars* apocalypse waged in the name of defense, or the absolute appropriation of women's bodies in a "masculinist orgy of war." Haraway notes, more accurately, that

> communication sciences and modern biologies are constructed by a common move—the translation of the world into *a problem of coding,* a search for a common language in which all resistance to instrumental control disappears and all heterogeneity can be submitted to disassembly, reassembly, investment, and exchange.[13]

Can we recognize the Marcellian person or Jaspers's "cipher of man" in such cultural cyborg imagery? No, not at all. Unlike Haraway, Marcel does not disassemble and reassemble the self in terms of race, sex, and class rooted in high-tech-facilitated social relations. Neither does he hold out a cyborg that rejoices in the illegitimate fusions of animals and machines, nor does he seek to collapse the Self/Transcendence dialectic into a "phallogocentrism" that codes all non-feminist texts as instantiating an "informatics of domination" over women. In the spirit of Marcel, however, it must be acknowledged that our dominant cultural cyborg is having some rather devastating social and economic impacts on communities, countries, and classes of persons who have not become socialized "borgs." Matriculation in the school of computer-mediated communications requires a computer, modem, available telephone line, and enough computer skill to set up the system and make it work. Literally millions of human beings cannot afford the Internet Service Provider (ISP) access fees and are technological lepers lacking access, connectivity, and, increasingly, the very conditions of possibility for livelihood, and thus dignity, in the new global information economy. Global cultural cyborgism still remains off limits to at least half of the human race who have never made a telephone call.

## Participation

For Marcel, the functionalization of the person works against participation or the affirmation of being that I *am* and that I *feel,* rather than that I utter or speak. Contemporary philosophers who shun the spirit of abstractionism should find in Marcel a ready ally insofar as he takes the body, rather than language, as the primary focus of his early reflection on existence. The body–subject does not so much assert being. Rather, being asserts itself through the body–subject and has primacy over knowledge, having, and technological doings. The question arises: Can being assert itself through the cyborg–subject? How much organic Dasein must remain? The person with a heart pacemaker is a modern cyborg. This technology is a human good, a value that can aid human health and well-being. And yet, how far may we go with biological and genetic technologies and still retain authentic personhood and an authentic cipher of human being? What is clear is that Marcel's transcending through cognitive and bodily participation is far less a matter of the will or volition than it is for Jaspers. Like Heidegger, Marcel is far more nonvoluntarist than Jaspers because the human being is governed by Being's call, which sets parameters within which authentic life is possible. Moreover, for Jaspers, the "if ... then" operations of the calculating "intellect" *(Bewußtsein überhaupt)*—no matter how sophisticated the logic of the algorithm—are ultimately encompassed by Being. Just as Existenz is for Jaspers the transcendent mode of the encompassing of subjectivity, *Das Umgreifende das wir sind,* that transcends technological orderings, doings, and makings, so too for Marcel participation in the realm of "mystery" remains distinct from the realm of the "problematic." Mystery "is a problem which encroaches upon its own data, invading them, as it were, and thereby transcending itself as a simple problem."[14]

## Mystery and metaproblematic thinking

Mystery and metaproblematic thinking go hand in hand. The problem of evil, the union of body and soul, profoundly liberating and freeing acts of fidelity, love, and hope all represent metaproblematic metaphors. Marcel rejects the view that metaproblematic thinking remains a content of thought or merely problematical to the "nth" degree. The epistemological critique of metaproblematic thinking fails to grasp the nature of metaproblematic thinking as "intermediary" thought. Metaproblematic thinking is neither purely objective nor purely subjective thinking, and is reminiscent of Jaspers's *transzendieren Denkens*, a *schwebend Denkens* between the poles of the subject–object split, only finally encompassed by the Encompassing of all encompassings. Similar to Jaspers's notion of the assurance of Being in "absolute consciousness,"[15] the metaproblematic "*is* certainty, it *is* the assurance of itself; it is, in this sense, something other and something more than an idea."[16]

Or again we might compare Marcel's metaproblematic realm with Jaspers's notion of "possible Existenz" as "the self-being that relates to itself and thereby also to transcendence from which it knows that it has been given to itself and upon which it is grounded."[17] Existenz and alterity are inextricably linked. "The test of the possibility of my Existenz," Jaspers stresses, "is the knowledge that it rests upon transcendence.... Existenz is only in relation to transcendence or not at all."[18] Whether we deny, directly oppose, or seek our way in the world with transcendence, transcendence remains the ceaseless question for possible Existenz. Marcel and Jaspers share a common epistemological insight here. Marcel's intermediary mode of metaproblematic participation on one hand, and Jaspers's experience of transcendence grounded in being's presentness for possible Existenz on the other hand, represent the respective cognitive loci wherein both philosophers hear the symphony of Transcendence: Marcel through the experience of embodied participation mediated through Christian faith, and Jaspers through the reading of ciphers mediated through philosophical faith and the absolute consciousness of possible Existenz.

We may also read the Marcellian self through a Jaspersian lens, via the language of modes of being. Thus, we may speak of the person's participation in the mode of *incarnation* actualized through sensation and bodily experience, the mode of *communion* actualized through transcending acts of love, fidelity and hope, and the mode of *transcendence* actualized through the primitive assurance of the mystery of being, the *Toi absolu*. Transcendence stands as the "irreducible [that] does not let itself be absorbed or reduced to an immanent process."[19] To use a different metaphor, we may say that Marcel's view is similar to Jaspers's admission that we are "touched" by transcendence and that we touch transcendence in turn "as the Other, the Encompassing of all encompassing,"[20] a touching that takes place within a "space of transcendence" where solutions to the problems of existence and approaches remain barely accessible yet inevitably open to metaphysical thinking.[21] To be sure, such a dissecting of the Marcellian self tends toward the negation of the unity of the concrete subject. As Marcel would remind us,

> it would certainly not be proper to deny the legitimacy of making distinctions of order within the unity of a living subject, who *thinks* and strives to *think of himself* ... [but] ... the ontological problem can only arise beyond such distinctions, and for the living being grasped in his full unity and vitality.[22]

Or again, consider Jaspers's notion of turning the transcending of temporal existence into "a *unity of presence and search*—a presence which is nothing but the search that has not been detached from what he is seeking."[23] The philosophies of Marcel and Jaspers both avoid optic metaphors for transcendence. We "hear," but never see, transcendence. And we hear only through ciphers that do not finally count as "scientific" knowledge.

## Incarnation versus ciphers

Marcel is far less hesitant to move straight in the direction of transcendental objectification than is Jaspers. This is the case despite Marcel's rejection of the appellation "Christian existentialist"[24] and his use of philosophically religious categories such as attestation, sin, faith, hope and love.[25] In contrast to the fundamental orientation and operation of Jaspers's philosophy, Marcel's neo-Socratism does not start from a systematization of the subject–object dichotomy and then move beyond it through a neo-Kantian formal transcending in world orientation, existential elucidation, and in the metaphysics of cipher reading. Marcel's philosophy moves, instead, from a "nonobjectifiable awareness of the finite to a nonobjectifiable discourse about the transphenomenal" convinced "that the transcendent cannot be identified with any conceptual point of view," and that "depersonalized thought cannot admit us to a realm worthy of being called metaphysical." "It is Karl Jaspers," Marcel writes,

> following in the steps of Kierkegaard and in all likelihood Heidegger too, who merits the important distinction of having shown that existence (and a fortiori transcendence) can only be apprehended or evoked in a realm beyond that of thought in general which must operate by means of signs on the contents of the objective world.[26]

For both Marcel and Jaspers there is no question of empirical demonstration of transcendence. Any "proof" for the immortality of the soul, even as a Kantian regulative idea, is a misguided undertaking, even though Marcel does not give up the experiential truth of immortality or swallow it up into an apophatic cipher. In opposition to Jaspers, Marcel views the self in its freedom as always "continuous with the man [or woman] of flesh and bone for whom immortality as a mere cipher would be a hollow consolation."[27] Despite their differing emphases on incarnation and the immorality of the soul, the "me" is not the last word for either thinker. The relation of Existenz to Transcendence is *l'âme commune* of Marcel's and Jaspers's philosophies in contrast to both Heidegger and Sartre.[28]

## Recollection

Marcel calls for transcending to the metaproblematical realm by means of *recollection*. Through Marcellian transcending, primary reflection is reborn as second-level reflection in a fashion similar to Jaspers's scientific world orientation being reborn to existential elucidation.[29] Second-level reflection for Marcel is a form of participatory faith rather than the primary act of the reflecting *cogito* that inhabits the world of *techne.* Marcel's second-level reflection, or recollection, is the road to the land of detachment, a journey that transcends the dualistic byways of being and action. In the journey of recollecting, one abandons oneself to, and "relaxes" in, the presence of being. Along the route of recollection we carry our belongings, both that which we are and that which our life is not. This recollective process brings out the gap between our being and our life, thus reinforcing the importance of the narrative unity of life. Recollection is not identical with the French *retour sur soi* or the *für sich sein* of German idealism. Why? Because "you are not your own." Marcel cites St. Paul in 1 Corinthians 6:19: "Do you not know that your body is a temple of the Holy Spirit within you, which you have from God, and that you are not your own?" The "I" (je) into which one withdraws is no longer an I but a Thou. Neither is recollection the same as "intuition." Recollection is about an assurance that can only be approached by a second order reflection that asks: "How and from what starting point I was able to proceed in my initial reflection, which itself postulated the ontological, but without knowing it."[30] The second-level reflection is a dramatic recollection whereby consciousness seeks to be conscious of itself. Marcel views "modern man" as at the mercy of technics and thus incapable of "controlling his own control." This problem of "controlling our own control" is the objective expression of second-level thought manifest in the artifacts and doings of human planning. Only through recollection can we combat the submerging of the individual by various technological functions that leave us not so much lost in the "many" in relation to the "one," as "empty" in relation to the "full" space of transcendence.

## Despair and hope

The shadow side of recollection is that the primal, ontological need can deny itself through despair, betrayal and suicide. Despair is essentially the negation of security or guarantee. Reality becomes insolvent. To change one of Marcel's favorite metaphors, the despairing self cannot obtain credit at The First Bank of Reality! In contrast, hope consists in asserting that there is

> at the heart of being, beyond all data, beyond all inventories and all calculations, a mysterious principle which is in connivance with me, which cannot but will that which I will, if what I will deserves to be willed and is, in fact, willed by the whole of my being.[31]

Marcel's nonvoluntarism and radical hope is something of a counterpoint to Jaspers's cipher of "being-in-foundering" and the courageous disinterest the shipwrecks of suffering and death require. Jaspers does not attest in the same way as Marcel to a radical hope that a loved one will recover from an incurable disease. Marcel, by way of contrast, believes it is impossible for Being to be hostile or indifferent to this very hoped-for good. Despite factual medical cases, other persons' experience, or even statistical probability, we hope that the health of the person we love will be restored. Marcel's dialectic of hope transcends to a plane that is opposed to "success" in any form in the same way that Jaspers's "cipher of being-in-foundering" requires a "leap from fear to serenity"[32] at the absolute limits of existence and Existenz. The origin of the rest that Marcel and Jaspers find beyond failure in their respective realms of the "problematic" or "foundering" before the being of nothingness has this key difference. Marcellian hope is a vociferous protest cry against death outside the City Hall of Being, a picket before the problematic, an affirmation of Christian tradition and its socio-historical instantiation of faith in the resurrection. Hope is primary for Marcel because hope is the most direct means of apprehending the meaning of transcendence. Hope provides the "spring," the "leaping" of a gulf to transcendence.[33] For Jaspers, reason is primary, the "glue" that holds the modes of the encompassing together and provides the condition of possibility for our being-with Transcendence as possible Existenz. Marcellian hope is highly untechnical. Technical thought never separates the means and the ends in its reflection. There can be no end for the technician if there is no vision of the means to achieve it. But hope is not ultimately concerned with means. Hope's personification is not the technical inventor or discoverer who says "there must be a way" and "I am going to find it." Hope simply asserts that "it will be found."[34]

## Moral degradation

We may now see how Marcel's understanding of transcendence, linked to second-order recollection and hope, fits with his view of technology; for the realm of the problematic is precisely the world of fear and desire, the realm of the functional, a "kingdom of technics." Marcel provided the following description of the metaphysical crisis of our age and the limits of technical efficiency and planning in 1933:

> Every technique serves, or can be made to serve, some desire or some fear; conversely, every desire as every fear tends to invent its appropriate technique. From this standpoint, despair consists in the recognition of the ultimate inefficacy of all technics, joined to the inability or the refusal to change over to a new ground—a ground where all technics are seen to be incompatible with the fundamental nature of being.... we seem nowadays to have entered upon the very era of despair; we have not ceased to believe in technics, that is to envisage reality as a complex of problems; yet at the same time the failure of technics *as a whole* is as discernible to us as its *partial* triumphs. To the question: What can man achieve? we continue to reply: He can achieve as much as his technics; yet we are obliged to admit that these technics are unable *to save man himself,* and even that they are apt to conclude the most sinister alliance with the enemy he bears within him.[35]

What Marcel means by the sinister alliance between technics and the enemy within becomes clear in *Les Hommes Contre L'humain* (*Men Against Humanity*) in 1951. Here he defines various techniques of degradation as a whole body of methods deliberately put into operation in order to attack and destroy in human persons belonging to some definite class or other their self-respect, and in order to transform them little by little into mere human waste products, conscious of themselves as such, and in the end forced to despair of themselves, not merely at an intellectual level, but in the very depths of their souls.[36]

With the horrors and killing techniques of the Nazi concentration camps in view, Marcel speaks of "moral degradation" as a type of technique. Techniques of moral degradation and humiliation turn human beings into garbage through the destruction of their dignity. Persons who retain a sense of their own value can always resist degradation, but the persecutor is able to strengthen a sense of his righteousness once the victim becomes degraded in their own eyes. Moral degradation is possible because of a systematic trampling underfoot of universal values. Marcel understands universal value not in an abstract Platonic sense of the idea of the good, but rather in the concrete relations of alterity that confer dignity. Marcel is far from Nietzsche's attempt to get beyond "good and evil." Nietzsche's "beyond" became a beneath, his way up, a way down, his thinking a transcendence not of ordinary moral categories, but a transcendence from them. When the universal is lost, values are created ex nihilo and technique gains the upper hand. Human beings deny createdness and claim self-dependence and are in turn led to consider themselves as beings who make themselves. This is a type of faith in which "man ... *is* only what he makes of himself; for if there is nobody who can destroy his self-sufficiency, similarly there is no gift which can be made to that sufficiency."[37] This Sartrean anthropology of non-giftedness leads willy-nilly to a view of human beings as excrement, shit; and, in turn, to self-abasement and ironic self-exaltation.

## Propaganda

Propaganda is another technique of degradation. It moves from a relative and subordinate recruitment of new adherents to some definite cause, to a seductive carrot-on-a-stick method in which benefits are held out to the novice for rallying to the cause. State propaganda simply reduces human beings to a condition in which they have no capacity for individual reaction. Drawing upon the Austrian writer, Joseph Roth, Marcel speaks rather demonically of the spiritual degradation that radio plays or will be seen to play in history. This view seems totally archaic to us today. And yet, Marcel asks an important question, one appropriate to the problem of "spamming"[38] on the Internet today and debates about the role of "Big Brother" in the regulation of computer-mediated communications: Why should someone be granted "the gift of being everywhere at once in return for the payment of an annual rent for radio time?"[39] Marcel is skeptical of the universality that radio confers and which he views as a peculiarly modern usurpation that may lead to evil ends. Why is this the case? It is the temptation to transcend our finite human condition and the "privilege of ubiquity" that Marcel finds so dangerous. To be sure, he insists that radio technology is not evil in itself or *malum in se.* If placed in the service of a genuinely universal mode of thinking, the radio could well serve the human good and have a wide-ranging positive impact.

May we say the same about Internet technology? Yes and no. The problem does not reside in the artifacts of technology as such, but rather in the will to technology or the volitional aspects of a given technology's use with respect to individual and societal goods and values. The fact that anyone can post hate literature or pornography on the Internet from a home computer given a modicum of technology and know-how raises the existential importance of computer-mediated communications. Questions concerning government regulation and the role of rights and responsibilities with respect to access, intellectual property, privacy, freedom of expression, data surveillance, and security on the information highway are vitally important and necessary. But they are insufficient if divorced from the prior question of existential communication—a question that is rarely if ever broached in current public policy debates on new information technologies, and insufficiently addressed in recent philosophical perspectives that explicitly attempt to bridge the gap between theory and practice in information ethics.[40]

Marcel's reflection on radio technology leads him to formulate an operative principle of technical progress. Every technical progress entails the payment of a heavy price for the individual "who takes advantage of it *without having had any share in the effort at overcoming difficulties of which such a progress is the culmination,*" and of which "a certain degradation at the spiritual level is the natural expression."[41] This principle commits Marcel to neither a Luddite worldview nor an extreme Romantic longing for some golden, pre-technological age. It does commit him, however, to the Bergsonian perspective that we need to balance outward, technical progress with an inner conquest of the self. Self-mastery must be greater than technological mastery. Stated negatively, there exists an inverse correlation between technological progress and the decline of self-mastery, between smooth functioning on the material level of existence and alienation from the spirit. Technique shifts the inner balance of the self. As we project ourselves into the realm of artifacts and into the technological apparati and objects necessary for life, the more we lose, or potentially lose, our center of gravity. We tend to substitute material satisfaction for spiritual joy, material dissatisfaction for spiritual inquietude.

Marcel's correlation between increasing functionalization and the decline of the spirit is not a strong one, however. His ambivalency toward technique is quite clear. Technological progress is both good in itself, and good because of the intelligibility and order it brings to reality. The problematical enters only when there is no personal appropriation. This, in turn, fosters resentment and envy of the Other who has or possesses technology as either knowledge or artifact. Marcel drives a wedge between any technical apparatus and its possessor. He gives an example of a peasant and the care he takes to cultivate a small vineyard. The peasant's relation of care in the harvesting of grapes is completely missing in a world where technique triumphs. What keeps the relation authentic is the lack of guarantees or security for the peasant farmer. A hailstorm could wipe out his crop. With mechanical technique, the ideal is to overcome what is contingent, unpredictable or insecure. It is the preoccupation with security, with the control of desire or fear, that diminishes life and leads to a spirit of revolution.

## Techno-slavery and techno-idolatry

Technique may also turn persons into "slaves" and "idolaters." Marcel is adamant that technical progress is not an expression of "sin" and to suggest as much is childishness. Technique, defined simply as "a group of procedures, methodically elaborated, and consequently capable of being taught and reproduced,"[42] is good in itself because it instantiates Being's gift of reason to reality. Techniques are analogous to possessions, however, because they are something we acquire, just as a habit is something that we acquire, and a kind of technique. As we can become slaves of habits, so too we can become slaves of our techniques. Technique, especially for the person who has to invent it, is not simply a means. It often becomes an end in itself, and, in the process, divorced from the purpose to which technique is subordinate. Technique's slippery slope toward "slavery" is thus tied to an inversion of end and means. Take the example of the automobile technician who, though unable to use the very cars he designs, nonetheless keeps the proper end of the car in view. The moral and intellectual investment of attention, ingenuity, and perseverance that the technician makes during the invention process naturally leads to feelings of power and pride in the invention. The problematic enters when there is no appropriation of the process of discovering and perfecting the car's means to its proper end. For a car aficionado, at least the one who regularly watches the Indianapolis 500, the car becomes an end in itself and no longer serves the goal of improved travel efficiency. This same problematic may function in the race for faster megahertz speed in CPU chips, though the drive for market shares and increasing profits probably sets the ultimate *telos* for mass-market computer developments. When this happens, we have passed from the technical realm, proper, to an occasion for idolatry: speed becomes an end in itself rather than a means. This development is one step away from *autolatry* or self-worship and, concomitantly, a loss of the sacred.

It was against the backdrop of the horrors of Auschwitz and Hiroshima that Marcel challenged those who decried the notion of technological idolatry as childish, savage, or superstitious. The emancipated person who prides him- or herself on believing in nothing fails to recognize that a person who believes in nothing does not really exist. What we "believe in" is not an opinion but that which we really cling to or hold on to. To believe is to have a "living link" with what we believe in. And the person who believes in nothing, who clings to nothing, really has no living links. The tissue of their body–self is dead in spirit. If I may metaphorically transform the image of Hypertext Markup Language (HTML) or Web page code, it is as if an existential text markup language, that is, human speech and action that links persons through authentic communication and alterity, has been erased due to lack of participation in being. The human intersubjective web is brought down, the communication drive crashes. Communication is ruptured not by hackers, but by the nothingness that undergirds the ultimate source code of life, now divorced from the spirit.

Marcel lived and wrote on the cusp of Marshall McLuhan's mass media revolution. Marcel's concern with global communication is relevant to contemporary debates concerning cultural protectionism and the putative global hegemony of Internet culture and language, which, at least in Canada, is often seen, wrongly I believe, as a tool of American cultural and economic–political control. The development of communications ought to produce greater possibilities for human communication, for example, greater access to education and health care information. But on both a global and national scale the development of global communications technologies may merely impose greater uniformity upon national cultures, thus negatively impacting individuality and the retention of local beliefs, customs, and traditions. The paradox is that greater communications technology may not, in fact, allow human beings to assimilate the universal. Instead, persons may be more than ever set on the path of parochial and aggressive loyalties. Technical and industrial progress combine to create the lowest common denominator of well-being, that is, "cash," which inspires both covetousness and envy. Although envy and greed have fueled quarrels and wars between local and national communities throughout history, there is something unique about the immediately past century stained by a variety of attempts at collective extermination.

Marcel may not have envisioned the pitfalls and possibilities of the Internet for contemporary global communications, but his philosophy has some perduring, perennial elements worth noting in our historical situation marked by the Infosphere. Consider, for example, the problem of leveling and competing moral visions of the Internet. The shift to the language of the "autoroute électronique" or information highway, with its exits to electronic shopping malls surely reflects a certain leveling effect. For it must be admitted by even the most devoted technophiles that the on-ramp to the Information Highway is paved with "cash" because online access depends not only on having the necessary hardware, which many persons cannot afford, but also being able to pay user fees charged by Internet Service Providers (ISPs). It remains to be seen whether Internet access will be universal, affordable, equitable, and gender inclusive, especially outside of North America and Europe. Will new information and communication technologies remain the domain of a small and demographically distinct segment of the population in First World countries?[43]

### Competing ethical visions: E-commerce versus e-commons

Indeed, even in information-rich countries we have seen two competing moral visions of the Internet: the Internet as a space for business and "e-commerce" versus a vision of the Internet as a space for communication or an "e-commons." The commerce vision or jobs-agenda policy has led a variety of individuals, organizations and special interest groups to criticize government information infrastructure policy as a market-driven, top-down approach, an information equivalent of trickle-down economics. On this view the global commodification of information is having negative implications for social control and power over community and cultural forms. By "commodification," I refer to the transformation of use value into exchange value in which objects and ideas are stripped of their intrinsic, moral, aesthetic, and utility values and replaced by market values, or what an object or idea will bring in trade.[44] According to some commentators, the e-commerce vision is really about increased transnational corporate power, the blurring of national identities, the breakdown of human solidarity and community, and a neoliberal policy agenda of privatization, deregulation, and user-pay that is leading more and more to the collapse of public space. The e-commerce vision encodes an information world of "in-your-face" capitalism where workers lose their jobs one week, the ex-employer's stock prices rise the next day, and the telecommunications CEO, already boasting a seven-digit salary, gets a fat stock option the third day. Technocratic polarization has been formulated by Jeremy Rivkin's research on the effect of technological destructuration on the progress and decline of work. He argues that information and communications technologies and global market forces are fast polarizing the world's population into two irreconcilable and potentially warring forces. His technological haves and have-nots are drawn in terms of

> a new cosmopolitian elite of 'symbolic analysts' who control the technologies and the forces of production, and the growing numbers of permanently displaced workers who have little hope and even fewer prospects for meaningful employment in the new high-tech global economy.[45]

In contrast, the electronic commons visions would instantiate a community-owned and financed communications tool linked to participatory democracy and creative social thinking that projects a flourishing of communication goods. The transformation to a knowledge-based society reflects the conflicting values of jobs and economic profit versus the values that concern self-identity in community. Justice in the ownership and regulation of the information highway requires a public broadcasting industry operating to reinforce culture, identity and linguistic duality. But because the private sector is establishing advanced networks, governments are increasingly relegated to a minor role of policy and regulation. In a different voice, Ursula Franklin and Sherry Turkle express heartfelt concerns that the Internet is being used not to serve the participation of persons in the electronic commons, but as a palliative for our vulnerable time of social and economic transition.[46] Policymakers have argued that information commercialization does not have to win the day on the information highway if a proper mix between public and private input can effect the interoperability and interconnectedness of networks and facilities, support infrastructure upgrade in northern and remote communities, contribute to the development of public access points, and assist community-based networks. But the marketplace alone will not lead to universal access to networks and facilities, services and contents, and could, in fact lead to an increase in the disparity between the information-rich and information-poor. We know that this disparity is already a reality when telephone and cable companies redline communities dominated by the poor and blacks.

## The problem of technological hubris

Marcel would perhaps remind us in view of our historical situation that a belief and hope in technical progress as the solution to societal and global problems in the areas of education,[47] environment, health, and the information economy does not obviate the need for recollection. Indeed, it is precisely our dominant cultural faith in technology that raises the question of a possible reconcilation between technological and ontological hope. Such a reconciliation or balance between the two is possible because Marcel affirms that technology is good in itself. But the reconciliation is possible only if hope is genuine, that is, when hope does not depend on ourselves but springs, instead, from inner humility. It follows that modern technology forcefully raises the problem of pride or hubris since technologies are tools, par excellence, for treating the realm of the problematic. Marcel stands very close to the ancient Skeptics in his view of *techne* when he invokes the problem of pride as the drawing of strength from oneself, and in direct contrast to Spinoza's *superbia,* which wrongly equated pride with vanity. Techniques potentially lead the mind into the temptation to abuse power. The exercise of any sort of power needs to be accompanied by the exercise of control over this power itself, a sort of "power at one remove."[48] What Marcel means by "power at one remove" corresponds closely to philosophical wisdom, which has traditionally stood guard against various forms of hubris.

## The logic of technocracy

Marcel speaks of the "logic of technocracy" that has about it an abstract character in which the intellect is the more at ease the more it is specialized.[49] There are serious dangers here, even if Marcel cannot naively avoid a role for specialization in human cognition and doing. The key danger is that a technical environment will become a paradigm for the universe by which we appropriate, wholesale, a technocratic representation of the world. The logic of technocracy is similar to Jaspers's concern in *Die Atombombe* book with technological panaceas resulting from the dominance of "departmentalized" thinking. Jaspers's interest is in a proper encompassing of culture by reason. When *Verstand* usurps its proper mode, technophiles devise a world of machines and mechanics that create a second world in which the masses assume an operative function. Culture becomes dominated by a form of technological positivism.[50]

Marcel goes a step further by stressing the inevitable way in which persons will consider themselves a natural part of a technocratic cosmos or, better, "a-cosmos." He wants to guard the view that technological information is not true *sophia;* technological artifacts are less real than natural objects. When a technocratic cosmos becomes natural, we become easy targets of manipulation through experimental psychology, psychiatry, and even education with its general pragmatization or functionalization of human beings and their relations. Devitalized rationality then turns human beings into robots, mere units of production in which life is no longer conceived except in sociobiological terms. Physicochemical conditions are then claimed to be objectively definable and placed in service of the collectivity. Marcel located this monstrous logic of dehumanization, of course, in totalitarian regimes—whether political or scientific—whose hallmark is the will to power, and who need a scapegoat in whom to focus that will. "It must not be forgotten," he writes during the heightened sabre-rattling of the early 1960s and Cuban missile crisis,

> that by now the [technocratic] system is endowed with its own power of self-consolidation, which incidentally is the crowning absurdity, since it may in the end involve us in the sacrifice of all, certainly not for all but precisely for no one—an immense holocaust to nothingness.[51]

What forms and impulses might the totalitarian spirit take in a post-1989 world less concerned now with nuclearism than with information warfare?[52] The danger of large, bureaucratic totalitarian regimes may have faded from recent history, but the will to power may yet take on a variety of avatars when endowed by a devitalized rationality and technological means. As Jaspers would rightly remind us, the fight against totalitarianism is as much an internal struggle as an external one.[53] Totalitarianism and fundamentalism can take on new and various forms depending on the historical situation, perhaps especially now in the global information age. We must not be historically naive to think that totalitarian impulses of recent history will never happen again.

## Technology and desacralization

In one of the final lectures Marcel gave in Heidelberg in 1963 he asks: "What are we to make of ourselves in face of the fact that we are gradually being thoroughly manipulated by a technology that we ourselves have devised?"[54] The sacral belongs to the realm of mystery, to the realm of the "meta-technical."[55] It is now the frail, defenseless person in prayer who is unabsorbed by a technical, dehumanized world who most fully symbolizes a meta-technical posture. In a technocratic culture, desacralization becomes a process directed against life and its manifestations, including the family. Totalitarian states have the same desacralizing effect when nationalism usurps the place of the sacral, thereby forming nothing more than a psuedosacral society. Libertarian contractual, voluntary modes of relation also reflect a naive view of the authentic human condition because agreement by its very nature requires an oath grounded on the sacral. Contractual agreements tend toward accords that each party tries to use to his or her own advantage. Or, only slightly less worse, they tend toward a mere bureaucratic fine-tuning to which nobody pays any attention. Marcel's critique of bureaucracy is reminiscent of Weber's iron cage, but now marked by another paradox. The men and women in the cage are statisticians because managers need the techne of statistics to solve the very problems posed by modern technology. Statistics is hardly sacral, for Marcel, since it entails an implicit indifference to the individual. Statistical methods are foreign to grace, and we search in vain for a total comprehension of human reality by such methods.

## Conversion

Marcel is led to the conclusion that in a technological Promethean age, with its attendant hubris, the sacral can only reveal itself by way of conversion.

> Conversion is first of all the movement by which the consciousness turns away from the oppressive and distressing spectacle that the technocratic view of the world offers, or—and this amounts to the same thing—by which consciousness transcends the obsession with numbers through the numberless. It is the inwardness we regain through an action which is not only free, but in fact is freedom *itself.* But we have to remember that in its essence inwardness is not tantamount to restriction, and it would be gravely deceptive to think so. And it is just as wrong to imagine the individual who becomes a unity all to himself on account of his conversion. The exact opposite is true; inwardness must be reciprocal; it is a relationship of one individual to another, of an "I" to a "Thou," as both Martin Buber and I have tried to prove in our writings.[56]

Similarly for Jaspers, human "conversion" (*Umkehr*) is the condition of possibility for a "new politics" in the context of the modern threats of political totalitarianism and nuclear war.[57] No technology can effect the human Umkehr that stands as the condition of possibility for the needed integration of philosophy, science and technology in the modern period. The Umkehr might effect the new reasonable politics imbued with a spirit of sacrifice that steers a course between *Realpolitik* and ethical idealism, a morality grounded in existential freedom and truth, and the transformation of dogmatic religion by the impulses of philosophical faith as a life lived out of the open horizon of the Encompassing.

## CRITICAL CONTRAST AND COMPARISON

Marcel's view of technology borders at times on a type of technological moralism. He often seems to have little positive to say about technology and its benefits for humanity. While there is a developmental ambivalence in his view of technology, Marcel nonetheless affirms that "the burden of technics has been assumed by man and he can no longer put it down because he finds it heavy."[58] We cannot return to a pretechnical age devoid of an ostensibly bloodless rationalism. In the 1960s Marcel admitted his initially hostile view of technology, but then tried to correct the impression he had given his followers:

> Today . . . I would no longer condemn a single instance of technology. I believe that technology is good in itself. On the other hand I think that technology can be put to the wrong use and that too little importance can be attached to the questions of what ends technology ought to serve.[59]

Marcel's main concern, then, is that techniques as means may become regarded as ends in themselves. When this happens they easily become an idol or an excuse for self-idolatry, often as the will to power. Participation in second-order reflection and recollection check the human will to power and hubris by creating a space for transcendence. Despite the ambivalency toward technology, Marcel's philosophy of concrete *engagement* and *disponibilité* deserves a new hearing lest it become a forgotten trace on our technological stage. It is the collapsing of transcendence into technology that worries Marcel's Christian faith:

> It is in gift, that is to say in grace, that there assuredly lies the only principle capable of breaking, I will not say the world of the techniques insofar as they are admirable means to be used for the good of all, but those superstructures which threaten in the long run to stifle their beneficent power, because they are ordained to the triumph of pride, which ultimately encompasses the destruction of the proud.[60]

For Jaspers, in comparison, technological making entails a form of "purposive action" aimed at accomplishment in the world (and in the transcending movement of world orientation). His philosophically relevant question concerns the limits of technological will and activity in relation to the realities of life, soul and the mind. There is a rather large intermediate area between the technological control of things in the world of facts and planning, and the free communication of one Existenz with another.[61] The point of philosophy is ultimately to transcend the intellect as a knowledge techne in the service of freedom or possible Existenz. Authentic communication remains inevitably tied to *Verstand* as a type of tool for the self in relation to the world, others and Transcendence. In sharp contrast to Marcel, Jaspers's non-divinized view of nature and culture can sustain a favorable view of secularization and a concomitant idea of the great benefits for science and culture through the use of technology. He rules out any kind of animism that might restrain technological investigations of nature because it is sacred. In short, the cipher of nature does not refrain human beings from intervening into its processes. What is crucial is the differentiation of power expressed in technological will, in the coercion of things, and the pure will to knowledge that exists apart from the aims of techne per se.[62]

Jaspers was concerned initially with the tension between "technical mass-order"[63] and the challenges it raises for truly human life. Later he denounced technological "total planning" that eliminates human freedom.[64] Despite his concern that modern science and technology have created the means for both virulent totalitarianism and nuclear annihilation,[65] the technological ordering of existence is an inevitable feature of modernity. In this respect, Jaspers stands very close to Max Weber, whereas Marcel stands close to Heidegger. Indeed, Jaspers would endorse the view espoused by some philosophers of technology that the solution to our problems is not less technology, but more technology.[66] He would add the important caveat, however, that all technological developments be placed under a suprapolitical ideal. Technology must be continually guided by a healthy heuristic of human finitude and the inevitable *Grenzen* or limits of cost–benefit prediction and planning.

Jaspers was far more insightful than Marcel in seeing that the inevitable technological ordering of existence effects not only the West, but also China and India who, a fortiori, have the problem of finding "new spiritual forms"[67] and intersubjective communication in a (post)modern technological tradition. With far greater emphasis than Marcel, Jaspers is as much concerned with the wrongful demonization of technology as he is with human misjudgment of its limits. "The demonism of technology," Jaspers writes,

> is only to be vanquished by following the road that leads to our penetration of it. Whatever mischief it gives rise to may perhaps be without our power to master. Organisation of the market, for example, can afford deliverance from temporary want then enable a free market to be re-established, instead of ending in annihilation, in which there is nothing more to distribute. But again every plan conceals the possibility of that 'demonism', of the unforeseen. Technological mastery of the mischief wrought by technology may add to this mischief. Absolute technocracy is, for its part, an impossibility.... The fate of man depends upon the fashion in which he masters the consequences of technology for his life (from the arrangement of its whole structure, as it presents itself at any particular time, to personal conduct at every hour of the day).[68]

It remains for us to ask what the technological drama means for us today in the information age. We are actors on a new electronic stage where Nietzsche has a modem and is rewriting the last pages of *The Will to Power* as *The Will to Virtuality*.[69] The gap between information haves and have-nots, between the virtual and unvirtualized classes grows larger, thus skewing economic and political power at a time when global unemployment has reached its highest level since the Great Depression.[70] While Marcel critiqued a functionalized world in which sleep, pleasure, and health were all on a vital schedule, Jaspers was far more concerned with the irony that technology has both saved and increased labor. For those not working "McJobs"[71] today, has technology actually freed us from work? Are we not more and more on the electronic leash of pagers, cell phones, fax machines, and laptop computers? Are not the natural dikes built between our work and family life leaking water, if not being washed away completely? Is not the pressure to work continually one symptom of the transformation of work life, even if we have not reached the "end of work"? Global markets demand instant availability and invisibility, an endocolonization of the unwired world of time, history, and human flesh by the electronic body, even as political pressures grow to dismantle workplace protections. Many persons find that their personal conversations and social lives are confined to e-mail, with social gatherings dominated by talk of work, investments and technology. Is this constant preoccupation with work and technology a symptom of the malaise that many people feel in our uncertain economic environment?

Marcel and Jaspers seriously challenge us to ask how the symbolics of technological desire and fear may dominate culture. More exploration is needed on how the archeology of the increasingly virtual subject is dialectically related to a teleology of desire and fear in its origins and ends. Will we ever have enough technics? When shall the will to technological power rest? Where is the Aristotelian mean between technological suspicion and technocratic faith, between a cipher of technology as "demon" on one hand, and a cipher of technology as having the capacity to reveal the sacred on the other hand? Marcel reminds us that the regeneration of thinking through technology cannot be grounded on anxiety or fear; regeneration is possible only by way of reflection on hope and joy.[72] Alterity in the form of existential communication is still the sine qua non for creatively living in the real world of technology where the burden of techniques cannot be put back into the genie's bottle. Philosophers have only begun to take up the ethical and metaphysical challenges posed by new information technologies. Second-person philosophers like Gabriel Marcel and Karl Jaspers have much to contribute to the current debates over science, technology, and society. At a minimum, humility is a precious if often forgotten virtue in approaching the challenges that new information technologies pose in the information age. We may at least hope and work toward a vision in which the functionalization of human beings via techniques gives way to an authentic and communicative alterity commensurate with our present information-age revolution.

## Concluding untechnological postscript: Philosophy and visual information processing

What do the philosophical discourses of Marcel and Jaspers contribute to the work of visual information processing, and how such processing relates to an understanding of intellectual functioning? Permit me to conclude with three brief observations.

First, it remains for the reader to tease out the implications of the foregoing philosophical frameworks for your field of technical specialization. One may only hope that philosophers and technologists can commit to a conversation across disciplines and field specializations. Visual information processing holds out new possibilities today, but the need for informational dialogue strikes me as all-important when the university (the *universitas* of knowledge) has become a pluriversity, or more accurately, a multiversity. The will to authentic multidisciplinary dialogue becomes increasingly difficult in the information age when pressures to publish for promotion and tenure reflect the transformation of means and ends that we need to guard against. In this context, the will to truth within the university, and, thus, intersubjective communication or authentic alterity, is all-important. Truth cannot be grasped if it remains incommunicable. As an educator I understand well that at least eighty percent of my students are visual learners. I appreciate the fact that the mind is inherently embodied, that ninety-five percent of thought is largely unconscious, and that abstract concepts are largely metaphorical. The goal or *telos* of information processes, therefore, must be methodologically and clearly defined in relation to internal educational values. Neither technical means nor basic scientific research can answer the question concerning the basic goals of science and technology. Philosophy is important here.

Second, information visualization is a process of transforming data and information that are not inherently spatial into a visual form that allows the user to observe and understand the information. The creation of technological means that provide useful and efficient visualizations of information is a human good and ethical value from the philosophical perspectives of both Marcel and Jaspers. Visual information processing seems to resonate directly with Marcel's concern for the concrete, embodiment, and avoidance of the spirit of abstraction. For Jaspers, too, the mind as intellect is never disembodied. His philosophical anthropology reminds us that the human intellect is inseparable from Dasein, Spirit, Reason, and Existenz; but these other modes of being are inevitably mediated through the intellect. As a type of techne, the intellect is the mediating mode of subjective being. In this context, Jaspers must give a legitimate weight to technological or instrumental rationality.

It is perhaps ironic to invoke the existential philosophies of Jaspers and Marcel in a volume whose concern is with visual information processing. Both thinkers part with the Western philosophical tradition's emphasis on optic metaphors in favor of auditory metaphors. The intellect cannot leave behind the shadowy images of Plato's Cave for .mpeg and audiocast files, but we may need to supplement visual information processing with an equal emphasis upon hearing and listening. Both Jaspers's metaphor about hearing ciphers of Transcendence and Marcel's dramatic and musical musings have much in common with cognitive science's understanding of the cognitive unconscious. Conceptual systems and our reason arise from our bodies. The cognitive also refers to our sensorimotor system that contributes to our abilities to conceptualize and to reason. If the cognitive unconscious accurately describes all unconscious mental operations concerned with conceptual systems, meaning, inference, and language, then perhaps auditory metaphors for cognitive science's metaphysical realism is more accurate than an optic metaphor. Indeed, color and color concepts make sense only in something like an embodied realism.[73] From an epistemological perspective, the philosophies of Jaspers and Marcel are expressions of cognitive interactionism that is neither purely objective nor purely subjective. But both emphasize the loving struggle for communication with other human beings, the I–Thou relation, as a primary value in the interactive unfolding of subject–object relations. The question, then, is how might visual information processing technologies better serve existential, communicative goals? The use of translation "bots" on the Internet may be an instantiation of such an ideal. And yet, language is polyvalent and even the most sophisticated translation technologies run up against limits. The same may be said for voice-recognition technologies, even though both technologies should be developed in conjunction with visual information processing techniques that aid human freedom and well-being.

One final caveat. Jaspers sharply distinguishes intellectual and rational thought. Intellectual thought is the mode of the technological inventor and maker. Its precepts can be carried out and can multiply the technological making by infinite repetition. One consequence of intellectual thinking is that it results in a world in which a few technologically sophisticated minds devise the technics—the technological operating system if you will. This then creates, as it were, a second world in which the masses of human beings assume an operative function in relation to some aspect of the technological system. Rational thought, in contrast, does not provide for the carrying-out of mass directives. It requires that each individual do his or her own thinking, original thinking. Reason believes that truth is not found by a machine reproducible at will, but by decision, resolve, and action whose self-willed performance, by each person on their own, is what creates a common spirit. When intellectual thought dominates the economic, political and social realms, and when it is untempered by the suprapolitical elements of reason, ethics and sacrifice, then humanity assumes an operative function in the state and society. Human will or volition becomes conditioned by means–end relationships instead of an unconditional volition and action grounded in the freedom of persons in authentic communication.[74] Culture becomes dominated by a form of technological positivism in which technology becomes the sole legitimate means for seeking truth, informing action, and governing society. Technology, including visual information processing, ought to serve alterity, however, because human beings are more important than things. We may then find that the true meaning of the information-age revolution concerns an evolutionary advancement in human intersubjective communication—a communication that relinquishes violent means in the struggle for human existence to the loving struggle for communicative Existenz.[75]

## NOTES

1. Gabriel Marcel, *Le mystère de l'être.* Tome I, *Réflexion et mystère.* Tome II, *Foi et réalité* (Paris: Aubier, 1951); *The Mystery of Being.* Vol. 1 *Reflection and Mystery,* trans. by G.S. Fraser, and Vol. 2 *Faith and Reality,* trans. by René Hague (Chicago: Henry Regnery Co., [1951] 1960).

2. Gabriel Marcel, *Les Hommes contre l'humain* (Paris: La Colombe, 1951); *Men Against Humanity,* trans. by G.S. Fraser (London: The Harvill Press, 1952).

3. Gabriel Marcel, *Le Décline de la sagesse* (Paris: Plon, 1954); *The Decline of Wisdom,* trans. by Manya Harari (London: The Harvill Press, 1954).

4. Gabriel Marcel, *L'Homme problématique* (Paris: Aubier, 1955); *Problematic Man,* trans. by Brian Thompson (New York: Herder and Herder, 1967), 17.

5. Marcel, *Problematic Man,* 28, 29.

6. Gabriel Marcel, *The Broken World and The Rebellious Heart,* edited by Francis J. Lescoe (West Hartford, CT: the McAuley Institute of Religious Studies, Saint Joseph College, 1974), 36.

7. Gabriel Marcel, *Being and Having [Être et Avoir].* The Fontana Library: Theology and Philosophy ed., trans. by Katharine Farrer (London: Collins, 1965), 109.

8. Marcel, "On the Ontological Mystery," in *The Philosophy of Existence [La Philosophie de l'existence],* trans. by Manya Harari (London: The Harvill Press, 1948), vii, 1–31.

9. Marcel, "On the Ontological Mystery," *The Philosophy of Existence,* 6.

10. *Webster's Third New International Dictionary of the English Language, Unabridged,* s.v. "cyborg."

11. Donna J. Haraway, "A Cyborg Manifesto: Science, Technology, and Socialist-Feminism in the Late Twentieth Century," *Simians, Cyborgs, and Women: The Reinvention of Nature* (New York: Routledge, 1991), 149–181, at 181.

12. Gabriel Marcel, *The Existential Background of Human Dignity: The William James Lectures, 1961–62 [La Dignité Humaine et Ses Assises Existentialles.* Collections Présence et Pensée] (Cambridge, MA: Harvard University Press, 1963), 123.

13. Haraway, *Simians, Cyborgs, and Women,* 154.

14. Marcel, "On the Ontological Mystery," *The Philosophy of Existence,* 8.

15. Karl Jaspers, *Philosophy,* Vol. 2 (Chicago: University of Chicago Press, 1969), 223.

16. Marcel, "On the Ontological Mystery," *The Philosophy of Existence,* 12.

17. Jaspers, *Philosophy,* Vol. 1, 56.

18. Jaspers, *Philosophy,* Vol. 3, 6,7.

19. Gabriel Marcel, "Je pense que le transcendant, c'est, avant tout, l'irréductible, c'est ce qui ne se laisse pas absorber ou résorber dans un processus immanent." Simonne Plourde, "Transcendance," in *Vocabulaire philosophique de Gabriel Marcel.* Collection Recherches; Nouv. Sér., 6 (Paris: Éditions du Cerf, 1985), 529–533.

20. Karl Jaspers, *Philosophical Faith and Revelation,* trans. by E.B. Ashton (New York: Harper & Row, 1967), 69.

21. Stephen A. Erickson, "The Space of Transcendence in Jaspers and Heidegger," in *Heidegger & Jaspers,* edited by Alan M. Olson (Philadelphia: Temple University Press, 1994), 126–137, at 137.

22. Gabriel Marcel, *Being and Having,* 17.

23. Jaspers, *Philosophy,* Vol. 3, 4.

24. Jean-Paul Sartre had distinguished between an atheistic and a Christian existentialism in *L'Existentialisme est un humanisme* (Paris: Nagel, 1946), 16–17. Marcel only tacitly accepted the title "existentialist" when he contributed a brief biographical statement to a collection of essays devoted to a discussion of his philosophy that was edited by Etienne Gilson, *Existentialisme chrétien: Gabriel Marcel* (Paris: Plon, 1947).

25. On this point, see Hans A. Fischer-Barnicol, "Systematic Motifs in the Thought of Gabriel Marcel: Toward a Philosophical Theory of Composition," in Schilpp, *The Philosophy of Gabriel Marcel,* edited by Paul A. Schilpp and Lewis E. Hahn (La Salle, IL: Open Court, 1984), 439.

26. Gabriel Marcel, *Creative Fidelity [Du Refus à L'Invocation],* trans. by Robert Rosthal (New York: The Noonday Press, 1964), 5.

27. Kenneth T. Gallagher, "Truth and Freedom in Marcel," in Schilpp and Hahn, eds., *The Philosophy of Gabriel Marcel,* 386.

28. Paul Ricouer, *Gabriel Marcel et Karl Jaspers: philosophie du mystère et philosophie du paradoxe* (Paris: Editions du Temps Prèsent, 1947), 265.

29. Jaspers, *Philosophy,* Vol. 1, 67ff.

30. Marcel, *The Philosophy of Existence,* 13, 14.

31. Marcel, *The Philosophy of Existence,* 16.

32. Jaspers, *Philosophy,* Vol. 3, 206.

33. Gabriel Marcel, l. *Being and Having [Être et Avoir].* The Fontana Library: Theology and Philosophy, edited and trans. by Katharine Farrer (London: Collins, 1965), 83–84.

34. Marcel, *Homo Viator: Introduction to a Metaphysic of Hope* (London: Victor Gollancz, Ltd., 1951), 51–52.

35. Marcel, *The Philosophy of Existence,* 18.

36. Marcel, *Men Against Humanity,* 30.

37. Marcel, *Men Against Humanity,* 31.

38. "Spamming" on the Internet, whether through e-mail, *Listserv* groups, mailing lists, and Usenet, has been seen as illegal since, at least under United States law, it is unlawful to use any telephone facsimile machine, computer, or other device to send an unsolicited advertisment to any equipment which has the capacity to transcribe text or images (or both) from an electronic signal received over a regular telephone line onto paper. The law allows individuals to sue the sender of such illegal "junk mail" for $500 per copy. Most states will permit such actions to be filed in small claims court. See Arlene Rinaldi, "The Net: User Guidelines and Netiquette,": http://www.fau.edu /rinaldi/netiquette.html.

39. Marcel, *Men Against Humanity,* 40.

40. Cf., Charles, Ess, ed. *Philosophical Perspectives on Computer-Mediated Communication* (Albany: State University of New York Press, 1996); Richard A. Spinello, *Ethical Aspects of Information Technology* (Englewood Cliffs, NJ: Prentice-Hall, 1995); Terry Bynum and Simon Rogerson, "Global Information Ethics," *Science and Engineering Ethics:* Special Issue on Global Information Ethics (April 1996).

41. Marcel, *Men Against Humanity,* 40.

42. Marcel, *Men Against Humanity,* 61.

43. A survey conducted by Nielsen Media Research for Commerce Net suggests that some 37 million adults in Canada and the United States (16.6% of all adults age 16 and older) have access either directly or indirectly to the Internet through a friend, a co-worker, or a commercial online service like America Online, Prodigy or Compuserve. Users tend to be highly educated (64% had at least four years of college) and affluent (25% had household incomes of $80,000 a year or higher), but only 35 percent are women, and more than half of the users were 35 years of age or younger. The Executive Summary of the survey is available at: http://www.commerce.net or from Nielsen Media's site at http://www.nielsenmedia.com.

44. Robert E. Babe, *Communication and the Transformation of Economics: Essays in Information, Public Policy, and Political Economy* (Boulder, CO: Westview Press, 1996); Vincent Mosco, *The Political Economy of Communication: Rethinking and Renewal* (London: Sage, 1995).

45. Jeremy Rifkin, *The End of Work: The Decline of the Global Labor Force and the Dawn of the Post-Market Era* (New York: G.P. Putnam's Sons, 1995), xvii. For empirical analysis and ethical critique of Rifkin and others, see Gregory J. Walters, "Information Technology, Work and Human Development: A Human Rights Perspective," *Canadian Journal of Development Studies/Revue canadienne d'études d développement,* Vol. 20, No. 1 (1999) 225–254.

46. Ursula Franklin, *The Real World of Technology* (Concord, Ontario: Anansi, 1994); Sherry Turkel, *The Second Self: Computers and the Human Spirit* (London/ Toronto/ Sydney/ New York: Granada, 1984).

47. For a discussion of the impact of science and technology on Jaspers's understanding of the university "idea," see Gregory J. Walters, ed., *The Tasks of Truth: Essays on Karl Jaspers's Idea of the University* (Frankfurt am Main: Peter Lang GmbH, 1996).

48. Marcel, *The Decline of Wisdom* [Le Décline De La Sagesse], 10.

49. Marcel, *The Decline of Wisdom,* 12.

50. Karl Jaspers, *The Atom Bomb and the Future of Man* (Chicago: The University of Chicago Press, 1963 [1961]), 7.

51. Marcel, *The Decline of Wisdom,* 18.

52. Gregory J. Walters, "A New Way of War in the Information Age," Argument & Observation, *The Ottawa Citizen* (March 14, 1998), B7. Available at:http://www.uottawa.ca/~hrrec/chair/walters.html

53. Karl Jaspers, "The Fight Against Totalitarianism," in *Philosophy and the World: Selected Essays and Lectures,* trans. by E.B. Ashton (Chicago: Regnery, 1963), 68–87.

54. Gabriel Marcel, "The Sacral in the Era of Technology," *Searchings* (New York: Newman Press, 1967), 41–53, at 41.

55. Marcel, *The Existential Background of Human Dignity,* 87.

56. Marcel, *Searchings,* 53.

57. Gregory J. Walters, *The Role of "Conversion" in the Nuclear Age* (Lanham and London: University Press of America, 1988).

58. Marcel, *The Decline of Wisdom,* 19.

59. Marcel, *Tragic Wisdom and Beyond,* 245–246.

60. Marcel, *The Decline of Wisdom,* 20.

61. Karl Jaspers, *Philosophy,* Vol. 1, 148.

62. Karl Jaspers, *The Origin and Goal of History,* trans. Michael Bullock (New Haven: Yale University Press, 1968 [1953]), 90; *Vom Ursprung und Ziel der Geschichte* (München: R. Piper, 1963), 119.

63. Karl Jaspers, *Man in the Modern Age,* trans. by Eden and Cedar Paul (Garden City, NY: Doubleday, 1957 [1951]), 41ff.

64. Jaspers, *The Origin and Goal of History,* 174; *Vom Ursprung und Ziel der Geschichte,* 217ff.

65. Jaspers, *The Atom Bomb and the Future of Man,* 282.

66. Joseph Agassi, *Technology: Philosophical and Social Aspects,* Vol. 11, "Episteme," edited by Mario Bunge (Dordrecht/Boston/Lancaster/Tokyo: D. Reidel, 1985), 259.

67. Karl Jaspers, *The European Spirit,* trans. by Ronald Gregor Smith (London: SCM Press, 1948), 47.

68. Jaspers, *The Origin and Goal of History,* 124.

69. Arthur Kroker and Michael A. Weinstein, *Data Trash: The Theory of the Virtual Class* (Montreal: New World Perspectives, 1994), 19.

70. Rifkin, *The End of Work,* xv.

71. Heather Menzies , *Whose Brave New World? The Information Highway and the New Economy* (Toronto: Between the Lines, 1996), 9. "McJobs" symbolize the turning of good full-time jobs into lousy part-time, shift, and temporary jobs, as well as the hollowing out of the middle ranks of administrative, sales, and service jobs, professional, management, and skilled-trades work.

72. Gabriel Marcel, *L'Homme Problématique,* 186.

73. George Lakoff and Mark Johnson, *Philosophy in the Flesh: The Embodied Mind and Its Challenge to Western Thought* (New York: Basic Books, 1999), 25.

74. Jaspers, *The Atom Bomb and the Future of Man, 7; Die Atombombe und die Zukunft des Menschen: Politisches Bewußtsein unserer Zeit.* 4th Edition (Munich: R. Piper, 1958 [1960]), 25.

75. See further, Gregory J. Walters, *Human Rights in an Information Age: A Philosophical Analysis* (Toronto, Buffalo, London: University of Toronto Press, 2002).

Chapter 12

# Alterity in Memory

*Kimiyo Murata-Soraci*

The uncanny (*daimonion, daimonic*) radiates in the field of memory (*anamnēsis*). Among many regions of scientific study, it is none other than this that is always summoned back to stand on the edge of an abyss. For ever since Plato spoke of memory as the "gift of Mnemosyne" (Theaetetus, 191d),[1] a science of memory seems to have inherited the problems of hermeneutics and nonbeing against which Socrates in the dialogues engaged in a constant battle of demarcation (e.g., the Ion, the Sophist). The pioneers and fieldhands who toil in the terrain of memory are obliged to gaze at and deal with these problems that may breed a vital threat to their scientific-methodological activities and presentation of the thing itself. Why are they a blockage?

Should memory be a dispensation from what is beyond man, our intimate relation with (*Mitsein*) memory is, strictly speaking, not ours alone. Although it begins on our part with a passive, or shall I say, a bit poetically, an enthusiastic (*en theos*) receptivity, we are left merely to find the given share afterwards in a resonant and savory wake of our "out of mind" or "out of this world" experience, because a course of sending and distributing memory from the goddess is beyond human comprehension, calculation, and will. And even during a state of being captured by the unknown, one gets enclosed so totally that there leaves no room for a self to be in a spectator's position from which to view and articulate the process of what is taking place. Having lost a spot of interpretation, it is impossible for us to make sense of "being outside of oneself" (*ecstasis*) in which an alterity, both of the divine and of man, intersects and co-shares awhile a common fate of "being outside of oneself." And the given share of memory that one is able to receive only by way of one's rapturous experience differs from one to another. Thus the grasping neither of an origination of memory with respect to its time and place nor of an interpretive generalization of memory is possible.[2] Here, then, in our relation with memory, we descry an enigmatic draw of nonbeing that conjoins magnetically belonging together of the opposite. In the nearest of near to us, the researchers of memory, as well as we, receive a signal of memory that resists a critical sighting, a hermeneutical explanation and generality as much as a heedless pursuit of bare facts, objectivity, and reportorial presentation of truth in the procedural production of memory studies. How shall we receive such an appeal of memory from within and beyond us?

If we trace the etymological roots of the Greek word *daimonion* in the verbs *daíō* and *theáō,* we can see that signaling and pointing are not foreign to the nature of the uncanny. The *daíō* designates "what presents itself in the ordinary and takes up its abode therein. To present oneself in the sense of pointing and showing is in the Greek *daíō* (*daíontes-daímones*)."[3] Far from implying a pervasive sense of deviation from normalcy, to the Greeks "uncanny" meant something that appears most natural and apparent "in the sense of physis."[4] Here, uncanny connotes the impersonal mode of proffering itself to a sighting that brings about an inevitable state of the concealedness (*lēthē*) in unconcealment (*alētheia*). As indicated by our inconspicuous familiarity with it, the self-showing appearance yields a certain aerial state of forgetfulness and envelops us in it. It is older and earlier than any intentional and acquisitive act of signaling[5] and points to a prehistorical dimension of human memory and recollection.

Furthermore, its way of nonintentional delivery (*daíō* means "in the middle voice," "to distribute or allot"[6]) demonstrates its sheer difference not only from our ordinary act of representation but also from our view of hermeneutics, either as a recovery of the previously hidden meaning by logical exploration and comprehensive elucidation, or as a methodological tool for research. Beyond an ordinary sense of the production and informing of meanings, the uncanny directs our gaze to a primordial dimension of hermeneutics, of that which our interpretive comportments of letting things appear in speech and deeds presuppose.[7]

With regard to the whence and how of such an impersonal dispensation of all that is ordinary and immediately familiar to us, the word uncanny takes us back to its hidden root in the word *alētheia,* by way of the verb *theáō* through *daíō.*[8] The English word "theater," which is also a classical metaphor for memory, bears a close linkage with this verb *theáō,* which designates an opening site of the diverse spectacles of the self-showing emergence of life. Naturally, the "look" (théa) is of no one's in the world and thereby is spotless of any type of beholding, including that of the metaphysical pure "look."[9] The neither subjective nor objective look (théa) is the theá of the goddess Alētheia.[10]

As the Greek *a-privativum* of a-lētheia indicates, the alētheic disclosure (*Entbergung*) is composed of a simultaneous double turning of the self-removal of concealment. First, it withdraws from concealment (lēthē) by a self-wrenching movement (Ent-bergen) and second, withdraws from unconcealment by a self-capturing in a manner of sheltering (Ent-bergen).[11] These countermovements of self-tearing withdrawal come to pass simultaneously without a purpose, and open a pathway to the presencing of life and set it on the way to be. The uncanny is the way of the self-opening of alētheia that lets the emergent self-presenting of all sorts of life come forth, and lets each life abide in a true nature by letting it be seen appropriately in the disclosed (*alèthés*) being. The truth, then, of alētheia precedes our logical assessments and truth-claims of things; for us to do so, it is necessary to count on the already given presence of beings and to take a look at a time and location of their happenstance.

We note here that the popular usage of the German conjunction *weil* ("because" or "since") has a temporal sense of "lingering" and "abiding" (*weilen*) and a logical sense of "true" (*wahr*) by way of etymological ties between *weilen, währen,* and *wesan* ("to be").[12] Since we commonly view time as a successive procession of now-points and intuit space as an enduring substance, we are inadvertently making a causal connection between the given things each time we use this word uncritically. Perhaps, the inconspicuous gap of a permanence-based esteeming and dis/con-joining of things is more pervasive in a theoretical construction of a house of memory. By intuitively relying on the assumption that time is a homogeneous series of now-points and space is an enclosed area, the "pastness" of our experience of the given beings is measured with respect to location and distance, and housed in an orderly manner by the lasting supports of truth as *homoiōsis* and *adaequatio*,[13] and the Platonic view of mimēsis. It seems that the habitual placing of a schema of the house of memory on a substantive view of the "whiling" of being blocks researchers' gaze to a self-opening of alētheia that makes possible a "whiling" of being.

Thus the uncanny is that which opens time and space in the self-sheltering unconcealment, and the uncanny is that which lets be seen the self-showing upsurgence that never avails itself to our gaze and experience, while at the same time holding sway over all presences of beings in the form of nonpresence and sustaining them together by its sublime power of self-shattering and self-disclaiming. Not only the researchers of memory themselves but also the object of their study are always already held sway by the ungraspable nonbeing at the very opening of the emergent life of memory. The uncanny is the truth (alētheia) of nonbeing and its noncausal enabling of a relation between beings, those of which are utterly different from the conventional view of nonbeing as annihilation, a deficiency, or a symmetrical opposite of presence.[14] And the uncanny pastness of its withdrawal is older than that of a past–present. How shall we call it to mind?

In what follows, I should like to address the alterity of dwelling and biding (i.e., whiling) in memory and of the meaning-bound production of memory studies, by circling around the Heideggerian notions of the Fourfold, *Raum,* and the Open. In outlining via Heidegger an opening of time and space, I shall show a possibility of reimagining and rethinking of memory from historicality and a possible direction of rebuilding the man-centered house of memory of beings by shedding light on a pre-ethical dimension of justice in memory studies. At the closure of the present corpus entitled *Visual Information Processing,* I request your time and attention to these issues and make an overture to another tracing of memory.

Let us first envisage an image of the house and find from the mundane an instance of the uncanny figure of abiding with a view to alterity. An image of the house constitutes many drawings of a border.[15] To make room for different uses and purposes for us to dwell, each borderline plays at once a double role of divider and junction of space and time. Opposite spheres and directions are conjoined without a reduction of one to the other by each borderline, and thereby the reciprocal touching and being touched by one another takes place. As the endpoint of one line becomes without respite a starting point of another on a border, the borderline as a midway between the two cannot be measured by a homogeneous yardstick of time (e.g., before/after, now/then, etc.); nor can it be appropriated from a perspectival position of either side. In a whole design of the house, each borderline abides awhile without becoming either an origin or an end. Each one moves in self-difference and self-postponement simply to meet, as it were, the "other than itself" without remaining at one spot and at one time. Here we have a trace of the nondialectical spacing of what Derrida calls "différance"[16] that makes possible the location and time of the meeting of phantasms in unity and separation without canceling out their unique individuality.

As Llewelyn points out in echoing the spatiality of *dasein* in Heidegger's *Being and Time,* the Greek noun "schemata" is a cognate of the verb "echō," which means "to have," "to dwell," and "to be on homeground."[17] Accordingly, an image of the house with a livable homey dimension as well as a moment of our imaging of it, are erected on the "alter base" of alētheia that opens time and space. We see a marked excess of its nonpresence in the abiding mode of the borderlines in the house that eludes a grasp of the common view of time and space.

And not only because memory is constituted of time (*OMR,* 449a24)[18] and images (*OMR,* 449b30) but also because absence (sterēsis) intrinsically belongs to memory's presence in and with us, the mundane image of the borderline reminds us of the necessity of reviewing our view of image and our way of allotment of location and distance to each phantasm in the housing project of memory studies. An examination of a base and a mode both of imagery and imagining is called for. As memory and imagination relate to both realms of sense and meaning and operate as a go-between of the opposite realms, turning our mind to imitate the borderline's simultaneous way of holding things in a way of neither the "either–or" nor the "both–and" may surprisingly open a way to the alterity of our having-been as well as that of a science of memory.

In the heritage of Greek philosophy, Judaeo-Christian spirituality, and humanism, we are accustomed to esteem values of permanence, purity, unity, and wholeness in being, based on having the power of self-sufficiency with regard to causality and freedom. The standard criteria for an apriority of being has repeatedly rested on this self-grounding power of letting-appear. Without relying on the Platonic concept of mimēsis, we cannot make things visible in an ordered way. By positing first of all an object of ideality in front of the mind's gaze, our mind proceeds to estimate the *sens* of the thing's being-present and its dis/as-sociable connections with one another in things' qualitative likenesses to ideal ones because the latter ones are believed to disclose meaning in advance of production. Homology and analogy are most familiar tools for our normal recognition of what things are and for our formations of things into worldly relations. However, in our ordinary way of reading the imagery of things,[19] we overlook not only a moment of being's ecstatic self-surpassing but also a uniqueness of each being's presence; we presuppose every existent being as an instantiation of some sort of the universal and take hold of it as a categorizable unit.

In our mundane way of beholding and housing things, the matter of "a singularity of existence"[20] (objective and subjective genitive) slips away from every twist and turn of our mentation. Thus a historicality of each thing remains forgotten and unthought of as we perceive and recollect things. Just as Heidegger has shown in *Being and Time* that the scientific world develops in and out of our daily practical world, the dominant lack of our regard for a historicality and a nonshareable dimension of each being infiltrates unbeknownst to the mindset of the researchers of memory, dictates from within and beyond their mind their daily course of information processing (e.g., design of experiments, collection and description of data, interpretation and storage of "facts," presentation of a hypothesis, etc.), and thus affects the theoretically built house of memory.

In his short treatise on memory, Aristotle broke his path from Plato's views of memory and recollection (e.g., *Phaedrus,* 249b–c; *Meno,* 73a–75d; *Phaedo,* 72e–73a), but kept the paradigmatic metaphors of memory (i.e., "a sort of waxen block," "a kind of aviary," *Theaetetus,* 197e) and the operation of mimēsis. For Aristotle, recollection is a privilege of human beings because it is a deliberative search that involves associations of ideas and syllogistic reasoning (*OMR,* 453a4).

In recollecting, the mind's gaze must try to speculate and read the portrayals of the mnēmic phantasms drawn on a panel in a hidden recess of the soul, just as, indicates Aristotle in recalling Plato, " ... people do who seal things with signet rings" (*OMR,* 450a25). And there, stamped picture-like imprints are quite ambiguous and muddled, because they manifest themselves in both an iconic and eidoletic likeness: "For the figure drawn on a panel is both a figure and a copy, and while being one and the same, it is both, even though the being of the two is not the same" (*OMR,* 450b20).

Thus they can be read either in their own right as objects of contemplation which would lead to the regathering of scientific knowledge, or as the imprints of something else that would function as reminders (*OMR,* 450b20). A moment of decidability and option in reading the equivocal imprint into the alternative types is brought about by the similarity of the motion of a sudden presencing of something else to which the drawn figure under speculation relates (*OMR,* 450b20). So then, the way of following another thing's presencing and its likeness to the drawn figure in question points back to the up- and downward swinging motion of time that allows for the revealing of the true being of the absent thing, out of concealment.[21] According to Krell (cf. *Of Memory*), the prefix "ana" of *anamnesis* bears such a motion of the rising up and falling down of time. It seems that a possibility of objectification of the imprint and of differentiation between contemplation and recollection are based on an originating fold of time.

In his *Physics,* Book 4.13, Aristotle speaks of a self-dispersing and overturning movement of time that comes about quite suddenly in and of itself:[22]

> A thing is said to depart suddenly if it does so in a time interval which is imperceptible because of its shortness; and every change is by nature a departure from an existing condition. . . . It is clear then that time in virtue of itself is a cause rather of destruction than of generation, as stated also earlier (for a change in virtue of itself is a departure from an existing condition), but that it is accidentally a cause of generation and of being. . . . And it is this [change] most of all that is usually said to be a destruction by time (*Physics,* Bk 4.13, 15–26).[23]

The sudden displacement and remotion of self (i.e., *metabolē*) is a very constitutive, and thus metabolic movement of "now," whose power of a total self-alteration opens at the same time a particular presence of "now." The "now" yields in and of itself both self-termination and self-openness. Since a rapturous self-displacement of "now" gives to itself the countability (i.e., a series of "nows" as a particular being and numerically countable in the ways of either addition or subtraction), the simultaneous way of the following of "now's" ontic/ontological difference is not a simultaneous movement of two things that happen in diachronic time.

Since each "now" arises in a form of alterity in and of itself, anamnēsis cannot be thought of appropriately on either a linear or circular basis of time because these alternative views start from the countable "now" as an objectively reliable unit of measurement of things. In Aristotle's thought, a metabolic mobility is tied to the key notions of *morphē, entelecheia,* and generation (*genesis*) of *physis.*[24] Then, rather than a familiar movement of diachronic retrogression to a starting point of occurrence, the "ana" of anamnēsis indicates an intrinsic way of the abiding of time and its immense power of undergoing a perpetual self-displacement in the course of its coming to presence. This hints that there is an impersonal and uncontrollable undercurrent in the Aristotelian anamnēsis.

What about the place upon which many graphic instances of mnēmic phantasms are drawn and the changes take place? Aristotle calls it "panel" (*OMR,*

450b20), by which he seems to translate Plato's metaphor of "signet rings" in a previous paragraph (*OMR*, 450a25). This translation appears to brood not only an entanglement of philosophical ideas with metaphors but also an internal linkage between recollection, metaphor, and mimēsis.

At first glance, the use of the word "panel" strikes us as odd because it means both an instrumental "surface for an oil painting" and the "painting [drawn] on a thin wooden board."[25] This choice of the word indicates the characteristics of a reciprocal belonging to the place, its resilient power to receive marking acts (e.g., inscription, furrowing, coloring, etc.) to let things become manifest, and its relation to a work of graphic art and a process of production (*poiēsis*). At this juncture, let us be reminded that the English word "character" is derived from the Greek word "*Kharassein*," which means "to engrave."[26] As we refer to the letters of the alphabet as "characters," not only the stock of living beings but also the written letters are grafted and represented on a "panel." Futhermore, according to Krell, "*[k]haractēr* (cf. *Kharagma*) is both engraver, engraving tool, die or stamp, and the mark engraved, the impression, in one."[27] So then, since the "*Kharactēr*" bears no sign of the binary division between agency/patient, active/passive, subject/object, etc., we may see in "panel" an erased trace of a middle-voiced occurrence.

The "signet rings" of which Aristotle appears to be making a coinage by the word "panel" refer to the mnēmic material of the "waxen block" upon which impressions are imprinted and recorded (*Theaetetus*, 191c9ff). The characteristic image of a slab of wax resembles an image of Khōra in Timaeus (50c).[28] Being derived from the verb "chōreō," which means "to make room for another, to give way or withdraw,"[29] Khōra is likened to a "type of mother" (*Timaeus*, 50d) or as "the receptacle, and in a manner the nurse, of all generation" (*Timaeus*, 49b).

However, as "chōreō" means also "to go forward, to be in motion or in flux,"[30] the Khōra in which the Platonic protopair of paradigm/copy is grafted is unlike a static matrix. It surges up into the *eidos* as its necessary a priori element and into the matter as their indirect homebase. It receives back all the images and names, both the nonsensuous and the sensuous types stamped (cf. *typtó*, "to strike")[31] on it, and shelters them in their decomposition.

The Khōra, the eternal bearer of impressions, is an alterity of both the philosophical ground and decidability of Platonism and its range of imagination. Conversely, the primal difference between the *eidos* and its other needs a matrix other than its own ground that would receive both, that would let them differentiate as they are, yet at the same time that would withdraw its identity from either of them and beyond both of them. As such, the Khōra is called "a third kind" (*Timaeus*, 49a), and "a third nature" (*Timaeus*, 52b). Doesn't the image of Khōra remind us of that of the borderline?

So then, when Aristotle creates a metaphor of "panel" and metamorphosizes "the signet ring" into "panel," he presupposes not only the Khōra-like space and its impersonal movement of self-showing presencing that houses magnetically all types of images as the space of "receptacle"—of that which the Platonic Khōra stands for—but also the Platonic Khōra that is already produced and available as a model for imitative reproduction. His reproduction of the "signet rings" operates on a resemblance and discloses in a duplication of a Khōra-like base both the unpresentable and the presentable. In this way, the "panel" bears a certain share of Khōra's characteristics. The streaks of her abyss-like hold, her neither active nor passive manner of reception, her duplicity, her graphicness, and her carefree setting of being-outside-of oneself are traceable in the texture of "panel." We have here an occasion to gather evidence of an internal linkage between metaphor, mimetic technique, and recollection, along with evidence of the entanglement between the so-called "science" of memory and metaphor.

As we have pointed out, Aristotle distinguishes recollection from remembering and crowns man at the center of dominion in the world on the basis of man's ability to reason deductively. In general, syllogistic reasoning operates on discernment of resemblance so that it cannot bypass the drawings of metaphor. During the course of the drawing of inference, mind must actively generate more or less similar images of those that stand for the phantasms, and by being entertained by their graphic demonstrations, mind reaches over the thing sought that has been hidden (*Poetics*, 1448b5–17).[32] Mind gets enlivened and expands in spite of a detoured path of the drawing of inference that is in contrast to the straight path of ideality. How is this so?

In this hidden, inner theater of "resembling double (*miméma*),"[33] mind plays at once a host of characters—a producer, a presenter, and a receiver. Mind moves soundly in tune with the engraving movement that bears a middle-voice-like character. In donning three masks at once, one's mind doesn't station itself in a single place of reception between these three different spots of taking a role; it is strewn in abiding in all these other places and simultaneously remains in touch with different roles. In dwelling neither in one spot nor in one time, it occupies everywhere else but its proper anchoring place.

And yet, mind appears to be removed from its customary position of receiving things transitively and unilaterally, and weaving them in a sequence of causal connections in order to make sense of the life of things. Also, let us not forget that the drawing of metaphor is guided by cultural propriety. As mind utilizes a common stock of cultural proprieties, it gets inevitably thrown back into a historical fold of common memory so as to recircle around the conventional forms and perceptions of life drawn by previous others. While reappropriating the anonymous footprints of the others' interpretations of the life-world in order to come to see the unrecognizable thing at hand, mind moves contemporaneously with personal and impersonal modes of repetition.

Here, mind surreptitiously glides a bit from its normal residence in the ego-base, the ego-oriented perception of things, and the surety of its footsteps of making sense, and moves its course by becoming increasingly bound to the other. The newly drawn metaphor, in which mind preserves the past forms and ways of life cultivated by the impersonal others, belongs neither to one's mind alone nor to the mind of the other alone.

So then, while drawing a metaphor, mind's experience of making its course and its experiencing the loss of its customary foothold are inseparably blended together. Thus mind's self-relation in the unfolding experience of the dividedness of self remains suspended before the mind's eye, and the impersonal self-relation stands before the mind's gaze, like the image of the borderlines in the house, undecidable to either side of an ordinary or extraordinary operation of mind.

And yet, despite a strewn setting, mind appears to be more gathered and moving more freely by removing a spot of its substantive foothold and regaining a deeper touch with its cultural heritage. Here, opposite moments of self-loss and self-gain are not held together in a neutral tension nor in a participation in the universal principle. Rather, in the midst of mind's forgetfulness of self-ground in grounding its course, those opposite moments leap over simultaneously to the other side of one and co-arise contemporaneously by thoroughly saturating each other's fold. Each side moves-and-being moved by the other at the same time so that each feels one's "self" directly in the feelings of the other.[34] Each one is immediately a mirror image of the other in the likeness of being arrested by the something that remains unfathomable to both sides of mind's self-experience, since mind has been carried away from its normal residence in will and reason. Accordingly, the imperceptible object that magnetically holds together the opposite sides of self-experience is wholly other than an absent object of ego's perception that was once present in the past before mind and can be retrievable for the sake of re-production.

Seen in the perspective of likeness, the affective touch of sympathy that arches over the self-difference of mind differs also from a common sense of sympathy as a property and a propriety of the free human individual. Customarily, the captured state of sympathy is determined by likeness. But it exceeds the common range of mind's receptivity of the other and responsiveness to the other that has been customarily measured by the yardstick of the likeness in kind, qualitative proportion (e.g., analogy) and predilection toward reciprocal responsiveness and responsibility.[35] These standard measurements of likeness, however, presuppose (a) the difference between meaning and sense, (b) man's ability to recognize the cardinal selfsame meaning (*huponoia*), (c) the two routes of seeing and presenting the truth either by homology, or by analogy, and (d) the Platonic sense of mimēsis that esteems the former (the ideal, the literal) as superior to the latter (the physical, the figural). In one broad stroke, we can say that the conventional pattern of imprinting an affinity between one and the other with an eye to an underlying identity has been monogenous and concentric, as well as anthropocentric and anthropomorphic.

Only the like feels the like in the feelings of the other, and the extension, in terms of near and far, of mind's responsiveness to the other starts from "me" to the likes of me and then stretches over afterward its list of minding and hospitality to the other others who and which are unlike me.[36] Not only freedom as a core property of personhood but also the other's existence as an alterego are presupposed by a human self. That is why at any time and at any place the making of a communal and communicable relation follows from "me" to between us human beings, to the beyond of human kinship. The farther the place the other occupies, the more the other's qualitative share with one with regard to the form of self-standing in one's own right (like me?) dissipates and becomes in need of the other's care and supplementation from outside. One's mind can either disregard or regard the need and appeal of the other at one's disposal and preference.

Conversely, if one's mind wills to lend its ear to those nonhuman others who have no voice of their own, and if one, like those of animal psychologists, needs to be mindful of their ways of being in his or her scientific experiments and productivity, or if one wants to become an active spokesman for a nonhuman creature kind, there is no other natural means than analogy for one's mind to use in order to not only refigure their right but also reappropriate one's right in taking the place of the other on the other's behalf. As Llewelyn points out, the repeated pattern of drawing an inference and making an emblem of a nonhuman right from an extension of the idea of human rights merely metamorphosizes the nonhuman other and excludes the nonsentient other (e.g., things, artifacts) even from the outskirts of mind's responsibility.[37] Not only the otherness of a free mind of the human individual but also the otherness of nonhuman others, including both the sentient and nonsentient, have been repeatedly overlooked in the conventional circles of communal affiliation between like minds and their communicative codes of agreement regarding the sharing of responsibility and obligation, charted by the benchmarks of likelihood and kindredhood.

In contrast, the affective state of mind in its undergoing of self-difference casts all kinds of the other out of the substantial storage of memory, concurrently with mind's being hollowed out of its substantive ego-base. Albeit in a strewn state of gathering, each figure stands on the verge of meeting the otherness of one's being and being met directly by one another in a singularity of a bare face under the removal of the categorical mask one has been assigned to wear. Although mind finds itself being cast in an overpowering climate of passivity beyond the regulatory alteration of an active/passive condition, mind's power of reception of the other becomes more extensive and penetrating than usual. It is the nonreflexive passivity of mind that makes room for it to meet a true face of each being that retains no spot of resemblance of either anyone or anything. In this respect, mind's affective state is more active than its usual state of active mentation to which the state of passivity is normally placed side by side, asymmetrically (e.g., active/passive, thought/feeling).

And yet, mind cannot cross over the line of "between" one and the other, drawn by it in its withdrawal from the privileged place of manhood and with its renunciation of the prestigious power of freedom. For the line crosses over both at once the otherness of a human individual whose mind is currently undergoing a stretch of self-difference and that of the others which happen to get involved in the happening of mind's self-surpassing journey. Here, the line of "between" is not composed of the now-points because the line of between crosses over both historical and nonhistorical aspects in and of every being. In effect, it makes it impossible for mind to see the other as an object in receiving the other and to operate upon it by recourse to the common property and probability. In this respect, mind is cast in the state of passivity that is more passive than its usual state of passivity.

The untraversable line of "between" is, then, a free space belonging jointly to each being in a singularity of its historical occurrence. It is an inoperable dimension of each life in which each being comes to stand in his, her, or its own right without a need of supplementation from outside. In the impersonal site of coming to presence, each one has no need of taking in a foreign thing first for the sake of historical transcendence. Each one is freed from the need of being a mirror-image of someone other than oneself for the sake of self-sustenance, and also freed from the need of a spokesman on behalf of one's nature. Any neighboring one that happens to stand alongside the other's upsurgence lets the other carry upward the other's own share of imprinting in a sheer absence of any model.[38]

This way of each one's letting of the self-showing imprinting that allows any one to be for the first time what one truly is opens a buoyant space of absolute freedom that uplifts one and the other equally.[39] In effect, any one's nonimage-bound trajectory of coming to presence carries forth not only that of the neighboring other's share but also the abundant force of absolute freedom. In the invisible dimension of historicity, one's free voice and that of the other are brewed from a common spring of absolute freedom so that each one bears inseparably and inherently the other's voice in and through which absolute freedom echoes forth.[40] To be sure, absolute freedom is, unlike the Hegelian absolute spirit, impersonal and nonsubstantial.[41] Everyone originates in and departs from a common *topos* of absolute freedom that belongs to no man, because mind has unlocked the power of freedom from the long-standing storehouse of thought and released it to a sphere of no man.

This common landing place of absolute freedom is a true place of alterity of all images that have been drawn by man's deeply seated supposition of a reciprocal cohesion between the power of freedom and thought and by the mimetic technique of reproducing things on the base of their presence and familiarity to man. The drawn line of "between" makes out of play mind's intuitive grasp of a relation between one and the other in terms of an adhesive bond between freedom and thought and puts to rest the accustomed measure of qualitative similarity. While suspending a moment of returning to the habitual position of mind, the line draws at the same time all types of being into a form of double transcendence (i.e., being simultaneously in and out of one's normal residence and one's proper home). And without a second's hesitation, it opens, beyond an outermost limit of man and objectivity, a space of absolute freedom along with the countless draws of such an ecstatic self-overturning of each and every participant and conjoins everyone in ecstasy.[42]

In lieu of the universal substance and egohood, ecstasy is a common bond. Ecstasy saturates the undividable and unownable sphere of in-between the historical and nonhistorical, and is the noncasual seam of their difference in terms of every one's singular and communal relation in a loss of self-identity. Like the image of the borderlines in a house, it abides in the middle of everything as at once both a divider and a uniter of each life's self-difference in extremity and sustains different lines of receiving the other, with or without mind, to the singularity of ex-istence.

The ecstatic space of "between" houses a nonqualitative quality of being free that belongs primordially to each being's dimension and that cannot be made into an object for any one's interest to be monopolized and reified.[43] Like the image of the borderlines in a house, the drawn line of "between" erects an absolutely irreducible dimension of "the free" that frees mind and things from a worn-down mold of thought, restores a fire of absolute freedom, and offers a possibility of a wholly other way of receiving and beholding life by starting from a common source of absolute freedom that belongs, primordially, equally, to every one and no one.[44] As such, the drawn space of "between" is an atopical "archi-trace" to which Derrida refers in *différance*,[45] and is a homeground of the unlimited surface space upon which a palette full of whole-world relations have been portrayed. It is a true home of all images, imagination, recollection, and the tapestry of homocentric memory.

Although no-self reigns and nothing subsists in this unlimited spacious home with a nonsymmetrical design, a thorough absence of identity pertaining to the homeground shares a blueprint genetically of neither a static *Urgrund* of general substance nor an *Abgrund* of absolute nothingness. For the space of "between" is not there first as a necessary foundation prior to and apart from the historical happenstances of beings. In this respect, its nonfoundational character differs from the likes of Buddhism and negative theology that end up, at the bottom line, redeeming essence by grasping nothingness as fullness. The drawing of "between" happens to emerge along with, in, and through a double transcendence of man and nonhuman beings. The crossing over of man and nonhuman beings to their nonhistorical root (i.e., the surpassing of the alterity of man and things) and the coming over of an utterly Other image of being toward man and things (i.e., the surpassing of the alterity of the other man and things) across from an imponderable depth underneath the continuum of a generic substance or an essence, and out of the immeasurable loss of homocentric images, open up the chiasmic zone of "between."[46] The "between" happens to be a restless zone of stamping off and on, in and out of images from all regions, and thus belonging to no one, it facilitates mindlessly as both their wasteland and matrix. Here the "between" reminds us of Khōra.

The crossing over of the double alterity of man and things draws time into the heart of time. It withdraws time from a monotonous continuum of ontic now-points and casts it back into an intrinsic flow of metabolē.

By metabolē, Aristotle means a change in and of a being that becomes all of a sudden wholly other than before by displacing itself completely in a violent stroke of inversion. As occasioned in beings' traversing into an alterity at the opposite ends (genesis, phthora) of their life (from nonbeing to being, from being to nonbeing), a metabolic mobility involves a complete change and loss of substance. Whereas an alteration of an intratemporal being can be measured in terms of quality, quantity, and place, and thus can be sensible and cognizable, a change in and of life's pure becoming leaves no permanent footprint of the same being so that the change in being with regard to its causal connection is absolutely improbable and illegible.[47] And since the dynamic power of metabolic mobility seems to spring from its inherent dispositional tendency to seek a total discharge of forces, the outbreak of change (metabolē) and the way of following the change demonstrate a mode of leaping. As a carefree urge for a complete self-dispersion and destruction (phthora) makes possible a generation (genesis) of a metabolic change in being, a stroke of nonpresence resides at the heart of metabolē and beats a syncopal (*sunkoptein* means to cut short: syn + *koptein* [to strike])[48] rhythm of abyssal destabilization. The character of metabolē is thus essentially ecstatic (*Physics,* Bk 4.13, 222 b16), and we see a trace of the middle-voiced *gignomai,* neither a man nor a thing-oriented becoming in Aristotle's notion of metabolē.

Time's retrieval into the metabolic mobility takes, along with its homeward journey, the fixed "now" into an innermost and outermost limit of "now," and unfixes it in the middle of time's pure becoming to clean off the strokes of punctuality, uniformity, and countability. Time gives back to the time of the present not only an intrinsic tie with nonpresence but also an infinitely self-disowning vitality that accommodates chance and non-sense. It draws forth an interval of self-difference regarding the metabolism and opens, in and through the drawn interval, a coming-forth of another time beyond humanism and essence.

Here, the taking and giving of time takes place in the midst of a no-self-and-nothing-oriented *polemos* of time as time recoils back upon an original polemos of time's self-withdrawing drawal. Nothing or no one initiates and directs a surpassing of the alterity of time. Time plays out its own coursing of genesis–phthora. The space between each time's coming to be passing away (*genesis–phthora*) is conjoined by nothing at all. Nonpresence saturates and keeps in touch, throughout and across time, not only each time's self-difference and self-différance in its singular occurrence, but also each time's communal relation with other times. In this intrinsic dimension of time, each time is ecstatic and in and out of a complete self-erasure leaps out, suddenly, over the yonder sides of one. Thus, there is no room for a dividing line between one another; nor is there a marker of provenance or providence; each ring of time shares equally a figure of "being outside of oneself;" and because the original power of ecstatic self-removal makes time resist the strokes of reduction and substantialization over and against common time, each time brings forth with it an outline of another time.

Here at last, beyond the time of man and thing-based space, time comes to land in the open space of "between" only to lose its home of origination eternally; as we have outlined, at the heart of time the ecstatic unfolding of temporalization and spatialization places constantly under erasure the causal "and" between time and space and rubs off the common sense of duration. And nothing but a mind-free "spacing" (espacement)[49] of time and space tears away, in a simultaneity of turning, the anchoring place of origin from each one. We restate here that time's self-retrieval to the sought place of origination is different from and older than not only the anamnēsis that seeks to retrieve a past that was once present but also the coincidental sense of simultaneity that comes across in time. In the anamnēsis, the directional change of the mind's gaze operates within a confrontational framework of position/opposition. In contrast, the self-retrieval of time never returns to itself but is destined to face no resting place of self. This homeground of self has been overtaken by the other, simultaneously with one's taking over the other's self-grounding place, so that time finds its destiny to be infinitely *unterwegs*.

In time's withdrawal from the present, both time and space come to be joined in a total exposure to each other and to share the seal of *Ursprung*. And yet, at the same time while bearing forth the seal of no self-presence, a carefree spacing of time and space yields an extent of spatiality whose spaciousness, belonging to neither time nor space but to nothing, is extended without qualification to all types of life at their root, and delivers each life's chance-ridden transcendence by letting each one take a proper share of absolute freedom. The "between" is of spatiality which hosts freely the transitional passage of the self-surpassing of man and nonhuman beings in the form of double alterity. As such, the "between" announces an *atopos* of interpretation.

It is then of necessity that the character "between" comes to don a mask, albeit behind or beneath of which nothing stands. The figure of Khōra and her choreography in Plato take over the mask and become a mask of this ever non(re)presentable ground (alētheia) of life.[50] And the Aristotelian "panel" replaces the mask of Khōra and further transmits the character of "being outside of oneself" in his treatise on memory. As we have stated, the human ability to draw a metaphor and chart a course of syllogistic inference by means of a metaphorical mirror-play separates the human race from nonhuman creature kinds so that the hidden threads of recollection, mimēsis, and metaphor underline a texture of this book, while still preserving an *aporeic* trace of nonpresence in the metabolic mobility of time and in his view of physis. Ironically, the drawing of metaphor as well as a play of metamorphosis are conditioned and supported by the inimitable figure of "being in and outside of oneself." The mask stands for an unthinkable and sublime power of nonpresence that delivers disclosive occurrences of all forms of life and mediates all lines of relation without ever availing itself as the archetype for mimetic duplication. The mask stands for the sublime vitality of opening that gives and delimits the origin of an image-making.[51]

The mask masks the simultaneous conjunction of time and space at their grounding, the unseverable tie with nonpresence that every disclosive presencing of life bears forth, and the illegible interconnection that occurs between the spacing of time and space and the coming-to-presence of each and every life. Also, it stands for the ecstasy of all beings that is destined to be effaced under and to slip over out of a speculative scope of memory and recollection, although the extraordinary uncanniness of this alterity filters thoroughly all disclosive beings and stands always already on our side. So then, the mask stands for the origination of our memory and for the inimitable nonhuman mode of repetition. It stands for an alterity of our memory and recollection.

Through (per) the mask, echoes (*"sonare"* of the Latin *persōna*) silently and freely the coming forth of time.[52] It continually inaugurates, in a tireless self-withdrawal, the present and the present scene of (re)production of the senses of life and the world. Yet at the same time, it extends the limit of our sensibility and cognizability by shaking off the common measurements of time and space. Thus, its ungraspable power brings to us all a time of relief by tearing the tissues of not only our obliviousness to the vital tie with nonpresence and the irreducible spatiality of each life but also our disposition to possess nothing but a bare truth and to erect the regions of determinations with no eye or ear to an alterity.

Through the mask of nonpresence vibrates a joyous sounding of absolute freedom that removes doubly, at once, the inertia of our memory. By continuing to return to the present scene of informing and producing the senses of memory and recollection, it gives back to the minds of scientists of memory and cognition a sound setting and a proper time to review the erasure of vital nonpresence and to reinscribe the roll of memory with what was never present, beyond the available techniques of kindredhood and likelihood. In this light, the ecstatic mask protects scientists from bare exposure to clarity and evidence-gathering, with rekindled awareness of the free spatiality of all existent, and lets them inform a coming of memory.

## NOTES

1. *Plato: The Collected Dialogues,* edited by Edith Hamilton and Huntington Cairns (Princeton University Press, Princeton, NJ: 1989), 897. The citations from Plato's Dialogues in the present paper are from this text.

2. I am indebted to Jean-Luc Nancy's reading of Plato's Ion in his article "Sharing Voices," in *Transforming the Hermeneutic Context,* edited by Gayle L. Ormiston and Alan D. Schrift (SUNY Press, Albany, NY: 1990), 211–259. Also, with regard to the intricate issues of interpretation and memory, see Charles E. Scott, *Boundaries in Mind* (Scholars Press, Chicago: 1982), and Charles E. Scott, *The Time of Memory* (SUNY Press, Albany, NY: 1999).

3. Martin Heidegger, *Parmenides,* trans. by Andre Schuwer and Richard Rojcewicz (Indiana University Press, Bloomington and Indianapolis: 1992), 102.

4. David Farrell Krell, *Daimon Life* (Indiana University Press, Bloomington and Indianapolis: 1992), 19.

5. Heidegger, *Parmenides,* 102–103.

6. Krell, *Daimon Life,* 20.

7. See Heidegger's accounts of hermeneutics and logos in *Being and Time,* section 7B and C. Martin Heidegger, *Being and Time,* trans. by John Macquarrie and Edward Robinson (Harper & Row, New York: 1962).

8. Heidegger, *Parmenides,* 107.

9. The Greek word *theōrein* ('pure looking") is derived from *theōros,* which is a compound of *théa* ("look") and *oraō* ("to see"). *Theōros* means the one who sees what is given to him in a sensuous immediacy, and also means a spectator at a Greek theater or festival. Aristotle coined the term "*theōretikos*" ("man of science," "theoretical man"), whose pursuit of pure knowledge "*sophia*" is done purely for the sake of knowledge alone from a quasi-spectator position to construct the ultimate ground of beings. This Greek notion of *theōria* was translated into the Latin *speculatio* ("speculation"). See Martin Heidegger, *Plato's Sophist,* trans. by Richard Rojcewicz and André Schuwer (Indiana University Press, Bloomington and Indianapolis: 1977), 44.

10. Heidegger, *Parmenides,* 108.

11. Heidegger, *Parmenides,* 133. In *Being and Time,* Heidegger describes this double turning of disclosure in *dasein's* anticipatory resoluteness in which a temporal mode of "repetition" and a spatial mode of forgetting are made manifest without a causal ground:

> The authentic coming-towards-oneself of anticipatory resoluteness is at the same time a coming-back to one's ownmost Self. . . . If Being-as-having-been is authentic, we call it "repetition." . . . The *ecstasis* (rapture) of forgetting has the character of backing away in the face of one's ownmost "been," and of doing so in a manner which is closed off from itself—in such a manner, indeed, that this backing-away closes off ecstatically that in the face of which one is backing away, and thereby closes itself off too. . . . Just as expecting is possible only on the basis of awaiting, remembering is possible only on that of forgetting, and not vice versa. . . . (*Being and Time,* 388–389)

12. Martin Heidegger, *The Principle of Reason,* trans. by Reginald Lilly (Indiana University Press, Bloomington and Indianapolis, 1991), xviii.

13. Martin Heidegger, *Basic Writings,* trans. by David Farrell Krell (Harper & Row, New York: 1977), 390.

14. Heidegger, *Basic Writings,* 107. Also, with regard to the vital intricacy of the connection between "enabling," "favoring," and "Being" in Heidegger's thought, see his "Letter on Humanism" in *Basic Writings:*

> Thought in a more original way such favoring [*Mögen*] means to bestow essence as a gift. Such favoring is the proper essence of enabling, which not only can achieve this or that but also can let something essentially unfold in its provenance, that is, let it be.This enabling is what is properly "possible" [*das Mögliche*], that whose essence resides in favoring." (*Basic Writings,* 196)

15. With regard to the image of the "borderlines" in the house, I am indebted to Andrew Benjamin's delineation of the experience of "spacing and distancing" in terms of a constitution of the house. Andrew Benjamin, *Art, Mimesis, and the Avant-Garde* (Routledge, London and New York: 1991).

16. Jacques Derrida, *Marges De La Philosophie* (Les Editions De Minuit, Paris: 1972). See his "La Différance" essay in this text.

17. John Llewelyn, "Imadgination," in *The Path of Archaic Thinking*, edited by Kenneth Maly (SUNY Press, Albany, NY: 1995), 83.

18. Richard Sorabji, *Aristotle On Memory* (Brown University Press, Providence, RI: 1972), 48. The citations from Aristotle's "De Memoria et Reminiscentia" are from this book, and cited with the abbreviation *OMR*.

19. It is noteworthy that according to Heidegger there is an essential nexus between reading (*lesen*, "to collect," "to gather"), writing (*graphein*), and saying (*legein*) in terms of letting things be manifest in their presencing out of hiddenness and letting them be seen as truly gathered in their presence, namely, in terms of the original meaning of logos. So, "logos" is not an assertoric statement. For instance, see *Parmenides*, pp. 84–85. See also Martin Heidegger, *What Is Called Thinking?*, trans. by J. Glenn Gray (Harper & Row, New York: 1968), 155, 208.

20. John Llewelyn, The *Middle Voice of Ecological Conscience* (St. Martin's Press, New York: 1991), 41–43, 265–268.

21. " ... For remembering is the presence within of the power which excites the changes, and this in such a way that the man moves of himself and because of changes that he possesses, as has been said." (*OMR*, 452a4). In Martin Heidegger, *The Basic Problems of Phenomenology*, trans. by Albert Hofstadter (Indiana University Press, Bloomington: 1982), Heidegger explores the Aristotelian *aporia* of time (time as both a particular countable being and a "continuum of the flux of time"), and by drawing attention to the unthought latter dimension of time, Heidegger reinterprets the now in terms of ecstatic self-"remotion" that opens up an expanse and keeps a horizon open (*BPP*, 267).

> The most general character of motion is *metabolē*, a turn or change or better a transition from something to something[17] ... Dimension expresses a general notion of stretch; extension in the sense of spatial dimension then represents a particular modification of stretch.... The determination of the *suneches*, being-held-together-within-itself, continuum, continuity, also belongs to stretch. Aristotle calls the dimensional character *megethos*. This determination *megethos*, extension or magnitude, also does not have a primarily spatial character, but that of stretch.... a stretching out that is closed within itself.... Aristotle expresses this set of circumstances in reverse order when he says that *akolouthei to megethei he kinesis*,[18] motion follows (comes in the wake of ) dimension (extension)." (*The Basic Problems of Phenomenology*, 242–243)

> Within itself, original time is outside itself; that is the nature of its temporalizing. It is this outside-itself itself. . . . As this ecstatic character is distinctive of temporality, each ecstasis, which temporalizes only in temporalizing unity with the others, contains within its own nature a carrying-away toward something in a formal sense. Every such remotion is intrinsically open. A peculiar openness, which is given with the outside-itself, belongs to ecstasis. That toward which each ecstasis is intrinsically open in a specific way we call the horizon of

> the ecstasis. The horizon is the open expanse toward which remotion as such is outside itself. The carrying-off opens up this horizon and keeps it open. As ecstatic unity of future, past, and present, temporality has a horizon determined by the ecstases. Temporality, as the original unity of future, past, and present, is ecstatically-horizonal intrinsically. "Horizonal" means "characterized by a horizon given with the ecstasis itself." Ecstatic-horizonal temporality makes possible not only the constitution of the Dasein's being, but also the temporalizing of the only time of which the common understanding of time is aware and which we designate generally as the irreversible sequence of nows. (*The Basic Problems of Phenomenology,* 267–268)

22. David Farrell Krell, *Intimations of Mortality* (Pennsylvania State University Press, University Park, PA, and London: 1986), 49–52.

23. Hippocrates G. Apostle, *Aristotle's Physics* (Peripatetic Press, Grinnell, IA: 1980), 87.

24. Martin Heidegger, *Pathmarks,* edited by William McNeill (Cambridge University Press, Cambridge, MA: 1998). See Heidegger's essay titled "On the Essence and Concept of Physis in Aristotle's Physics B, I" in this text.

25. *The American Heritage College Dictionary,* 3rd edition (Houghton Mifflin, Boston: 1997), 986.

26. James Hillman, *The Force of Character* (Random House, New York: 1999), xxvii.

27. David Farrell Krell, *Of Memory, Reminiscence, and Writing* (Indiana University Press, Blomington and Indianapolis: 1990), 24.

28.

> she is the natural recipient of all impressions, and is stirred and informed by them, and appears different from time to time by reason of them. But the forms which enter into and go out of her are the likenesses of eternal realities modeled after their patterns in a wonderful and mysterious manner, which we will hereafter investigate. For the present we have only to conceive of three natures: first, that which is in the process of generation; secondly, that in which the generation takes place; and thirdly, that of which the thing generated is a resemblance naturally produced. And we may liken the receiving principle to a mother, and the source or spring to a father, and the intermediate nature to a child, and may remark further that if the model is to take every variety of form, then the matter in which the model is fashioned will not be duly prepared unless it is formless and free from the impress of any of those shapes which it is hereafter to receive from without. (*Plato,* [Timaeus, 50c–d])

29. John Sallis, "Of the Khōra" in *Epoché,* Brigham Young University, Provo, UT: 1994 Vol. 2, No. 1, (1994), 4.

30. Sallis, *Epoché,* 4–5.

31. Krell, *Of Memory,* 23.

32. H.G. Apostle, E.A. Dobbs, and M.A. Parslow, *Aristotle's Poetics* (Peripatetic Press, Grinnell, IA: 1990), 4.

33. Derrida, *Marges De La Philosophie,* 285.

34. Rudolf Bernet, "The Other in Myself" in *Deconstructive Subjectivities,* edited by Simon Critchley and Peter Dews (SUNY Press, Albany, NY: 1996), 181–183.

35. Llewelyn, *The Middle Voice,* 255–256.

36. Llewelyn, *The Middle Voice,* 255–256.

37. Llewelyn, *The Middle Voice,* 260–261.

38.

Earth and sky, divinities and mortals—being at one with another of their own accord—belong together by way of the simpleness of the united fourfold. Each of the four mirrors in its own way the presence of the others. Each therewith reflects itself in its own way into its own, within the simpleness of the four. This mirroring does not portray a likeness. The mirroring, lightening each of the four, appropriates their own presencing into simple belonging to one another. Mirroring in this appropriating-lightening way, each of the four plays to each of the others. The appropriative mirroring sets each of the four free into its own, but it binds these free ones into the simplicity of their essential being toward one another. [Martin Heidegger, *Poetry, Language, Thought,* trans. by Albert Hofstadter (Harper & Row, New York: 1971), 179.]

39.

How is this essence of freedom to be thought? That which is opened up, that to which a presentative statement as correct corresponds, are beings opened up in an open comportment. Freedom for what is opened up in an open region lets beings be the beings they are. Freedom now reveals itself as letting beings be. ... To let be—that is, to let beings be as the beings which they are—means to engage oneself with the open region and its openness into which every being comes to stand, bringing that openness, as it were, along with itself. Western thinking in its beginning conceived this open region as *ta alēthea,* the unconcealed." (Heidegger, *Basic Writings,* 127)

40. Jean-Luc Nancy, *The Experience of Freedom,* trans. by Bridget McDonald (Stanford University Press, Stanford, CA: 1993).

For this reason, disclosure also offers itself—this is the logic of *alētheia* in Heidegger—as the renewed concealment of the very being that discloses itself, and of the being of disclosure itself: in other words, as the concealment of the being of being, and of the being of freedom, of the freedom of being, and of being as freedom. Freedom: what is concealed in disclosure, if we can understand this not as a remainder that stays concealed in disclosure, but as the very movement of disclosure, or as its aspect or tone (its intensity): what is "veiled" in a voice, for example. In this way existence is exposed: *Dasein* is exposed to the surprise of the disclosure of beings, because this surprise happens in the *da* of *Sein* and as this *da*—as "being's being-the-there"—whereas the being-there of *Dasein* does not belong to it as its own before this surprise. (94–95)

41.

But neither are they merely coupled together. For world and things do not subsist alongside one another. They penetrate each other. Thus the two traverse a middle. In it, they are at one. Thus at one they are intimate. The middle of the two is intimacy—in Latin, inter. The corresponding German word is unter, the English inter-. The intimacy of world and thing is not a fusion. . . . In the midst of the two, in the between of world and thing, in their inter, division prevails: a dif-ference. . . . What it now names is not a generic concept for various kinds of differences. It exists only as this single difference. It is unique. (Heidegger, *Poetry, Language, Thought,* 202)

42.

The dif-ference is not abstracted from world and thing as their relationship after the fact. The dif-ference for world and thing disclosingly appropriates things into bearing a world; it disclosingly appropriates world into granting of things. . . . The dif-ference is, at most, dimension for world and thing. . . . The difference is the dimension, insofar as it measures out, apportions, world and thing, each to its own. (Heidegger, *Poetry, Language, Thought,* 202–203)

43.

Raum, Rum, means a place cleared or freed for settlement and lodging. A space is something that has been made room for, something that is cleared and free, namely, within a boundary, Greek peras. A boundary is not that at which something stops but, as the Greeks recognized, the boundary is that from which something begins its essential unfolding. That is why the concept is that of horismos, that is, the horizon, the boundary. Space is in essence that for which room has been made, that which is let into its bounds. That for which room is made is always granted and hence is joined, that is, gathered, by virtue of a location, that is, by such a thing as the bridge. Accordingly, spaces receive their essential being from locations and not from "space." (Heidegger, *Basic Writings,* 332)

44. Martin Heidegger, *Early Greek Thinking,* trans. by David Farrell Krell and Frank A. Capuzzi (Harper & Row, New York: 1984).

"How should what is present as such give the jointure of its presencing? The giving designated here can only consist in its manner of presencing. Giving is not only giving-away; originally, giving has the sense of acceding or giving to. Such giving lets something belong to another which properly belong to him. What belongs to that which is present is the jointure of its while, which it articulates in its approach and withdrawal. In the jointure whatever lingers awhile keeps to its while. It does not incline toward the disjunction of sheer persistence. The jointure belongs to whatever lingers awhile, which in turn belongs in the jointure. The jointure is order." (43)

45. Derrida, *Marges De La Philosophie,* 13.

46. Llewelyn, *The Middle Voice,* 228.

47. Françoise Dastur, *La mort* (Hatier, Paris: 1994), 32–33.

48. *American Heritage Dictionary,* 1376.

49. Derrida, *Marges De La Philosophie,* 13–14.

50.

Or does it happen because self-concealing, concealment, lēthē, belongs to alētheia, not just as an addition, not as shadow to light, but rather as the heart of alētheia? And does not even a sheltering and preserving rule in this self-concealing of the opening of presence, from which unconcealment can be granted to begin with, so that what is present can appear in its presence?

If this were so, then the opening would not be the mere opening of presence, but the opening of presence concealing itself, the opening of a self-concealing sheltering. (Heidegger, *Basic Writings,* 390–391)

51. In *Being and Time,* the notion of "Vorlaufen" in dasein's "being-towards-death" plays this role of delimitation (see, e.g., Heidegger, *Being and Time,* 262, pp. 306–307).

52. Nancy, *The Experience of Freedom,* 112.

# Index

# About the Contributors

LYN M. BALSAMO is Graduate Research Associate at American University. Her research interests include neuropsychology, neural plasticity, and the use of fMRI to investigate language reorganization in children with seizure disorder.

ROMARIZ DA SILVA BARROS is Assistant Professor of Psychology at the Universidade Federal do Para. He is interested in the cognitive and perceptual capacities of *Cebus Apella* and in the application of state-of-the-art teaching methods to improve the efficiency of training for nonhuman primates.

MICHAEL T. CARLIN is Assistant Professor at the University of Massachusetts Medical School, Shriver Center. He is interested in stimulus factors that influence attention and how these variables relate to intelligence. He is also interested in generative learning and memory processes.

ROBERT G. COOK is Professor in the Department of Psychology at Tufts University. His research interests are in the general area of comparative animal cognition. In particular, he has been involved in investigating the mechanisms of visual perception and discrimination learning in pigeons, and their comparative relations to our own perception of the world.

ALICE CRONIN-GOLOMB is Associate Professor of Psychology and the Director of the Center for Clinical Biopsychology at Boston University. The focus of her work is the relation between sensory function and cognitive function in neurodegenerative disease. In particular, she is interested in how lower-level visual impairments impact upon higher-order visual cognition in Alzheimer's disease and Parkinson's disease.

WILLIAM V. DUBE is Associate Professor in Psychiatry at the University of Massachusetts Medical School, Shriver Center. His primary research interests are related to stimulus control and discrimination learning in individuals with mental retardation. Current areas of emphasis include the analysis and remediation of learning problems related to overselective attention and restricted stimulus control and the use of computer-mediated programmed instruction.

KEVIN M. FARREN is a teacher at the Varnum Middle School and is interested in stimulus control methodologies.

MICHAEL J. FLORY is Research Scientist and Head, Research Design and Analysis Laboratory, at the New York State Institute for Basic Research in Developmental Disabilities. His research interests are in the statistical methodologies used to analyze longitudinal data from epidemiological and psychological studies of development and health, and he also explores data visualization techniques.

DEBRA S. FLUSSER is Project Coordinator at the University of Massachusetts Medical School, Shriver Center. Her work involves the translation of basic behavioral research findings into improved teaching methods for special education classrooms.

THOMAS R. FOWLER is a researcher at the University of Hartford. His experience with developmentally disordered children led to an interest in childhood psychopathology.

OLAVO DE FARIA GALVÃO is Professor of Psychology at the Universidade Federal do Para. He is interested in the behavior of nonhuman primates, with a special emphasis on stimulus control processes and relational learning.

JUDITH M. GARDNER is Senior Research Scientist and Head, Neurobehavioral Laboratory and Infant Development Follow-up Program, New York State Institute for Basic Research in Developmental Disabilities. Her primary research interest is the modulation of arousal and attention in normal and high-risk neonates; how this early arousal/attention modulation may influence the development of perceptual, cognitive, and autoregulatory control systems through the first years of life; and how differentiation of difficulties in modulation in high-risk neonates may lead to increased understanding of the mechanisms underlying both normal and atypical development.

GROVER C. GILMORE is Professor of Psychology and Social Work and the Dean of the Mandel School of Applied Social Sciences at Case Western Reserve

University. He is interested in individual differences in the amount and quality of information that can be captured in a brief glance. He is also interested in the extent to which the information capture rate influences higher-order processing.

ROBERT M. HODAPP is Professor at the UCLA Graduate School of Education and Information Studies. He examines how children with different mental retardation syndromes develop and interact with their families. A leading proponent of developmental approaches to children with disabilities, his interests relate to the so-called "indirect effects" of genetic mental retardation disorders on others, intellectual strengths and weaknesses in Down syndrome, and the possibilities of etiology-based educational interventions.

DON IHDE is Distinguished Professor of Philosophy at SUNY, Stony Brook. He examines the historical development of interpretation theory, focusing on how the practice of science relates to interpretive processes. He is interested in how the visual hermeneutics of science produce images, visual displays, and so on, and depend upon visually based processes.

BERNARD Z. KARMEL is Senior Research Scientist and Head, Neurophysiological Development Laboratory, New York State Institute for Basic Research in Developmental Disabilities. His research interests involve understanding normal and abnormal neurofunctional behavior as the basis for development of early perception, attention, and cognition in infants. Development of models to understand trajectories in brain-behavior relationships is also explored to predict differential outcomes across populations.

JEFFREY S. KATZ is Assistant Professor in the Department of Psychology at Auburn University. His research program is focused on the comparative mechanisms of learning, cognition, and perception. Current research involves avian visual search, same/different concept learning, and the processes and conditions by which pigeons learn matching to sample.

KRISTIN M. LOMBARD is Research Associate at the University of Massachusetts Medical School, Shriver Center. She is interested in studying the learning processes in children with developmental disabilities and disseminating the results in a way that will lead to more effective tools for special education teachers.

TRAN M. LY is a graduate student at the UCLA Graduate School of Education and Information Studies. She is interested in applying an attributional framework to issues pertinent to families with children with different mental retardation syndrome and is currently investigating parental attributions of intellectual strengths and weaknesses of children with Williams syndrome and Prader-Willi syndrome.

WILLIAM J. McILVANE is Professor of Psychiatry at the University of Massachusetts Medical School and Director of the E.K. Shriver Center. His broad research program concerns the analysis of intellectual disabilities from multiple perspectives and the modeling of human intellectual functioning in nonhumans.

KIMIYO MURATA-SORACI is a lecturer at Boston University and Emerson College. She has taught in the areas of 19th and 20th Century Philosophy, Ethics, Philosophy of Religion, and Eastern Philosophy at the University of Alabama, Union Theological Seminary, Emerson College, and Boston University. Her special areas of research interests include the thought of Martin Heidegger, Jacques Derrida, and Jean-Luc Nancy. She is currently working on a book reinterpreting Heraclitus through the thought of Heidegger and Dogen.

STEPHEN OROSS III is Assistant Professor in the Department of Psychology at Kutztown University. He examines the development of perceptual abilities by assessing low-level sensory/perceptual abilities in populations varying in age and clinical status. When impairments are noted in individuals and populations, he seeks to explore correlations between sensory/perceptual abilities and skills typically thought of as higher-level, issues of etiology, and questions concerning underlying neural foundations.

SAL SORACI, JR. is Associate Professor of Psychology and Director of the Human Factors Program in Liberal Arts at Tufts University and a Research Scientist at the University of Massachusetts Medical School, Shriver Center. He conducts research on the relationship between perceptual and cognitive processes and how these relate to individual differences. His research also concerns generative learning and its implications for theories of memory.

CHRISTINA STRAWBRIDGE is a research assistant at McMaster University. She studies developmental factors impacting upon learning in visual and conceptual domains and is interested in intelligence-related differences.

GERSON Y. TOMANARI is Assistant Professor of Psychology at the University of Sao Paulo. His current research interests are focused on the experimental investigation of conditioned reinforcement and observing responses in human and nonhuman participants.

GREGORY J. WALTERS is Associate Professor of Ethics on the Faculty of Philosophy and Faculty of Theology at Saint Paul University. His interests include the relationship between technological advancement and human rights, and he writes about contemporary issues of ethics.

EDWARD A. WASSERMAN is Stuit Professor of Experimental Psychology, University of Iowa. His research activities have centered on the study of learning, memory, and cognition in nonhuman animals. Additional research areas include conceptual behavior, visual object recognition, foraging behavior, temporal discrimination, higher-order conditioning, observing behavior, response differentiation, and contingency discrimination.

CHARLES B. WOODS is Assistant Professor in the Department of Psychology at Austin Peay State University. He is currently conducting research to evaluate the visual temporal processing abilities of individuals with mental retardation. The specific goal is to determine the origin of observed impairments in temporal processing by assessing whether they reflect inefficient functioning in one or more of the stages of the flow of visual information.

MICHAEL E. YOUNG is Assistant Professor and Director of Brain and Cognitive Sciences Program, Southern Illinois University at Carbondale. His research interests include predictive learning in people and visual variability in people and pigeons. His research on predictive learning focuses on the acquisition of information about causal interactions and the temporal relationships among events in learning tasks.